ALIEN & INVASIVE ANIMALS

a South African perspective

Mike Picker
Charles Griffiths

Published by Struik Nature
(an imprint of Random House Struik (Pty) Ltd)
Reg. No. 1966/003153/07
80 McKenzie Street, Cape Town, 8001
PO Box 1144, Cape Town, 8000 South Africa

www.randomstruik.co.za

Log on to our photographic website **www.imagesofafrica.co.za**
for an African experience.

First published in 2011 by Struik Publishers

1 3 5 7 9 10 8 6 4 2

Publisher: Pippa Parker
Managing editor: Helen de Villiers
Editor: Julia Casciola
Concept and cover designer: Janice Evans
Typesetter and designer: Tessa Fortuin
Proofreader: Glynne Newlands
Indexer: Cora Ovens

Reproduction by Hirt & Carter Cape (Pty) Ltd
Printed and bound by Tien Wah Press (Pte) Limited, Singapore

ISBN 978 1 77007 823 9

Front cover: Flock of Common starlings in flight
(Nature Picture Library/Digitalsource)
Back cover: Orchid snail (D. Herbert)
Title page: Shell of Black sea urchin (C. Griffiths)
Imprint page: Potato tuber moth parasitoid wasp (G. Goergen, IITA)
Affiliations page: Bark-eating earthworm (B. Muller)
Contents page: Northern pine weevil (C. Ernst)
Page 221: Mallard (C. Griffiths)

AUTHOR AFFILIATIONS

Graham Alexander
Animal, Plant and Environmental Sciences, University of the Witwatersrand, Johannesburg, South Africa
e-mail: graham@gecko.biol.wits.ac.za

Chris Appleton
School of Biological and Conservation Sciences, University of KwaZulu-Natal, Durban, South Africa
e-mail: appletonc@ukzn.ac.za

Gary Bronner
Zoology Department, University of Cape Town, Rondebosch, South Africa
e-mail: gary.bronner@uct.ac.za

Kevin Christison
Department of Agriculture, Forestry and Fisheries, Cape Town, South Africa
e-mail: kevinch@nda.agric.za

Ansie Dippenaar-Schoeman
Agricultural Research Council (ARC)-Plant Protection Research Institute (PPRI), Pretoria, South Africa
e-mail: dippenaara@arc.agric.za

Charles Griffiths
Zoology Department, University of Cape Town, Rondebosch, South Africa
e-mail: charles.griffiths@uct.ac.za

Fabian Haas
ICIPE – African Insect Science for Food and Health, Nairobi, Kenya
e-mail: fhaas@icipe.org

Dai Herbert
Natal Museum, Pietermaritzburg, South Africa
e-mail: dherbert@nmsa.org.za

Phil Hockey
Percy FitzPatrick Institute and DST/NRF Centre of Excellence, University of Cape Town, Cape Town, South Africa
e-mail: phil.hockey@uct.ac.za

Charlene Janion
DST-NRF Centre of Excellence for Invasion Biology, Faculty of Science, Stellenbosch University, Stellenbosch, South Africa
e-mail: cjanion@sun.ac.za

Ashley Kirk-Spriggs
National Museum, Bloemfontein, South Africa
e-mail: ashley.kirk-spriggs@nasmus.co.za

Ian Millar
South African National Collection of Insects, Biosystematics Division, ARC-PPRI, Pretoria, South Africa
e-mail: millarl@arc.agric.za

Mike Picker
Zoology Department, University of Cape Town, Rondebosch, South Africa
e-mail: mike.picker@uct.ac.za

Danuta Plisko
Natal Museum, Pietermaritzburg, South Africa
e-mail: jdplisko@saol.com

Marc Rius
Zoology Department, University of Cape Town, Rondebosch, South Africa
e-mail: marcriusvil@gmail.com

Paul Skelton
South African Institute for Aquatic Biodiversity, Grahamstown, South Africa
e-mail: p.skelton@saiab.ac.za

Riaan Stals
South African National Collection of Insects, Biosystematics Division, ARC-PPRI, Pretoria, South Africa
e-mail: stalsr@arc.agric.za

Michael Stiller
South African National Collection of Insects, Biosystematics Division, ARC-PPRI, Pretoria, South Africa
e-mail: stillerm@arc.agric.za

Eddie Ueckermann
Agricultural Research Council (ARC)-Plant Protection Research Institute (PPRI), Pretoria, South Africa
e-mail: ueckermanne@arc.agric.za

Simon van Noort
Natural History Department, Iziko South African Museum, Cape Town, South Africa
e-mail: svannoort@iziko.org.za

Olaf Weyl
South African Institute for Aquatic Biodiversity, Grahamstown, South Africa
e-mail: o.weyl@saiab.ac.za

CONTENTS

INTRODUCTION 6

SPECIES ACCOUNTS

ACKNOWLEDGEMENTS

The authors are grateful to the Centre for Invasion Biology (University of Stellenbosch) for a grant in support of the preparation of this volume. The generosity of those persons who permitted us to use their images is gratefully acknowledged; photo credits are given alongside individual images. A special thanks to D. Visser, P. Ryan and G. Goergen for providing a large number of images. The distribution map for *Linepithema humile* in South Africa was kindly provided by M. McGeogh and N. Luruli, and P. Brock provided the unpublished Australian locality for *Carausius morosus*. W. Lane, K. Packer, C. Nichol, L. de Vos and T. Parker all assisted with retrieval of locality data, map drawing and curation of the image library. For the Diptera chapter, the following individuals kindly provided information and/or literature: M. Ackland, M. de Meyer, A. Pont, F.C. Thompson, M. Villet and N. Woodley. The following are thanked for their contribution to the fishes chapter: A. Chakona, D. Impson, E. Swartz, M. Rudman and E. Truter (information on alien fishes in South Africa), and W. Coetzer (assistance in assembling distribution data). Louis Deharveng provided useful comments on the Collembola chapter. Thanks to Janice Evans (design concept and cover design) and Tessa Fortuin (typesetting and design) for their work that resulted in such a visually pleasing product. Lastly, thanks to our tireless and meticulous editor, Julia Casciola.

The Freshwater jellyfish feeds on small planktonic organisms and invades standing waters, and is the most conspicuous stage of the life cycle, especially when it forms swarms.

C. Griffiths

INTRODUCTION

'A hundred years of faster and bigger transport has kept up and intensified this bombardment of every country by foreign species, brought accidentally or on purpose, by vessel and by air, and also overland from places that used to be isolated ... make no mistake: we are seeing one of the great historical convulsions in the world's fauna and flora.'

Charles Elton, 1958. *The Ecology of Invasions by Animals and Plants*

ALIEN AND INVASIVE ANIMALS DEFINED

For the purposes of this book we have considered alien animal species in South Africa to be those that have been deliberately or accidentally introduced through human intervention and that have subsequently established self-sustaining wild populations. Where aliens have spread significantly beyond their point of introduction, they are considered to be invasive.

Animal species that have established self-sustaining populations 'in the wild' fall into two distinct categories. These are (a) deliberately introduced species that have either been released directly into natural or disturbed habitats (e.g. Rainbow and Brown trout (pp. 61 and 62), Grey squirrel (p. 28) and Common starling (p. 41)), or that have escaped from culture and subsequently invaded natural habitats (e.g. Feral domestic cat, p. 23), and (b) accidentally imported species, such as rats, cockroaches and crop pests.

Both groups may cause a variety of problems, out-competing, consuming, parasitising or otherwise altering native animal and plant communities in the invaded habitat. To make matters worse, introduced species have often escaped from their natural predators, competitors and pathogens (which would control their population numbers in their home environment), allowing them to reach unnaturally high densities in their invaded habitat.

The scope of alien and invasive species in this book necessarily excludes: (1) animals native to South Africa that have undergone a human-facilitated range expansion (apart from a few fish species included due to their important ecological impacts); (2) the very large number of species that are only found in captivity; (3) farmed or cultured animals; (4) those that occasionally escape from captivity, but have not established breeding populations in the wild, e.g. a number of alien snake species kept as pets escape continuously into the wild, but none appear to have established; and (5) cryptogenic species, which are those whose native range is unknown or debatable, but whose very wide global distribution patterns suggest that they have been dispersed widely by humans and are therefore likely aliens.

Typical invasive species tend to share characteristics that facilitate their establishment in new habitats – these include wide physiological tolerances, and life-history traits such as high levels of fecundity and fast growth rates, which make them both good competitors and tolerant of human-altered habitats. Many such species, including mice, rats, cockroaches, land snails and silverfish, have become virtually global in distribution, resulting in what is sometimes referred to as the 'MacDonaldisation' of the global biota.

In general, one might expect that alien species introduced a long time ago would have the widest geographical ranges, and while this may often be true, there are certainly exceptions, such as the Giant garden slug (p. 198) and

M. Picker

The Giant garden slug reaches lengths of 20 cm.

Black River (Cape Town) choked by Water hyancinth, an aggressive invader that flourishes in nutrient-rich freshwater.

Common chaffinch (p. 38), which have persisted within very confined areas for a century or more since their original arrival.

Tens of thousands of deliberately or accidentally introduced plant and animal species have now established themselves in new habitats around the world, as the result of human activities. The impacts of these on the native biota is now recognised as the most serious global environmental problem after habitat destruction, and numerous academic and popular books have been written on this topic, some of which are listed in the section on 'Further reading' (p. 209).

No previous attempt has been made to list the alien and invasive animals known from South Africa, and although the 601 alien species listed here (Appendix, p. 222) represent the most comprehensive current listing to date, this still certainly remains an underestimate of the true number of alien animals actually present. This is both because many animal groups and potential introduction sites have not been adequately sampled, and because we lack the taxonomic knowledge to identify species accurately for many invertebrate groups. In contrast, the given inventory of alien vertebrates in South Africa is probably complete.

HISTORICAL PERSPECTIVE

Before human intervention, all species had geographically finite distributions: each 'region' on Earth was characterised by its own particular fauna and flora. These biogeographic differences were maintained by natural barriers that prevented the movement of species from one region to another. In this regard, freshwater species are particularly isolated in 'islands' of water, separated by impassable 'seas' of land, and hence often have very restricted ranges. For terrestrial species, barriers to dispersal include rivers, mountain ranges, deserts and oceans. Marine species might appear to live in a single, continuous ocean, but their distributions are also restricted; coastal species, for instance, cannot cross the open ocean between continents or islands, and oceanic species may be restricted to specific regions by thermal barriers, or the flow pattern of ocean currents.

However, ever since humans first developed modes of transportation that allowed them to cross natural biological barriers, particularly oceans, they have carried other species with them to regions that those species could never have reached without human intervention.

Redistribution of species would have started with the agricultural revolution and early spread of humans. When Europeans began to explore the Americas, large numbers of species began to be introduced in both directions. The number of non-native species introduced from Europe to North America, and vice versa, peaked in the mid-19th century, mostly due to global commerce.

Because of its colonial history and its strategic location on major trade routes, South Africa has been particularly prone to invasion by introduced species. In terms of animal introductions, there is little doubt that the first of these date back to the initial European explorers, who not only introduced domestic stock, but who would inadvertently also have carried rats, cockroaches, crop pests, human and livestock parasites and many other animal 'hitchhikers' to each of their ports of call.

More than 8 750 plant species are known to have been introduced to South Africa, of which 161 rank as serious pests, causing problems that include loss of biodiversity, loss of crop production, soil erosion and reduced freshwater runoff. Both deliberate and accidental animal introductions have continued to occur up to the present day – indeed, the rate of introduction is thought to be accelerating. The ecological and economic costs of these impacts have been estimated to exceed ZAR100 billion annually.

Many of these introductions have been deliberate and beneficial, at least to humans. For example, most of our food plants and domestic animals have been spread around the world to such an extent that we often forget that they are relatively newly introduced over

C. Griffiths

Felling of pine plantations on Table Mountain will enable the restoration of granite fynbos vegetation.

C. Griffiths

Importation of agricultural produce without proper phytosanitary control has allowed for the establishment of a wide range of alien agricultural pests.

C. Griffiths

The grape-like galls of the imported Acacia gall wasp reduce the seed set of these invasive woody trees.

much of their current range. It is difficult, now, to imagine Europe without potatoes, or Africa without maize, yet both are relatively recent deliberate human introductions from the New World. Similar introductions of domesticated plants and animals blanket the entire planet and underpin the dramatic recent growth of human populations, with its resulting environmental impacts.

A SOUTH AFRICAN PERSPECTIVE

Terrestrial introductions and vectors

The majority of invasive vertebrates have purposely been introduced into South Africa. Early explorers deliberately released various species, such as European rabbits (p. 29) on Robben Island, hoping that they would breed and be available for later use as food. However, the vast majority of early invertebrate introductions would have been accidental hitchhikers. These included soil species introduced with potted plants; crevice-inhabiting snails and woodlice carried in freight, food, timber and other trade goods; and pest species like rats, mice and cockroaches, introduced via ship. The majority of crop pests were introduced with seed, seedlings, fruit and potted plants; while domestic stock, and indeed human immigrants, would also have brought parasites and pathogens along as unwanted guests.

During the colonial era, numerous 'decorative' animal species were also deliberately introduced in order to 'enhance' the appearance of the colonies and make them more like 'home'. Examples include Grey squirrels (p. 28), European fallow deer (p. 32), Common starlings (p. 41) and Common chaffinch (p. 38), which were all imported from Europe by Cecil Rhodes. A much more recent development has been the release of many biocontrol agents, deliberately introduced to stem the spread of alien plant species, or less frequently to control alien animals.

Freshwater introductions and vectors

There are few opportunities for freshwater species to be transported accidentally between countries or continents, unless they are capable of surviving prolonged periods out of water. The majority of freshwater animal introductions have therefore been deliberate, or were accidentally introduced in association with deliberate introductions, as in the case of fish parasites. More recently, the aquarium or aquaculture trade has evolved as a new vector and several freshwater species have been deliberately or accidentally released into the wild by aquarists, or have escaped or spread from aquaculture facilities.

The earliest freshwater fish to be introduced, the Common carp (p. 52), was

SAIAB, O. Weyl

Brown trout have been introduced to most parts of South Africa for recreational angling.

C. Griffiths

Gribbles are small crustaceans that erode wooden pilings of piers and jetties as they feed on and burrow into the wood.

C. Griffiths

Shipworms are highly modified bivalve molluscs that can severely damage wooden piers and ship hulls.

imported as early as the 1700s by European settlers as a food species. Later it became the official policy of nature conservation authorities to enhance freshwater fisheries, resulting in a number of species, including trout and bass, being widely propagated for this purpose – a policy that was reversed only in the 1980s. Additional smaller fishes were imported as forage for these larger predatory species, or in the case of mosquito fish, for pest control. Accidental introductions include aquarium fishes and their associated parasites (some of which have spread to indigenous fish species) and a few free-living freshwater invertebrates, mostly snails originally imported by the aquarium trade.

Marine introductions and vectors

Ever since European explorers first landed in South Africa in the late 15th century, marine species have been continuously introduced into the region, along with international shipping activities. A few of these, such as oysters, were brought in intentionally, but the vast majority arrived by accident.

The mechanisms, or vectors, by which importation has taken place have varied greatly over time. The wooden vessels of early seafarers travelled slowly and were ideal habitats for external 'fouling' organisms, as well as for specialised species, such as Shipworms (p. 183) and Gribbles (p. 165), that tunnelled into the wood itself. Many early ships were also stabilised by dry ballast, in the form of sand and rocks. This was loaded at the port of origin and then dumped on arrival, along with any drift-line fauna and flora that had survived the journey. Modern steel vessels have become much larger, more numerous and traverse the world's oceans more rapidly. Although these vessels are impenetrable to borers and are painted with anti-fouling paint to discourage fouling organisms, they are often stabilised with ballast water. This has provided a new mechanism for the transport of an additional wave of alien species, particularly those that float or swim in the ballast water. It has been estimated that at any one time there are around 10 000 marine species being transported in ballast water around the world! Modern aquaculture has also been implicated as a further vector for a few recent introductions.

C. Griffiths

The Mediterranean mussel is South Africa's most widespread and prolific marine invader, smothering rocky shores along much of the coastline.

Some 80 marine species are currently confirmed to have been introduced to the region by these means, with an additional 40 suspected to be so, although even this is certainly a considerable underestimate. The most commonly introduced groups are attached fouling species, such as hydrozoans, tube-dwelling worms, small crustaceans, molluscs, bryozoans and sea squirts. Fortunately, most marine introductions remain restricted to sheltered harbours and estuaries, probably because conditions there most closely resemble those in their areas of origin. The open wave-exposed coast has been invaded by only a handful of species, although some of these now dominate the seashore, particularly that of the west coast.

COSTS AND CONSEQUENCES

Terrestrial invertebrate invaders

South Africa is home to many alien invertebrates including insects, snails and slugs, millipedes, worms, spiders and other less well-known groups. However, the impacts of only a few of these have been studied. The number of introduced animal species is high, although probably not as great as the 2 300 and 1 500 species of alien invertebrates known to have established in the USA and New Zealand respectively. Most of the accidentally introduced alien invertebrates in South Africa were introduced more than 100 years ago, although new introductions still take place constantly. Biocontrol agents have been purposefully introduced in more recent years in an attempt to control other alien species (mainly plants). Indeed, about half of all major weed species have been successfully impacted by biocontrol agents.

Alien animal species typically establish themselves in habitats dominated by alien plants (e.g. plantations, croplands and orchards, or weedy wastelands). Only a small proportion manage to establish in native vegetation. Narrow feeding preferences of herbivorous insects restricts their spread to plantations or croplands, where they may assume pest proportions and require management through the introduction of biocontrol agents. As an example, two parasitic biocontrol agents, the Woodwasp nematode worm (p. 173) and the Sirex ibaliid wasp (p. 141), have been introduced to control the Sirex woodwasp (p. 141) which impacts pine plantations by inoculating trees with fungal spores. Typically, biocontrol agents are small parasitic insects known to be very host-specific. After laboratory trials to confirm their host specificity, these can be released with a fair measure of confidence that they will not begin parasitising non-target, native hosts. Only very rarely have biocontrol agents attacked native (non-target) insect hosts, or insects introduced to control weeds have moved onto native plant species.

The most obvious economic costs associated with animal introductions are losses caused by crop pests. Indeed, of the 40 top crop pests in the country, 42% are introduced. The economic costs associated with introduced arthropods alone in South Africa are considerable, as most crops are attacked by these alien pests. Economically important examples include Red scale (p. 96), Codling moth (p. 120), and Potato tuber moth (p. 121), among many others.

B. Hurley

Female Sirex woodwasps threaten the commercial pine industry by injecting a fungus into pine trees that then kills affected trees.

C. Griffiths, M. Picker

Red scale is one of the most economically important plant pests globally, best controlled using natural enemies and parasites rather than insecticides.

Alien invertebrates with non-specialised diets (e.g. generalist carnivores and detritus feeders) tend to be more capable of establishing in native vegetation. For example, at least 17 species of alien terrestrial molluscs have been recorded from the Cape Peninsula alone. One of the most conspicuous of these is the Dune snail (p. 192), which established in the fynbos biome in 1881 and now forms dense clusters on plants along roadsides up the west and eastern coastal strips.

The ecological impact of most alien invertebrates in South Africa has not been evaluated. A recent introduction from Europe to the fynbos biome is the German wasp (p. 142), which in New Zealand has upset the ecological balance by competing with nectar-feeding birds for honeydew. Although climatic conditions in the fynbos biome are not ideal for this species, eco-climatic modelling indicates that wetter and cooler parts of South Africa offer ideal habitat for its establishment. The economic cost of eradication measures for

C. Griffiths

Along the west coast the Dune snail attains huge densities, aggregating on fence posts and plants.

this and other invasive aliens are considerable, but the impact of the German wasp has not yet been adequately estimated.

Arguably one of the most harmful alien insects worldwide, the Argentine ant (p. 140) was accidentally introduced into South Africa in about 1898. Although the ant has penetrated undisturbed fynbos and Afrotemperate forest in the Western Cape, it is more typically associated with paved roads, other forms of disturbance and moist drainage lines. The reliance of the Argentine ant on a constant supply of moisture may limit its spread into parts of the Western Cape, where soils become very dry in summer. Argentine ants are known to disrupt a number of mutualisms (mutually beneficial interactions between two species), including some in the Western Cape. They have negative impacts on crop production and crop quality by protecting harmful crop pests such as mealybugs, aphids and scale insects from predators and parasites. In the Western Cape this association with ants results in vineyards becoming heavily infested with mealybugs, negatively impacting the plants.

The Varroa mite (p. 158) recently crossed from its Asian honeybee (*Apis cerana*) host onto the Western honeybee (*Apis mellifera*). It is possibly the most serious pest to plague bees, often resulting in total destruction of the hive. In most parts of the world where it is invasive, it has resulted in the complete elimination of natural honeybee colonies. In the Cape Peninsula, it has extended its range well beyond the fynbos biome, although it appears as if there is some resistance to infection by native Cape honeybee (*Apis mellifera capensis*) colonies, as few have yet succumbed to infestations of the mite. It is therefore not yet known what level of threat this mite will pose to honeybee activity in the fynbos. Cape honeybees are among the most important pollinators of fynbos, and the possible extinction of wild honeybees would have major implications for pollination of many fynbos plants. It has been estimated that around 60% of all African plants are pollinated by African honeybees. There are few currently available control measures against the mite (apart from chemical treatment of domestic hives) and the selective breeding of resistant strains of honeybee might prove the only viable control method.

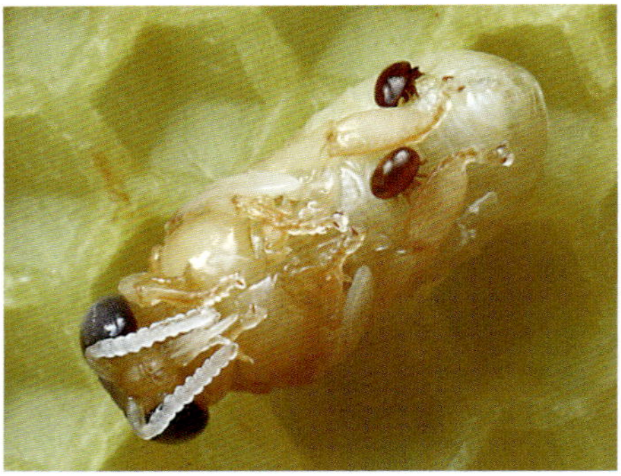

Waugsberg, Wikimedia Commons

In other parts of the world the Varroa mite has been implicated in the collapse of honeybee colonies.

Of course not all terrestrial animal introductions are harmful, indeed many were deliberately introduced for good reason and continue to perform their original function, for example, as biocontrol agents for crop pests. After careful screening, a few hundred 'agents' have been released – most to regulate their original host plants, which had become invasive in South Africa, and a smaller number to regulate alien insects that are crop pests. Even accidentally introduced species may be useful, such as earthworms and woodlice, which play an important function as decomposers of organic plant material. These positive functions, however, still need to be balanced against the negative impact such species may have on their indigenous competitors.

C. Griffiths

Although the Composting earthworm is commonly used in worm farms, it is also widespread in urban soils, but cannot survive in dry soils lacking adequate humus.

Freshwater invaders

The majority of freshwater introductions are fishes: either large predatory species such as carp, bass and trout, deliberately introduced to enhance angling, or smaller species introduced as forage fish, to control mosquitoes or for ornamental purposes. The main impact of these species has been to prey upon, and sometimes eliminate, smaller indigenous fish species. This is particularly a problem in the Western Cape, where almost all native fish species are range-restricted endemics. Only 19 native species of fish are currently recognised from the Cape Floristic Region, and of these 70% are listed as either Critically Endangered or Endangered, mainly due to predation by introduced species such as bass. Some introduced species may also indirectly cause an increase in the spread of diseases. Alien freshwater snails may serve as hosts for various parasitic worms (e.g. liver and intestinal flukes) that cause disease in humans. The Fish louse (p. 161) is an external parasite that can impact survival of both farmed and native fish populations.

C. Griffiths

The Vase tunicate is an important pest on west coast mussel farms, coating ropes and smothering the mussels.

C. Griffiths

Redfin minnows are small native fish with a narrow distribution range, and are very vulnerable to predation by alien fishes.

Marine invaders

Fortunately, relatively few South African marine introductions are known to have significant ecological or economic impacts, although the impacts of most species remain to be investigated. The Vase tunicate (p. 72) is an important fouling species that causes economic damage by growing over and smothering farmed mussels, themselves an introduced species. European shore crabs (p. 170) are voracious predators currently common only

C. Griffiths

High-pressure hoses are used to clean fouling organisms such as the Vase tunicate from culture ropes of the Mediterranean mussel.

C. Griffiths

Although a recent invader, the Pacific barnacle is already the most common barnacle on west coast rocky shores.

in harbour areas on the Cape Peninsula, but could decimate shellfish stocks if they spread to other sheltered sites, such as Saldanha Bay. The Pacific barnacle (p. 160) has become the dominant barnacle on the upper intertidal zone along much of the west coast, displacing some intertidal species, but providing habitat for others. The most significant introductions to date are those of the Mediterranean mussel (p. 181) and Bisexual mussel (p. 182), which together now blanket large sections of the west and, in the case of the Mediterranean mussel, south coasts. They have a multitude of ecological impacts, such as greatly increasing intertidal biomass, providing habitat for many small invertebrates, displacing large limpets and providing additional food resources for predatory species, including whelks, starfish and African black oystercatchers (*Haematopus moquini*). The Mediterranean mussel invasion also has direct economic benefits, as it forms the basis of a significant mussel culture industry, as well as a recreational harvest.

SOUTH AFRICA AS AN EXPORTER OF PROBLEM ANIMALS

Although the main text of this book deals only with invasions into South Africa, the region has, of course, also acted as an exporter of species to the rest of the world. In particular, the diversity and beauty of the endemic flora of the region has resulted in many native South African flowering plants being deliberately transported around the world as ornamental species. Not surprisingly, many of these have become invasive in their introduced habitats, particularly in Australia, where similar climatic conditions favour their establishment. Well-known examples of southern African plant invasives include Sour fig (*Carpobrotus edulis*), Arum lily (*Zantedeschia aethiopica*), Wild dagga (*Leonotis leonurus*), Tick berry (*Chrysanthemoides monilifera*) and Oxalis (*Oxalis* spp.). Among water plants, Waterblommetjie (*Aponogeton distachyos*) has also become a problem plant in Australia, New Zealand and parts of both Europe and South America, where it is disparagingly known as 'Cape pond weed'.

Far less has been written about African animals that have become problematic elsewhere. One of the most notorious of these is the East African land snail (*Achatina fulica*), which is recognised as one of the worst invasive species in the world and has devastated crops in introduced sites in Asia, various Pacific islands, the Caribbean, Florida and Queensland. Other terrestrial examples include the Big-headed ant (*Pheidole megacephala*), which is considered to originate from southern Africa, and is ranked as one of the world's worst invasive species, impacting native invertebrate communities. The South African Geranium bronze butterfly (*Cacyreus marshalli*) feeds on cultivated *Pelargonium* species in the UK and Mediterranean. The Brown widow spider (*Latrodectus geometricus*) is thought to have invaded North America via Australia and South Africa, while the common South African mantid (*Miomantis caffra*) was first reported in New Zealand in 1978. The Helmeted guineafowl (*Numida meleagris*) has been spread from central or southern Africa to the West Indies, Hawaii, Brazil and France, while other small South African finches, waxbills and the African Sacred ibis (*Threskiornis aethiopicus*) have also

M. Picker

Miomantis praying mantid, a South African native first recorded in New Zealand in 1978.

spread to other tropical parts of the world. More recently various species of South African plant-feeding insects have been introduced into Australia as biocontrol measures for South African plants that have established there.

Among aquatic species, the African clawed frog (*Xenopus laevis*) was dispersed worldwide during the period when it was used in pregnancy testing, and later as a model for developmental biology. Escaped populations naturalised in Europe and North and South America. The primary impact is not through the frogs themselves, although they are opportunistic carnivores, and therefore

C. Griffiths

The African clawed frog ('platanna') is thought to have carried the devastating chytrid fungus to many parts of the world, where it has resulted in the decline of various amphibian populations.

have the potential to impact other species (including other amphibians and fish). Far more devastating is the likelihood that they are responsible for the spread of the devastating fungus (*Batrachochytrium dendrobatidis*), which is now considered the major cause of decline in amphibian species around the world. In spite of this, regulated export of African clawed frogs from South Africa continues.

Several South African marine species have also spread elsewhere. These include the Four-tone nudibranch (*Godiva quadricolor*), which

G. Branch

The Four-tone nudibranch is a South African export that has established in Australia.

preys on anemones, and has been introduced to the Mediterranean and Western Australia, and the tiny bristleworm, *Terebrasabella heterouncinata*, which was introduced into abalone farms in California, and subsequently spread into the adjoining sea. Infestations of this burrowing worm in the shells of commercial abalone stocks causes shell deformation and slowed growth, resulting in substantial losses to the industry. The Californian population has now been eliminated, but only through destruction of the infected abalone stock and a labour-intensive programme to remove susceptible shellfish from the adjoining marine environment – one of just a few examples of successful eradications of marine invasive species

PREVENTION AND CONTROL

Some introduced species, notably biocontrol agents, have been introduced to perform specific tasks and are therefore deliberately dispersed widely in order to best perform that function. Many other species have established naturalised populations, but appear relatively benign. Such species would not be priorities for control or eradication, but their presence, at least in untransformed habitats, should still be discouraged, as most introduced species have at least some subtle effects on native species and community structure within invaded habitats. Prevention and control efforts therefore tend to be focused on known problem species.

The first rule of control is that it is much cheaper and easier to prevent alien species from arriving in the country than to eliminate them once they have become established – indeed the latter can be a hugely expensive and complex operation, and in many cases, impossible. The most widely used methods of preventing deliberate introductions are regulations prohibiting the importation of non-indigenous species. These are relatively easily enforced in island nations, where there are few ports of entry. Australia and New Zealand are particularly well known for their current tight restrictions prohibiting importation of animals and plant products, although unfortunately numerous introductions took place before these regulations were put into place. Preventing introductions into countries that have extensive and relatively porous land borders, such as South Africa, is much more problematic, but importation, culture and translocation of many potential problem species is restricted under the Alien and Invasive Species Regulations 2009 of the National Environmental Management: Biodiversity Act of 2004.

Accidental introductions are by their nature more difficult to control and require the establishment of inspection and eradication programmes. Regulations can therefore demand that imported goods, particularly live produce such as flowers and fruit, be fumigated, inspected or certified to be pest-free at the port of origin, or of arrival, or indeed both. Similar stringent regulations apply to importation of wildlife. Strict enforcement of such 'phytosanitary' regulations is an essential requirement in retarding the influx of alien species. Pest control measures, such as rat or insect traps, can also be concentrated around ports, border posts and other introduction sites. The fumigation of aircraft cabins, cleaning of ship hulls and mid-ocean exchange of ballast water are additional familiar examples imposed to restrict accidental movement of animal species.

Once a species has established, a number of management strategies can be imposed and these are conventionally divided into four groupings – eradication, containment, control and mitigation.

S. Gollasch

Ballast water release has initiated a new wave of planktonic invaders, as well as other marine species that have planktonic larvae.

D. Gudlindlu, NCC Environmental Services

House crows gather at an evening roost in eucalyptus trees at Nyanga, Cape Flats. Targeting roosts for eradication is much more effective than efforts directed at scattered individuals.

Eradication is generally the most desirable option, but it is also the most difficult to achieve, particularly if the species has spread significantly from its point of origin, as populations can often recover quickly from a mere handful of survivors. As an example, efforts to control 20-odd House crows (p. 40) when they were initially detected in the vicinity of Cape Town International Airport proved unsuccessful, and subsequent local spread and increase in numbers in the Cape Town metropole have resulted in increasingly expensive and complex, but as yet still unsuccessful, efforts to achieve complete eradication. Methods of eradication vary greatly depending on target species, but can include hand picking, trapping, hunting, poisoning and various forms of biocontrol, such as introduction of predators, parasites, pathogens or sterile males. Rapid detection and swift, decisive action are, however, important indicators of success. A major drawback of eradication programmes is that they can easily fail, in which case considerable cost and effort may have been wasted. This option should therefore not be attempted unless adequate manpower and resources are available and the chances of success are good. Successful eradication efforts in South Africa include the elimination of Feral domestic cats (p. 23) from Marion Island.

Containment refers to the restriction of a species within a particular region and can be effective against less mobile species. For example, invasive freshwater species can be relatively easily contained within particular catchments or water bodies. An extreme form of containment is exclusion, where a widespread species is specifically excluded from important conservation sites by fencing areas off and either eliminating the alien species from inside the fence, or preventing its entry into the fenced area. This is the preferred option for freshwater fishes in South Africa, where the conflict between the economic and recreational value of alien angling fishes and the risk they pose to native species can be resolved by allocating some catchments for conservation and allowing sport fishes to be introduced to others.

The control option involves suppression of the population below a threshold at which the damage caused is deemed acceptable. This is a widespread and pragmatic approach, widely used to manage agricultural and household pests. The downside is that it requires continuous effort and expense. The relentless battle against imported household rats, mice, cockroaches and booklice are familiar examples of this approach.

The last option, mitigation, involves learning to live with an invasion by minimising its impacts on natural species and ecosystems. One extreme mitigation measure is to translocate Endangered native species to new sites that are not invaded. For example, in Australia some mammals and birds now only survive because they have been established on rat-free islands off the coast. The majority of alien species are restricted to habitats impacted by humans, a further factor arguing for minimal environmental alteration.

HOW TO USE THIS BOOK

The aim of this book is to document the alien animals found within the political boundaries of South Africa, although the ranges of many of the species mentioned extend into adjoining countries, particularly Namibia, Zimbabwe and Mozambique. Since many of the species covered in the book have also been widely introduced around the world, these species accounts will also be useful to readers across a much wider geographic range.

This book deals with all groups of multicellular animals, from the most advanced (mammals) to the simplest (sponges). Bacteria, viruses and single-celled animals are excluded, although many of these have been introduced and have caused major economic, ecological and health problems. Plants and fungi are also excluded, but are covered in several excellent alien plant guides (see Further reading, p. 209). Species are arranged by major taxonomic group, starting with the most advanced and proceeding to those with simpler body plans. Apart from the vertebrates, it has not been possible to cover more than a handful of the most significant alien and invasive species in many of the groups, particularly for the large insect orders such as flies, beetles, moths and bugs, which include many alien species. The Appendix (p. 222) gives a more complete listing of known alien species in South Africa.

Common and scientific names

Both common and scientific ('Latin') names are given for each species, as well as taxonomic position (listed as Phylum, Class or Order and then Family). Many users find common names easier to pronounce and remember, but these names lack consistency – most obviously they are different in each language. Even within a single language the same species is frequently known by a variety of common names in different countries, or even regions within a country (for example, the pilchard of the west coast and the sardine of KwaZulu-Natal are one and the same fish). Moreover, groups of more obscure species often bear the same common name, while many smaller species have no common name at all. Scientific names, on the other hand, are unique to each and every species on earth and are globally consistent (even in Chinese, Arabic or Russian).

Each species is allocated to a Phylum (a group of organisms with a particular body plan), and within each phylum species are further divided into classes, orders and families, grouping them in progressively more detailed and more closely related structures. Finally, each species is given a unique two-part name or 'binomial', equivalent to the family and first name of a person; both words are written in italics. The first part of the binomial is the name of the genus, and the second, that of the species.

M. Picker

The assignment of a unique scientific name to all known species is essential to avoid ambiguity, and to facilitate archiving the diversity of life.

As an example, the full classification of the domestic dog is:

Phylum: Chordata
Class: Mammalia
Order: Carnivora
Family: Canidae
Genus: *Canis*
Species: *Canis lupus*

M. Picker

Structure of species entries

After the common and scientific name and classification of each species, and the **group** to which it belongs, each entry is divided into sections that deal with the **habitat** occupied by the species, notes on its **identification and biology**, its probable mode and date of **introduction and spread** and its ecological and economic **impacts**; notes on possible **control** measures are included, should these be known. Finally, **key references** are provided for each species.

Each of the 260-odd species entries is accompanied by two distribution maps. The **world map** shows the native range of the species in green, and also indicates other countries to which it has been introduced in tan-brown. These maps indicate presence/absence within a country only, as it is not possible to portray more accurate distributions within countries, given the scale of the world maps used. The **South African map** shows the known distribution range within this country. Few introduced species (other than vertebrates) have been adequately surveyed, therefore these maps need to be treated with some caution. They reflect the known/likely distribution at the time of publication of this book. Most species have fluctuating distribution ranges, and this is particularly true for alien species, where a single colonisation event usually deposits species in a very localised area from which point of establishment they spread progressively. However, certain alien species are known to have been introduced multiple times into a country. Ranges within South Africa are often based on a relatively small number of published or museum records, many of them collected many years ago. The maps are therefore most likely to under-report current distribution ranges. Inadequate surveying and ongoing spread (in some cases facilitated by climate change) impact the distribution of all alien species, which is more likely to be in a state of flux, compared with native species.

AUTHOR: P. HOCKEY

BIRDS Class Aves

As birds are widely kept as pets and semi-captive ornamentals, their potential for escape or release is high. Several species have been recorded as escapees in South Africa, e.g. Mandarin duck (*Aix galericulata*), Emerald starling (*Coccycolius iris*) and Red-crested cardinal (*Paroaria coronata*). There have also been deliberate introductions that have failed, including the Common pheasant (*Phasianus colchicus*) and Northern bobwhite (*Colinus virginianus*). Cecil Rhodes was particularly persistent in attempting to introduce aliens, favouring those that reminded him of England, but his attempted introductions of Rooks (*Corvus frugilegus*), Eurasian blackbirds (*Turdus merula*), Song thrushes (*T. philomelos*) and Common nightingales (*Luscinia megarhynchos*) all failed. He did, however, leave one major alien bird legacy, the Common starling (p. 41), and a minor one, the Common chaffinch (p. 38). Of the other eight aliens with self-sustaining wild populations, four were deliberately introduced, three escaped from captivity, and one, the House crow (p. 40), probably arrived by ship.

There have also been a few deliberate translocations of native species that have resulted in range extensions. Historically, for example, Helmeted guineafowl (*Numida meleagris*) did not occur in the western half of South Africa. Today, due to translocations and spread of agriculture, the species is widespread and abundant in the west. In 1933, the nominate race of Red-eyed dove (*Streptopelia semitorquata*) was introduced to Elgin, Western Cape, from Beira. The race *australis* was already present in the Elgin area, with the result that the Western Cape population is now a hybrid swarm. Not all such translocations have been successful, for example, Red-winged francolin (*Scleroptila levaillantii*) were introduced to the Stormberg in the Eastern Cape by wing-shooters, but died out soon afterwards.

Many species have also expanded their ranges due to indirect interactions with humans, some dramatically, such as Hadeda ibis (*Bostrychia hagedash*), Glossy ibis (*Plegadis falcinellus*) and Acacia pied barbet (*Tricholaema leucomelas*). Two factors driving these expansions are an increase in availability of water through provision of farm dams and irrigation, which has allowed the expansion of water-associated species, and establishment of plantations and wooded gardens, along with invasions of alien invasive trees, which have facilitated the spread of woodland species ranging from top predators like Black sparrowhawks (*Accipiter melanoleucus*) to nectar- and insect-feeding Amethyst sunbirds (*Chalcomitra amethystina*).

A family of Mallards; female left front, male right back

36 BIRDS Class Aves

Chukar partridge

Alectoris chukar

Group: Galliformes: Phasianidae

Habitat: Natural habitat sparsely vegetated, rocky mountain slopes. Local habitat stands of alien trees and adjacent strandveld. **Identification and biology:** Combination of plain, brown-grey upperparts, a white throat patch bordered by a bold, black line, and strongly barred, black-and-chestnut flanks diagnostic. Lives in groups of 5–15 birds, eating mostly plant material. Nests on the ground. **Introduction and spread:** Introduced from Eurasia to the Western Cape, Northern Cape and KwaZulu-Natal. The only surviving population is on Robben Island near Cape Town where six birds, confiscated by customs officials, were released in 1964. By the early 1970s the population had increased to about 500, falling to c. 300 by the early 1980s. **Impacts:** Probably negligible. **Control:** No control measures in place. There is no evidence that these birds have ever dispersed to the mainland.

Key reference: Hockey, P.A.R., Dean, W.R.J. & Ryan, P.G. (eds) 2005. *Roberts birds of southern Africa*. 7th ed. Cape Town: Trustees of the John Voelcker Bird Book Fund.

Common peacock

Pavo cristatus

Group: Galliformes: Phasianidae

Habitat: Natural habitat near water sources in dry, deciduous forests. Local habitat alien tree infestations, wooded suburbia and farmsteads. **Identification and biology:** Large size, head plumes, blue body and massive 'eyed' tail of male unmistakable. Female has whitish face, green neck, mottled upper breast and white lower breast and belly. Upperparts brownish-grey. Omnivorous. Largely sedentary, roosting communally in trees at night. **Introduction and spread:** Native to Indian subcontinent. Popular ornamental bird with many small, semi-feral populations. **Impacts:** On Heron Island, Australia, peacocks eat eggs and chicks of seabirds; unknown whether this occurs on Robben Island. **Control:** No control measures in place. Sedentary behaviour suggests that this species is unlikely to become a significant problem, and numbers could be contained easily.

Key reference: Hockey, P.A.R., Dean, W.R.J. & Ryan, P.G. (eds) 2005. *Roberts birds of southern Africa*. 7th ed. Cape Town: Trustees of the John Voelcker Bird Book Fund.

C. Griffiths

Class Aves BIRDS 37

MAMMALS

Class Mammalia

Compared with invertebrates, and even some vertebrates (such as birds), there are relatively few alien invasive mammal species in South Africa – a trend also evident globally. However, the species that have become invasive have economic, agricultural, health and environmental impacts that are probably more severe and wide-ranging than those of any other animal group. All are closely associated with humans, most being domesticated agricultural animals, pets, biocontrol agents or aesthetically pleasing species deliberately introduced outside their native habitats as humans colonised the far corners of the world. Others, particularly rats and mice, are commensal hitchhikers that are now as widespread as humans, and much more numerous.

Like other invasive animals, invasive mammals are ecological generalists able to adapt to a wide range of habitats and diets, with a high reproductive capacity allowing them to establish independent, self-sustaining populations very quickly that spread by natural dispersal and are difficult to control. Their impacts vary geographically but are often greatest in areas without natural competitors and predators, particularly on oceanic islands, but also human-transformed habitats where indigenous equivalents and potential predators have been largely reduced or eradicated.

The accounts that follow detail invasive mammals introduced to South Africa either deliberately or accidentally by humans, and that have established economically, agriculturally, environmentally or aesthetically significant feral (i.e. self-sustaining) populations in the country. Many of these species are included in the Top 100 Worst Invasive Species, and more detailed information on most local invasive mammals is available on the IUCN (International Union for Conservation of Nature) Global Invasive Species Programme database (www.issg.org/database).

Perhaps the greatest current threat to local biodiversity, however, is the ongoing translocation of indigenous ungulates outside of their natural ranges, to stock private nature reserves, eco-tourism resorts and game farms. Such translocations contaminate local gene pools through hybridisation, result in the homogenisation of local and regional communities, alter the habitat characteristics and functioning of local ecosystems, and impact negatively on ecologically sensitive habitats (such as the Cape Floristic Region) and their constituent species. Although this problem was recognised as far back as 1986, regulatory measures are only being formulated now. For more about this important problem, see Chown *et al.* (2009) in the 'Further reading' section (p. 209).

F. Theunisson

Resident herd of feral horses at Botrivier Lagoon in the Western Cape.

•Feral domestic cat

Felis catus

Group: Carnivora: Felidae

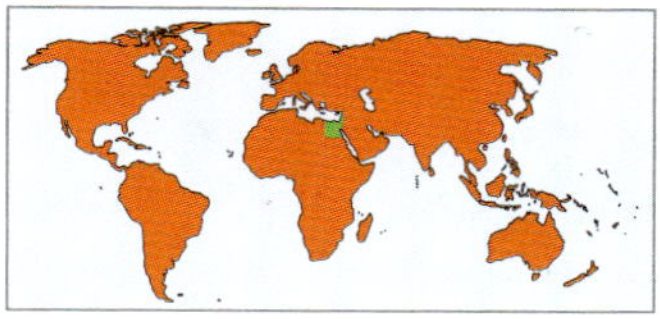

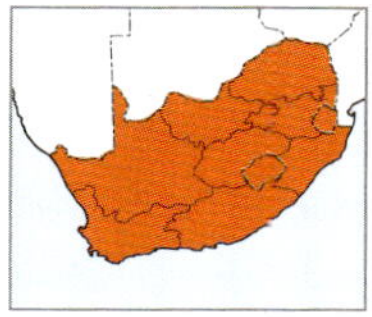

Habitat: Terrestrial, cosmopolitan. **Identification and biology:** A small carnivore weighing 1.5–3.0 kg when feral, but often heavier when domesticated. Body length approximately 46 cm. Unlike variable coloration of domestic breeds, feral cats commonly have black, ginger, tabby or tortoiseshell coats. Possibly confused with African wild cats (*Felis sylvestris lybica*), but these have distinctive reddish fur on the ears, hindlegs and abdomen, and black tail stripes. **Introduction and spread:** Domesticated in Egypt *c.* 3 000 years ago, from African wild cats. Translocated throughout Europe during Roman times, subsequently introduced to most parts of the world accidentally via ship, or deliberately by European colonisers. Occur in association with human settlements throughout South Africa. A large population emanating from five cats introduced onto the sub-Antarctic Marion Island to control house mice had devastating effects on resident breeding petrels, but was successfully eliminated by 1991. **Impacts:** Rated among the Top 100 Worst Invasive Species (IUCN Global Invasive Species Programme). Primary impact is predation on indigenous small vertebrates (birds, reptiles and mammals), causing localised decline or extinction of native populations, including several Endangered species. Predation impacts are greatest on populations of endemic and insular birds on oceanic islands, with the extinction of at least 33 species globally. On Dassen Island, predation on chicks severely impacted resident groundnesting seabirds in the 1980s, prompting an eradication programme that successfully exterminated resident cats by 2008. The only other offshore island with feral cats is Robben Island, where a control programme has been effective, but has not yet eliminated all individuals. Secondary impacts include competition for prey with indigenous predators, transmission of diseases (e.g. felid HIV) to other feline species, and contamination of African wild cat gene pools through interbreeding. **Contro :** Eradication programmes involving shooting, trapping, poisoning and/or inoculation with diseases (e.g. feline panleukopaenia (FLP) virus) have proved effective, especially on islands. More humane procedures based on trapping and subsequent release of sterilised cats, with regular supplementary feeding, are underway in some South African towns. These may control numbers and reduce hunting of indigenous prey, but sustainability of such programmes is tenuous given the expenses and labour involved. Non-surgical contraception is the most promising future control method.

Key reference: Robertson, S.A. 2007. A review of feral cat control. *Journal of Feline Medicine and Surgery*, 10:366-375.

A. Jones

House mouse

Mus musculus

Group: Rodentia: Muridae

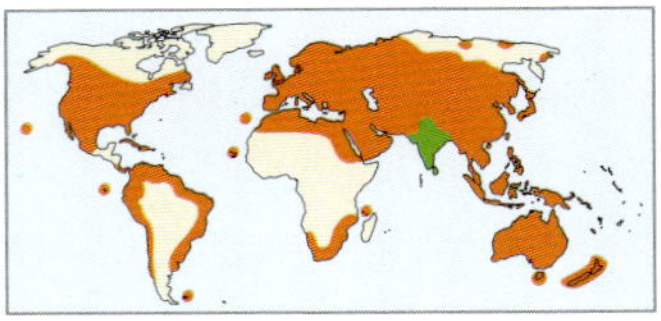

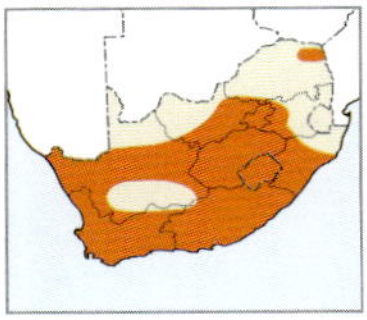

Habitat: Terrestrial, commensal with humans and the most widespread mammal species after *Homo sapiens*. Largely confined to human settlements, normally indoors, but also adjacent to buildings, in disturbed areas and cultivated lands. **Identification and biology:** A small rodent, body 65–95 mm long, tail similar length; adults weigh 12–30 g, usually about 20 g. Upperparts dark brownish-grey, underparts light buffy-brown. Size varies geographically, often larger on oceanic islands. Possibly confused with the Southern African pygmy mouse (*Mus minutoides*), which is smaller (length 70 mm, mass <10 g) with reddish-brown upperparts and white underparts. **Introduction and spread:** Originated on Indian subcontinent, spread throughout Eurasia to reach Europe 4 000 years ago, then transported worldwide by European colonisers. Spread facilitated by ships, vehicles, aircraft and freight carriers, with subsequent natural dispersal into local habitats. **Impacts:** Included by the IUCN in the Top 100 Worst Invasive Species. Major agricultural and economic pest, consuming and despoiling crops and human foodstuffs. In areas with few indigenous rodent competitors, such as southern Australia, sporadic population irruptions (>1 000/ha) may cause crop damages exceeding ZAR420 million. Impact on indigenous rodent species through competition for resources; vegetation communities as a result of foliage and seed consumption; and invertebrates through predation and habitat modification. Direct ecological impacts accentuated on oceanic islands; on Gough Island, House mice, which grow to three times the size of mainland counterparts, eat chicks of two Endangered seabirds. They host several pathogens and parasites that may infect humans, notably bubonic plague and salmonella, but are generally not important disease vectors. Domesticated House mice are used widely in biomedical research. **Control:** Population reduction methods (trapping, fumigation and poisoning) are used widely and effectively. Poisoning based on anti-coagulant rodenticides, preferably using bait boxes designed to exclude non-target species. Zinc phosphide applied aerially over large areas is used to counter imminent outbreaks of House mice in Australia. Integrated pest management programmes employ preventive measures (sanitation, repellents and mouse-proof construction) and population reduction techniques. Use of a genetically modified virus for immunocontraception of House mice in Australia is also being developed.

C. Griffiths

Key references: Berry, R.J. 1981. Town mouse, country mouse: adaptation and adaptability in *Mus domesticus (M. musculus domesticus). Mammal Review*, 11:91-136.

Wanless, R.M., Angel, A., Cuthbert, R.J., Hilton, G.M. & Ryan, P.G. 2007. Can predation by invasive mice drive seabird extinctions? *Biology Letters*, 3:241-244.

Brown rat

Rattus norvegicus

Group: Rodentia: Muridae

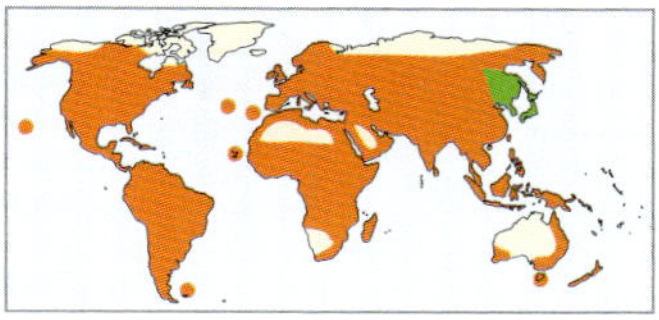

Habitat: Terrestrial, cosmopolitan, commensal with humans. Like the House rat (p. 26), it is associated with urban settlements and adjacent habitats, but tends to prefer more mesic habitats. Where both species occur together, Brown rats dominate in more built-up areas, and displace House rats to more peripheral urban habitats. **Identification and biology:** Similar to the House rat, but bulkier (150–300 g, sometimes up to 500 g) with relatively smaller ears, and a shorter tail (about 90% of head-body length) that is bicoloured (dark above and pale below) with a generous covering of bristly hairs. Body length 25 cm, tail about 22 cm long. Upperparts greyish-brown, flanks sometimes tinged with yellow, underparts paler with throat and chin being pale grey. Stronger and more aggressive than the House rat, which it tends to exclude. Nocturnal and often territorial. Omnivorous and opportunistic. Rarely climbs, unlike House rat. **Introduction and spread:** Native to north-eastern China, present in Europe by 1700, became the dominant rodent on ships which then transported the species globally. May swim up to 2 km to colonise islands. Date of introduction to South Africa uncertain, but documented in Cape Town by 1830. Found mainly in coastal cities and ports, especially associated with dockyards. Recent molecular studies have also detected this species in three localities in Gauteng near O.R. Tambo International Airport, suggesting that it has been introduced inland via airfreight. **Impacts:** See the House rat. Damage and impacts of these species are indistinguishable (especially where they occur together), and commensurate. Numbers of Brown rats are reaching significant pest proportions in the larger ports (e.g. Durban and Cape Town), where they pose a significant health risk in informal settlements, not only by despoiling foodstuffs and spreading diseases, but also directly by biting sleeping people. In one such incident in 2004, a physically handicapped child at a care centre on the Cape Flats died after Brown rats inflicted hundreds of bites on her body and consumed most of her face. 'White' rats (albino Brown rats) are sold widely as pets and used throughout the world for biomedical research. **Control:** See the House rat.

Key reference: Jackson, W.B. 1990. Rats and mice. In: Mallis, A. (ed.). *Handbook of pest control*. Cleveland, Ohio: Franzak & Foster, pp. 9-85.

Reg McKenna, Wikimedia Commons

House rat/Black rat/ Ship rat

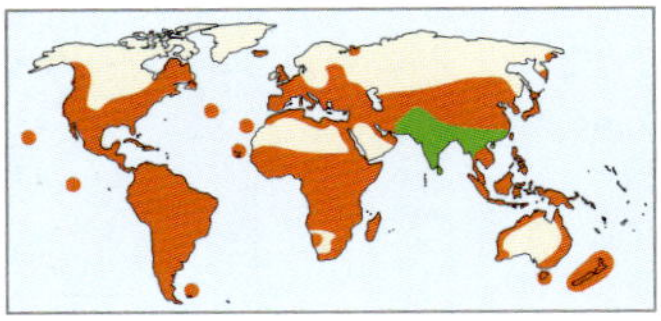

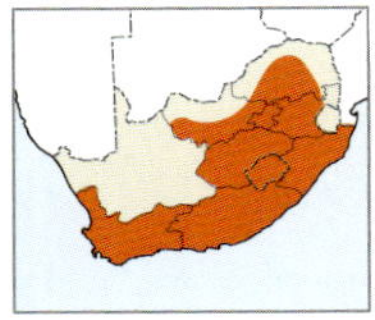

Rattus rattus

Group: Rodentia: Muridae

Habitat: Terrestrial, cosmopolitan, commensal with humans. Largely confined to human settlements and adjacent areas in wetter parts of South Africa. They have established some field populations where abundant vegetative or rocky cover is present, but seldom far from human dwellings and farm outbuildings. **Identification and biology:** A medium-sized rat, total length 34–40 cm, mass 105–184 g, tail 18–21 cm long (longer than head and body combined, and uniformly dark, hairless and coarsely scaled). Fur harsh, black to blackish-brown or greyish-brown above, underparts vary from slate-grey tinged with sepia to white or yellowish. Nocturnal and predominantly herbivorous, but opportunistic feeder, taking a wide variety of small animal prey. An agile climber and swimmer. Externally indistinguishable from the Asian house rat (*Rattus tanezumi*) which was detected in South Africa by molecular studies in 2005, and is currently known from seven localities in KwaZulu-Natal, Gauteng and Limpopo. The two species co-exist at several localities in north-eastern South Africa. The introductory pathway of this newly recognised species in South Africa is not yet clear, but probably similar to that of the House rat. The House rat may be confused in coastal areas with the Brown rat (p. 25), which is bulkier with a bristly tail that is shorter than head-body length and bicoloured (black above, whitish below), and relatively smaller ears. **Introduction and spread:** Six ancestral lineages, native to India and the Far East, have been identified but only two became invasive. The most widespread lineage originated in India, spread to the Middle East about 20 000 years ago, colonised the Mediterranean by 400 BC, and was later transported to the Americas, Africa and Australia by European colonisers. Another lineage, originating in Taiwan, colonised Japan, the Philippines and Indonesia. Introduced to South Africa at least twice, first

C. & T. Stuart

by AD 800, probably via Arab traders along the east coast of Africa; and from the 17th century onwards via European ships. Main introduction pathway via ship cargo, with movement inland via road freight, and spread to local habitats by natural dispersal. May swim distances of up to 500 m to colonise islands. **Impacts:** Included by the IUCN in the Top 100 Worst Invasive Species. Causes extensive damage (amounting to billions of US dollars annually) to crops and stored foodstuffs (particularly grains and nuts), often despoiling as much as they consume. Competition with humans for food may lead to malnutrition in rural subsistence communities. Responsible for infrastructural damage by gnawing (especially through cables), also pose a public health threat by contaminating foodstuffs with pathogens (such as salmonella), and are vectors of various pathogens causing bubonic plague, Weil's disease and toxoplasmosis. House rats can have marked impacts on plant communities through feeding (and also restrict regeneration by eating seeds and seedlings), and have contributed to range reduction or the local extinction of invertebrate and small vertebrate species. Like the House mouse (p. 24), the environmental impact of this species is greatest on oceanic islands, where they have had catastrophic effects on populations of endemic birds and other fauna. **Control:** Predominantly by poisoning, second-generation anti-coagulant rodenticides (e.g. bromadiolone and difethialone) are most effective; first-generation rodenticides (e.g. warfarin and diphacinone) still widely used but resistance has been noted. Poisons are best administered via bait boxes to reduce toxic effects on non-target species. Efficacy of chemical repellents is unclear. Fumigation is of limited use as House rats seldom dig burrows. Integrated pest management strategies (based on physical exclusion, early detection and monitoring of increasing rat populations), eradication programmes and management of refuge areas are effective in heavily infested areas. Preventive measures to halt colonisation remain the most effective way to protect uninvaded and cleared islands. Oral immunisation with contraceptive formulations is still experimental but shows promising results.

Key references: Jackson, W.B. 1990. Rats and mice. In: Mallis, A. (ed.). *Handbook of pest control*. Cleveland, Ohio: Franzak & Foster, pp. 9-85.

Taylor, P.J., Arntzen, L., Hayter, M., Iles, M., Frean, J. & Belmain, S. 2008. Understanding and managing sanitary risks due to rodent zoonoses in an African city: beyond the Boston Model. *Integrative Zoology*, 3:38-50.

A. Parker

•Grey squirrel

Sciurus carolinensis

Group: Rodentia: Sciuridae

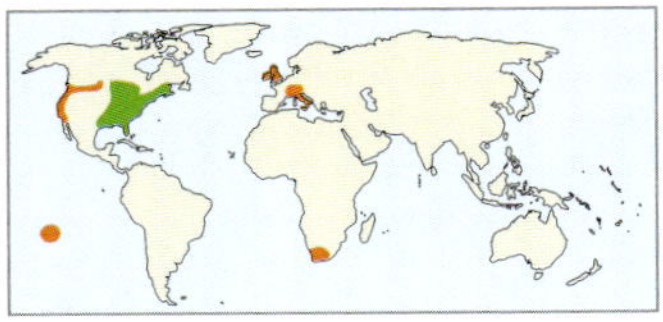

Habitat: Terrestrial, tree-dwelling. Present only in parts of the Western Cape. **Identification and biology:** A medium-sized tree squirrel (body length 38–50 cm, tail 15–25 cm, mass 480–650 g). Upperparts pale grey washed with cinnamon, underparts whitish. Agile climber. Restricted to areas with sufficient oak trees (*Quercus robur*) and/or pines (*Pinus* species), upon which it depends primarily for food. Avoids plantations and fynbos. Feeds mainly on acorns, pine nuts and other seeds, which it scatter-hoards during the summer. Also eats fruit, shoots, flowers, insects, fungi, eggs and rarely small vertebrates. Solitary, except for females with young. Active mostly in early mornings and late afternoons. During winter emerges from nest only every few days to feed. **Introduction and spread:** Native to deciduous forests of the USA, introduced to the UK, Ireland, Italy and South Africa. Cecil Rhodes introduced Grey squirrels to Groote Schuur Estate in the 1890s; by 1920 had spread across the Cape Flats, and reached the Hottentots Holland Mountains by 1930. Subsequently introduced to Swellendam (1957) and Ceres (1968). Distribution patchy, limited to gardens, parks, vineyards and farms where there is sufficient food-plant cover. **Impacts:** Included by the IUCN in the Top 100 Worst Invasive Species. In temperate European woodlands has caused dramatic declines in numbers of indigenous Red squirrels (*Sciurus vulgaris*) through competition and by transmitting the squirrel poxvirus. Other impacts include damage to woodlands by bark-stripping and declines in woodland bird populations. Impacts in South Africa are less severe and include damage to young trees, seedlings and ornamental plants; consumption of vegetables and fruit; and spoiling of thatched roofs. Considered a tourist attraction in Cape Town's Company Gardens although handling of Grey squirrels is discouraged as they are temperamental and can inflict severe bites. **Control:** Poisoning using anti-coagulants (especially warfarin) is the only cost-effective measure when densities are high. Other methods include shooting, bounty payments and trapping. In Britain, immunocontraception methods have thus far failed to reduce fertility of free-ranging squirrels.

Key reference: Gurnell, J., Wauters, L.A., Lurz, P.W.W. & Tosi, G. 2004. Alien species and interspecific competition: effects of introduced eastern grey squirrels on red squirrel population dynamics. *Journal of Animal Ecology*, 73:26-35.

C. Griffiths

European rabbit

Oryctolagus cuniculus

Group: Lagomorpha: Leporidae

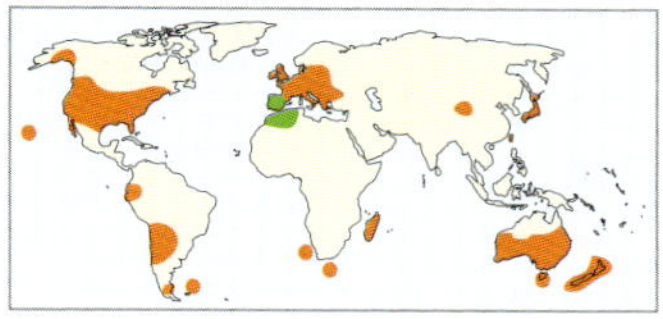

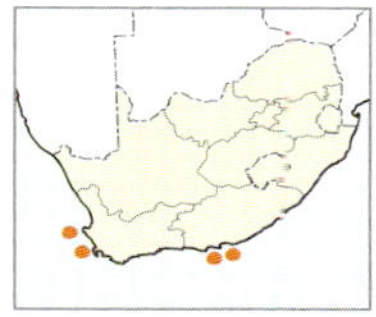

Habitat: Terrestrial, cosmopolitan. In South Africa, populations are restricted to seven offshore islands. **Identification and biology:** Smaller than indigenous hares (genus *Lepus*), body 34–45 cm long and weighs 1.0–2.2 kg. In domesticated forms colour varies from white through sandy to black, or mixtures thereof, but in feral populations it is commonly grey-brown above and whitish-grey below. Relative to the forelimbs, hindlegs are slightly shorter than in hares. Thrives in areas with annual rainfall of <1 000 mm, light soils and adequate plant cover. Nocturnal and crepuscular. Feeds on grasses and other herbaceous vegetation, which it crops closely, thus keeping feeding grounds heavily grazed. Highly gregarious and prolific breeder. **Introduction and spread:** Native to Iberian Peninsula, and domesticated in French monasteries between AD 600 and AD 1000. Colonies were established on most continents and many islands during the 17th–18th centuries, either deliberately as food for new settlements, or accidentally by escapees from stocks kept on ships. Introduced to Robben Island by the Dutch East India Company in 1654, as a source of meat for European settlers, later introduced to 10 other offshore islands. Accidental introductions on the South African mainland have surely taken place, but no feral populations have become established, probably because of depredation by indigenous predators. **Impacts:** Included by the IUCN in the Top 100 Worst Invasive Species. May cause severe damage by competing with indigenous small herbivores for food and shelter; reducing plant abundance and diversity through feeding; removing plant cover, thereby affecting communities of small animals; and facilitating soil erosion. In Australia it has directly contributed to local extinctions of some marsupials. At high densities may also damage crops and compete with livestock for food. **Control:** Methods include fencing, warren ripping, baiting, fumigating and biocontrol using fleas, myxomatosis and rabbit haemorrhagic disease virus. Acute poisons (e.g. cyanide, strychnine, sodium monofluoracetate) have been effective but adversely affect non-target predators. Shooting is an inefficient control method, but used on Robben Island so as not to impact on non-target species (nesting seabirds). Rabbits were also removed by shooting from Bird Island, Algoa Bay, in the 1980s.

Key reference: Flux, J.E.C. & Fullager, P.J. 1992. World distribution of the rabbit *Oryctolagus cuniculus* on islands. *Mammal Review*, 22:151-205.

K. Dunn

Feral domestic pig

Sus scrofa

Group: Artiodactyla: Suidae

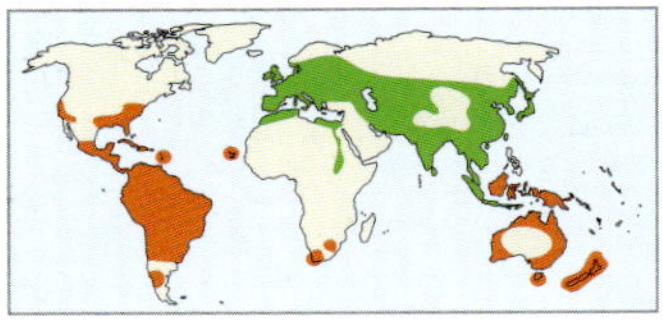

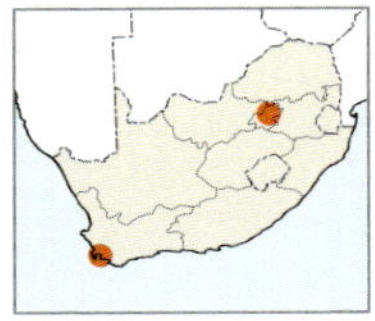

Habitat: Terrestrial, cosmopolitan. In South Africa, significant feral populations are restricted to the Boland, Overberg and west coast of the Western Cape. **Identification and biology:** Medium-large (body length 90–180 cm, shoulder height 55–110 cm, mass 60–200 kg), hoofed mammal with wedge-shaped head, pointed ears, flattened mobile snout and powerful body. Males generally larger than females, with four continually growing tusks. Body covered with a scant coat of coarse, bristly hair, adults often also have a scruffy mane. Tail not coiled, as in domestic pigs. Considerable variation in size and colour as a result of domestication and cross-breeding with Wild boars (the ancestral subspecies). Coat colour and patterns range from black through grey, brown, to whitish, often with spotted and belted combinations thereof. Their mobile snouts and sharp tusks are used to uproot prey or plant material. Omnivorous and aggressive. Unlikely to be confused with indigenous bushpigs (*Potamochoerus porcus*) as these species seldom occur together. **Introduction and spread:** The natural range extends throughout Eurasia as far south-east as the Malaysian Peninsula, Sumatra and Java. Domesticated from Wild boars independently at least seven times, starting about 9 000 years ago. Transported by Neolithic migrants, and later by seafarers who released them at many settlements, or on islands as potential food for marooned sailors. Expansion into new areas by natural dispersal from wild populations or from free-ranging commercial farms. Most populations in the Western Cape originated from domestic pigs that strayed from farms, but two populations were deliberately introduced by the Department of Forestry into pine plantations (Tulbagh 1926 and Franschhoek, 1941) to control severe infestations of the Pine emperor moth (*Nudaurelia cytherea cytherea*). Domestic pigs released at Tulbagh were subsequently augmented by four Austrian Wild boars from Groote Schuur Zoo released in 1930, and genetic throwbacks bearing longitudinal back stripes typical of Wild boars are sometimes encountered. Small groups of feral pigs undoubtedly occur on some farms

Richard Bartz, Munich Makro Freak, Wikimedia Commons

at other sites in South Africa, but usually do not persist, probably because of competitive exclusion by bushpigs and their susceptibility to swine fever. A small feral pig population originating from two Wild boars that escaped from a wildlife refuge has become established in the Pienaars River Valley near Pretoria, but long-term viability of this population remains uncertain. **Impacts:** Included by the IUCN in the Top 100 Worst Invasive Species. Cause extensive agricultural damage by feeding on crops, damaging pastures, preying on lambs, destroying infrastructure (especially fences), fouling water bodies, competing with livestock, and soil erosion associated with their rooting and wallowing. Rooting and trampling by pigs alters soil characteristics and seed banks, and creates open patches colonised by alien plants. This, together with their habit of feeding on tubers, has led to a decline or local extinction of some rare and Endangered plant species, changes in the diversity, composition and dynamics of natural plant communities and altering of nutrient cycles. Impacts on animal communities directly by preying on bird eggs, small mammals, reptiles, amphibians and soil invertebrates; and indirectly by modifying habitats and vegetation communities. Severity of their impacts varies spatially and seasonally, but is often greatest in wetlands, near water bodies and on islands. Pigs also pose a health risk, as they are vectors of various human and domestic livestock diseases, including brucellosis, leptospirosis, foot-and-mouth disease and swine fever.

In the Western Cape, Feral domestic pigs impact negatively on the few remaining patches of Critically Endangered lowland renosterveld by retarding succession on old fields, facilitating the spread of alien plants, feeding on geophytes (including some rare bulb species), and competing with small mammals (such as molerats and porcupines) that also feed on tubers. May also consume the Geometric tortoise (*Psammobates geometricus*), an Endangered species endemic to Renosterveld. **Control:** Poisoning with sodium monofluoracetate (1080), yellow phosphorus and warfarin is the most popular control method, but poses significant risks to non-target species. Hunting is effective but may be expensive on a large scale. Hunting and poisoning used together have successfully eradicated pigs on some large islands. Trapping is most useful when numbers are low, or in proximity to populated areas where baiting and shooting are too risky.

Key reference: Joubert, L. 2009. The invasion sounder: Eurasian wild pigs. In: Joubert, L. *Invaded: the biological invasion of South Africa*. Johannesburg: University of the Witwatersrand Press, pp. 153-158.

R. van der Walt

European fallow deer

Dama dama

Group: Artiodactyla: Cervidae

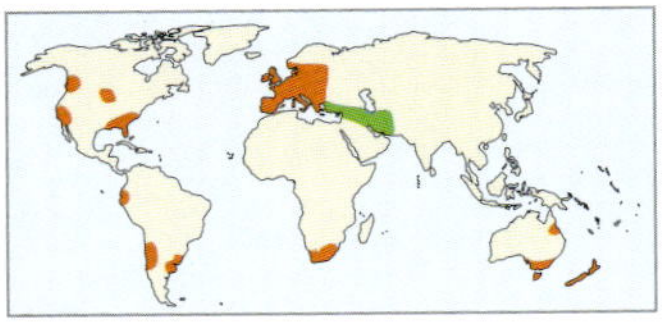

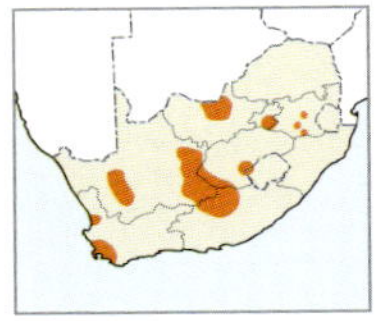

Habitat: Terrestrial, cosmopolitan. **Identification and biology:** Large deer, buck 140–160 cm long, 90–95 cm at shoulder and weighs 90–110 kg; doe smaller, 130–150 cm long, shoulder height about 80 cm and mass 80 kg. Many colour varieties, commonly tan or fawn with white stripes and spotting on the flanks and thighs, a broad black stripe down the mid-back, and a white rump patch. Male bears flattened antlers up to 70 cm long with numerous points. Catholic habitat requirements, typically oak or pine woodlands and open grasslands, but adapt to a wide range of vegetation types. Usually found in groups of three or four in dense habitats, but in larger groups in open country. Predominantly nocturnal, some activity at dawn and dusk. **Introduction and spread:** Native to Iran and Iraq, introduced widely to Europe during Roman times. Introduced several times from England into South Africa; first introduction uncertain, but present in Cape Town by 1869, and subsequently re-introduced by Cecil Rhodes in 1897. Animals from these stocks supplied many farms and estates in all provinces except Limpopo; feral populations are now widespread in these areas. They are expanding into dry, winter-rainfall habitats, which are similar to their native Mediterranean ecotype. Introduced to Robben Island in 1963. **Impacts:** As opportunistic feeders that both browse and graze, but also feed selectively, their impacts on natural vegetation can be severe. Sustained feeding changes vegetation communities, reducing diversity and abundance of plants when deer densities are high. On Table Mountain the proliferation of deer (in the absence of natural predators) impacted on flowering fynbos plants largely through trampling. Other impacts include fouling of waterholes, dispersal of weeds, ring-barking of trees and erosion resulting from overgrazing. They sometimes also damage crops. On Robben Island, numbers reached about 500 in 2009, and deer (together with about 25 000 European rabbits) had stripped nearly all edible vegetation. **Control:** Legislation now prohibits further introductions and regulates translocations of fallow deer within South Africa. Culling programmes currently underway in Table Mountain National Park, and on Robben Island, had eliminated 170 deer by December 2009.

Key reference: Feldhamer, G.A., Farris-Renner, K.C. & Barker, C.M. 1988. *Dama dama. Mammalian Species*, 317:1-8.

C. Griffiths

Feral goat

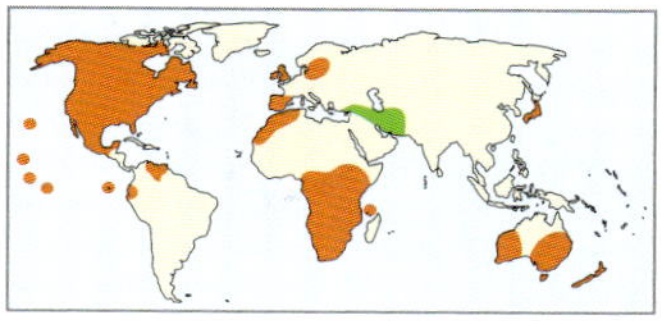

Capra hircus

Group: Artiodactyla: Caprinae

Habitat: Terrestrial, cosmopolitan. **Identification and biology:** Indistinguishable from domesticated goats, males 45–55 kg with beards, females 25–35 kg and beardless. Unicoloured (mostly black, to various shades of brown) or multi-coloured. Wide habitat and dietary tolerances. **Introduction and spread:** Domesticated in Iran highlands 10 000 years ago, subsequently spread throughout the world for meat and milk. **Impacts:** Included in IUCN Top 100 World's Worst Invasive Species. Degrade ecosystems through overgrazing and trampling, impacting fauna and flora. Size and impact of feral populations in South Africa largely undocumented and difficult to distinguish from those of free-ranging goats kept by pastoralists. Goat herbivory most damaging in Succulent Thicket (Eastern Cape), and still evident in areas now under conservation (such as Addo Elephant National Park). **Control:** Variety of hunting methods, including use of 'Judas goats' to detect feral herds. No co-ordinated control programme in South Africa.

Key reference: Parkes, J., Henzell, R. & Pickles, G. 1999. *Managing vertebrate pests: feral goats.* Canberra: Australian Government Publishing Service.

G. Hempson

Sambar deer

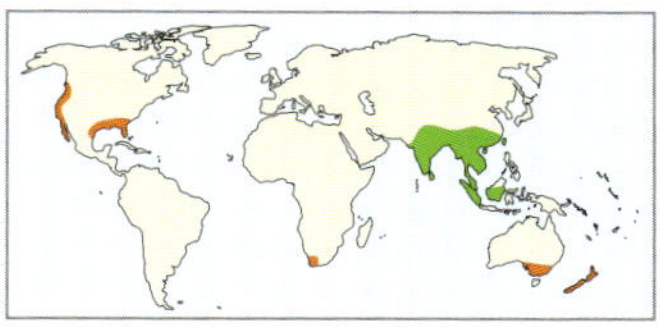

Rusa unicolor

Group: Artiodactyla: Cervidae

Habitat: Forests. **Identification and biology:** A large deer, 102–160 cm at the shoulder, weighing 160–260 kg, with large, forked antlers and a short, dense mane. Coat is dark brown with chestnut marks on the rump and underparts. **Introduction and spread:** Native to forests of south-eastern Asia. Introduced to Groote Schuur estate in the late 1800s, now confined to forested habitats (including pine plantations) on Table Mountain in the Orange Kloof area and above Hout Bay. Also introduced to various hunting and safari parks. **Impacts:** Unknown, but probably negligible as numbers (currently about 30) on Table Mountain are low. **Control:** A culling and relocation programme is underway in Table Mountain National Park.

Key reference: Timmins, R.J., Steinmetz, R., Sagar Baral, H., *et al.* 2008. *Rusa unicolor.* In: *IUCN Red List of Threatened Species.* Version 2010.4. www.iucnredlist.org

Paul Asman and Jill Lenoble, Wikimedia Commons

•Himalayan tahr

Hemitragus jemlahicus

Group: Artiodactyla: Caprinae

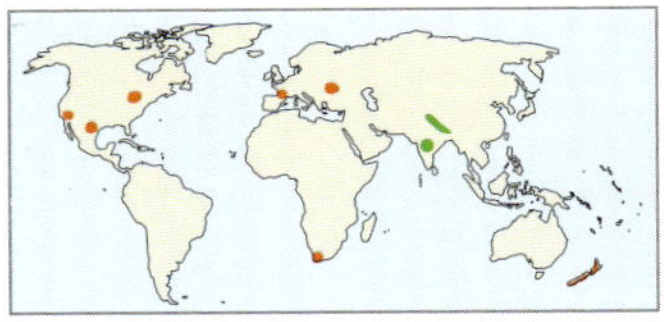

Habitat: Steep mountain slopes. **Identification and biology:** Goat-like; weight 135–180 kg, length 120–170 cm, height 60–90 cm, with dense reddish to dark brown fur. Both sexes horned. **Introduction and spread:** Native to rocky slopes of Himalayas. Introduced to New Zealand, USA and Canada for recreational hunting. In South Africa, Table Mountain population founded in the 1930s by a pair of zoo escapees. **Impacts:** Numbers increased to over 600 by mid-1970s, damaging montane fynbos, and causing soil and rock erosion, impacting resident Klipspringer (*Oreotragus oreotragus*). **Control:** South African National Parks Board initiated controversial culling programme, as translocation was not feasible given the rugged, inaccessible habitat. Culling ceased in 2005, and klipspringers have been successfully reintroduced, but sporadic anecdotal reports suggest that a few tahrs have survived.

Christian Jansky (User:Tschaensky), Wikimedia Commons

Key reference: Bhatnagar, Y.V. & Lovari, S. 2008. *Hemitragus jemlahicus*. In: *IUCN Red List of Threatened Species*. Version 2010.4. www.iucnredlist.org

Donkey

Equus asinus

Group: Perissodactyla: Equidae

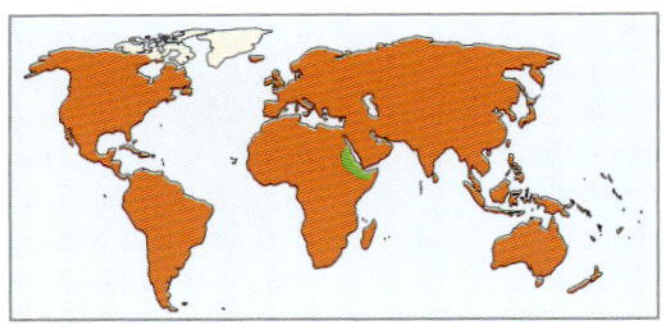

Habitat: Arid areas. **Identification and biology:** Resembles horse, with large head, long ears and cow-like tail; length about 200 cm, height 125 cm, weight variable, up to 250 kg. Colour black, through browns and grey, to white. **Introduction and spread:** Native to North African deserts and Arabian peninsula, transported widely by humans. **Impacts:** In drier regions, overgrazing, trampling of vegetation and soil erosion (Succulent Karoo), and competition with native herbivores. Feral populations likely exist in drier parts of South Africa, but poorly documented. Crossbreed with Cape mountain zebra (*Equus zebra zebra*), producing infertile mules. **Control:** Hunting, translocation and fertility control used worldwide. Feral population in the Ai-Ais/Richtersveld Transfrontier Park not considered a significant threat, but eradication programmes planned for the Greater Cederberg Biodiversity Corridor (Western Cape) and Baviaanskloof Cluster of nature reserves (Eastern Cape).

C. Griffiths

Key reference: Global Invasive Species Database, 2005. *Equus asinus*. www.issg.org/database/species/ecology.asp?si=639&fr=1&sts=&lang=EN

Feral horse

Equus ferus caballus

Group: Perissodactyla: Equidae

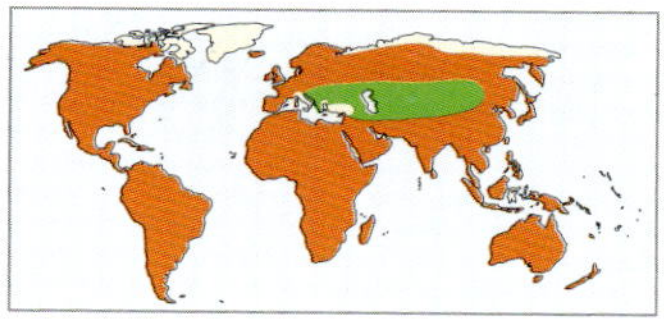

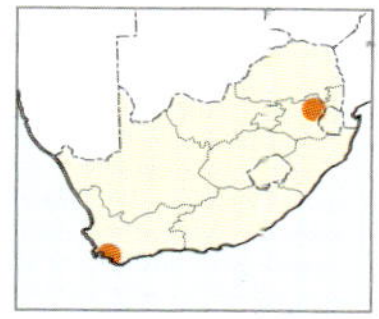

Habitat: Terrestrial, cosmopolitan. **Identification and biology:** Resembles the domestic horse, shoulder height 1–1.6 m, weight 350–450 kg. Coat short and fine, with a forelock and neck mane; colour varies from black, brown, tan and white through white with orange or brown patches. **Introduction and spread:** Originated in the open grasslands and steppes of Eurasia, transported worldwide by humans. Two known feral populations occur in South Africa, one in the Kaapsehoop district (Mpumalanga), and a smaller herd at the Botrivier Lagoon in the Kogelberg Biosphere Reserve near Kleinmond (Western Cape). The Kaapsehoop population numbers about 200 individuals, occurring in small bachelor herds (3–4 stallions) and structured herds (>20 individuals) that have roamed free over an area of about 17 000 hectares for many decades. Provenance of these horses is unclear, local legends being that they are descendants of mounts used by British or Boer forces stationed in the area during the Anglo-Boer War, or of horses released by local police when motorised transport became available. However, it is more probable that they are derived from a few escapee animals that evaded a roundup of work horses in the late 1960s and early 1970s. **Impacts:** The Kaapsehoop horses have not become invasive and numbers are regulated naturally by horse sickness. The Botriver herd grew to over 400 individuals by the 1930s and caused extensive agricultural damage, leading to an intensive eradication programme. The current Botriver herd of about 16 horses is descended from three individuals that survived the culling, and seldom leaves the reedbeds and vlei at the Botrivier Lagoon. **Control:** Control methods have included fertility control, relocation and (mostly) shooting.

Key reference: van der Merwe, F.J. 2007. *The Kleinmond wild horses*. Cape Town: Whale Printing.

F. van der Merwe

BIRDS

Class Aves

As birds are widely kept as pets and semi-captive ornamentals, their potential for escape or release is high. Several species have been recorded as escapees in South Africa, e.g. Mandarin duck (*Aix galericulata*), Emerald starling (*Coccycolius iris*) and Red-crested cardinal (*Paroaria coronata*). There have also been deliberate introductions that have failed, including the Common pheasant (*Phasianus colchicus*) and Northern bobwhite (*Colinus virginianus*). Cecil Rhodes was particularly persistent in attempting to introduce aliens, favouring those that reminded him of England, but his attempted introductions of Rooks (*Corvus frugilegus*), Eurasian blackbirds (*Turdus merula*), Song thrushes (*T. philomelos*) and Common nightingales (*Luscinia megarhynchos*) all failed. He did, however, leave one major alien bird legacy, the Common starling (p. 41), and a minor one, the Common chaffinch (p. 38). Of the other eight aliens with self-sustaining wild populations, four were deliberately introduced, three escaped from captivity, and one, the House crow (p. 40), probably arrived by ship.

There have also been a few deliberate translocations of native species that have resulted in range extensions. Historically, for example, Helmeted guineafowl (*Numida meleagris*) did not occur in the western half of South Africa. Today, due to translocations and spread of agriculture, the species is widespread and abundant in the west. In 1933, the nominate race of Red-eyed dove (*Streptopelia semitorquata*) was introduced to Elgin, Western Cape, from Beira. The race *australis* was already present in the Elgin area, with the result that the Western Cape population is now a hybrid swarm. Not all such translocations have been successful, for example, Red-winged francolin (*Scleroptila levaillantii*) were introduced to the Stormberg in the Eastern Cape by wing-shooters, but died out soon afterwards.

Many species have also expanded their ranges due to indirect interactions with humans, some dramatically, such as Hadeda ibis (*Bostrychia hagedash*), Glossy ibis (*Plegadis falcinellus*) and Acacia pied barbet (*Tricholaema leucomelas*). Two factors driving these expansions are an increase in availability of water through provision of farm dams and irrigation, which has allowed the expansion of water-associated species, and establishment of plantations and wooded gardens, along with invasions of alien invasive trees, which have facilitated the spread of woodland species ranging from top predators like Black sparrowhawks (*Accipiter melanoleucus*) to nectar- and insect-feeding Amethyst sunbirds (*Chalcomitra amethystina*).

C. Griffiths

A family of Mallards; female left front, male right back.

Chukar partridge

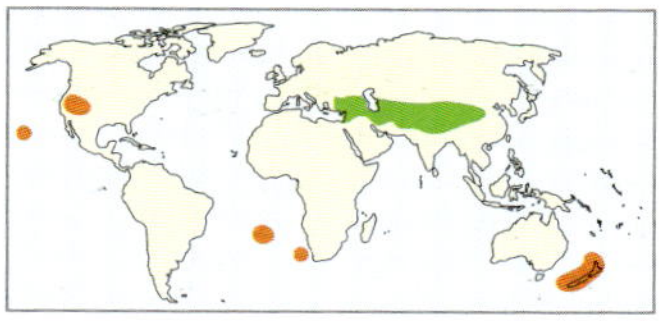

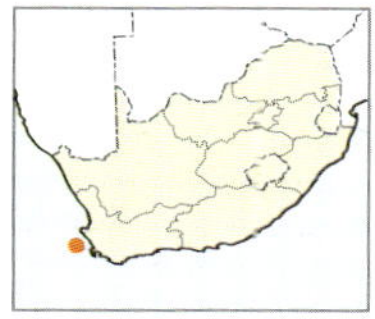

Alectoris chukar

Group: Galliformes: Phasianidae

Habitat: Natural habitat sparsely vegetated, rocky mountain slopes. Local habitat stands of alien trees and adjacent strandveld. **Identification and biology:** Combination of plain, brown-grey upperparts, a white throat patch bordered by a bold, black line, and strongly barred, black-and-chestnut flanks diagnostic. Lives in groups of 5–15 birds, eating mostly plant material. Nests on the ground. **Introduction and spread:** Introduced from Eurasia to the Western Cape, Northern Cape and KwaZulu-Natal. The only surviving population is on Robben Island near Cape Town where six birds, confiscated by customs officials, were released in 1964. By the early 1970s the population had increased to about 500, falling to *c.* 300 by the early 1980s. **Impacts:** Probably negligible. **Control:** No control measures in place. There is no evidence that these birds have ever dispersed to the mainland.

Karunakar Rayker from Occupied Kashmir, Wikimedia Commons

Key reference: Hockey, P.A.R., Dean, W.R.J. & Ryan, P.G. (eds). 2005. *Roberts birds of southern Africa.* 7th ed. Cape Town: Trustees of the John Voelcker Bird Book Fund.

•Common peacock

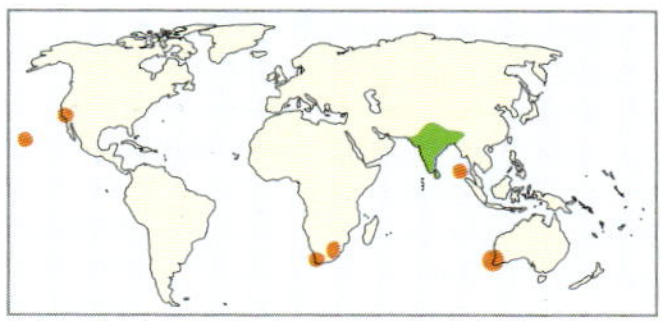

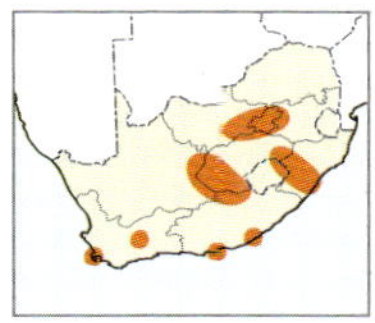

Pavo cristatus

Group: Galliformes: Phasianidae

Habitat: Natural habitat near water sources in dry, deciduous forests. Local habitat alien tree infestations, wooded suburbia and farmsteads. **Identification and biology:** Large size, head plumes, blue body and massive 'eyed' tail of male unmistakable. Female has whitish face, green neck, mottled upper breast and white lower breast and belly. Upperparts brownish-grey. Omnivorous. Largely sedentary, roosting communally in trees at night. **Introduction and spread:** Native to Indian subcontinent. Popular ornamental bird with many small, semi-feral populations. **Impacts:** On Heron Island, Australia, peacocks eat eggs and chicks of seabirds; unknown whether this occurs on Robben Island. **Control:** No control measures in place. Sedentary behaviour suggests that this species is unlikely to become a significant problem, and numbers could be contained easily.

C. Griffiths

Key reference: Hockey, P.A.R., Dean, W.R.J. & Ryan, P.G. (eds). 2005. *Roberts birds of southern Africa.* 7th ed. Cape Town: Trustees of the John Voelcker Bird Book Fund.

Rose-ringed parakeet

Psittacula krameri

Group: Psittaciformes: Psittacidae

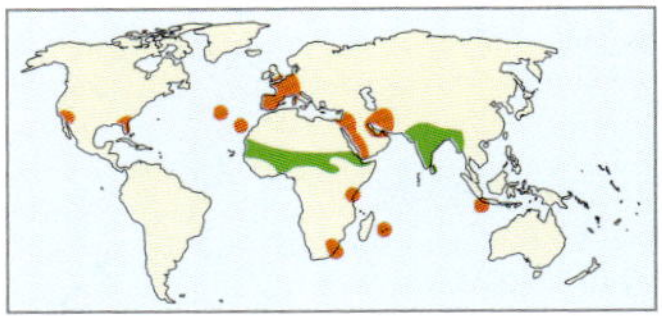

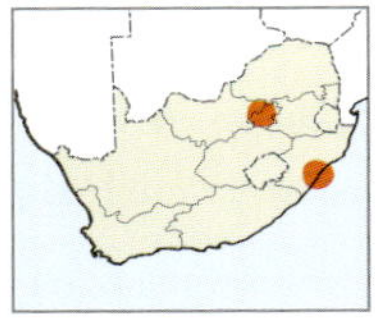

Habitat: Native habitat desert to riparian woodland and mangroves. Local habitat suburban parks and gardens. **Identification and biology:** The only long-tailed, green parrot in South Africa. Both sexes have red bills and bright yellow underwing coverts. Gregarious and noisy. Eats flowers, seeds and fruits. Nests in natural tree cavities, buildings, rock crevices and nest boxes. Usually lays 3–4 eggs. Roosts communally in large trees. **Introduction and spread:** Various subspecies native to central and North Africa, Indian subcontinent and Asia. Regular escapee from captivity (breeding recorded in Cape Town as early as the 1850s). Founder population for KwaZulu-Natal birds, now centred around Durban, believed to be birds released by campers in the province. Common in Durban by the 1970s, estimated 700–1 000 birds in early 2000s. Recorded breeding in Gauteng (probably a separate introduction) by mid-1980s. **Impacts:** May compete for nest sites with other hole-nesting bird species. **Control:** None.

C. Griffiths

Key reference: Hockey, P.A.R., Dean, W.R.J. & Ryan, P.G. (eds). 2005. *Roberts birds of southern Africa*. 7th ed. Cape Town: Trustees of the John Voelcker Bird Book Fund.

Common chaffinch

Fringilla coelebs

Group: Passeriformes: Fringillidae

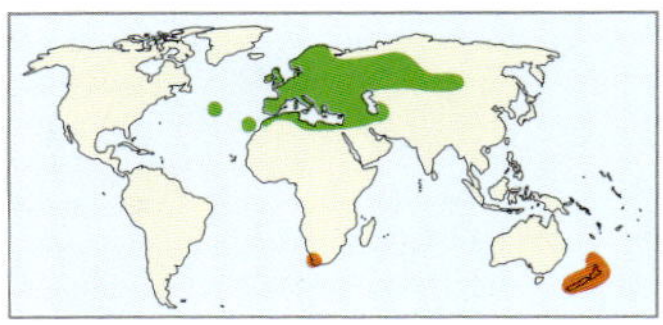

Habitat: Parks, gardens and alien woodlands. **Identification and biology:** A small species; male combination of pinkish face and underparts, grey crown and nape, chestnut back and white shoulder patch and wing bar diagnostic. Female plainer, paler and browner, but has the conspicuous shoulder patch and wing bar. Eats seeds and invertebrates. Occurs in pairs or small family groups; territorial when breeding in spring and early summer. Population estimated at 1 700–2 000 individuals. **Introduction and spread:** Native to Europe, North Africa and western Asia, it was introduced by Cecil Rhodes to Rondebosch, Cape Peninsula, in the 1890s. Remains largely confined to the Peninsula, mostly along eastern slopes of Table Mountain and adjacent lowlands. Recent felling of alien conifer plantations has caused a slight outward spread of the range into adjacent suburbs. **Impacts:** No evidence of negative impacts on native species. **Control:** No measures in place. Felling of alien plantations within the range expected to reduce population size.

P. Ryan

Key reference: Hockey, P.A.R., Dean, W.R.J. & Ryan, P.G. (eds). 2005. *Roberts birds of southern Africa*. 7th ed. Cape Town: Trustees of the John Voelcker Bird Book Fund.

• Rock dove/Feral pigeon

Columba livia

Group: Columbiformes: Columbidae

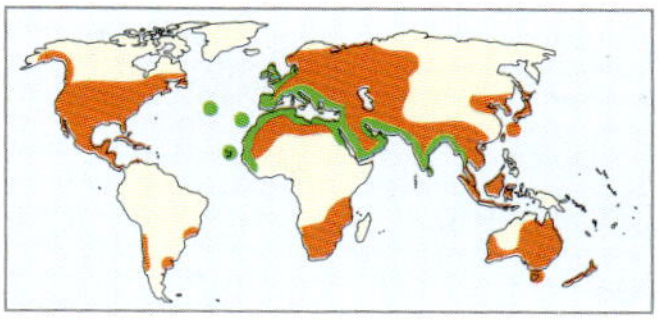

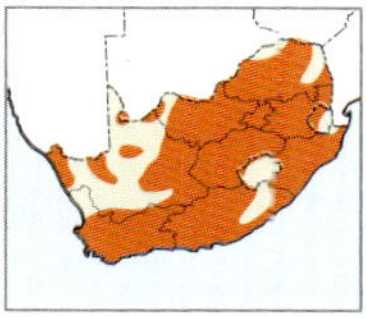

Habitat: Urban areas and cultivated fields. **Identification and biology:** Highly variable in appearance due to hybridisation with both domestic pigeons and native Speckled pigeons (*Columba guinea*). Original wild type has iridescent purple-green head, neck and upper breast, pale blue-grey upperparts and a white band across the upper rump. Two black bars visible on folded wing. Underparts slate grey, legs and feet deep red. Hybrids range from largely black to largely white. Social, often in flocks of hundreds, regularly with Speckled pigeons. Breeds year-round. Nests mostly on ledges of buildings, rarely on cliffs; clutch size 1–2. Feeds mostly on seeds and grain, but much food derived from human discards, such as bread. Probably has closer contact with humans than any other bird species in South Africa. **Introduction and spread:** Native to parts of Europe, North Africa and South Asia. Many introductions documented worldwide, from southern Canada and the Americas east to New Zealand. Date of introduction in South Africa uncertain, but present by mid-1800s. Now in towns and villages throughout the country, with lowest densities in arid north-west. **Impacts:** Diverse, including fouling of buildings, vehicles, grain stores, etc. with their droppings, and potential transmission of diseases, including histoplasmosis and possibly encephalitis. Regularly forages in animal feed lots, causing competition with agricultural sector, and significant consumers of grain stored for human consumption. In Cape Town consume 2–3 tonnes of stored grain daily. Also eat fresh green shoots of seed crops, placing them in further competition with agriculture. **Control:** No official control measures implemented to date, although losses from grain stores in Cape Town may lead to such measures being taken in future. However, given the species' extensive South African range and regular supplementation of the population with escaped domestic birds, any control measures are likely to become ongoing operations. One of the major challenges to control efforts is its close association with humans.

Key reference: Hockey, P.A.R., Dean, W.R.J. & Ryan, P.G. (eds). 2005. *Roberts birds of southern Africa*. 7th ed. Cape Town: Trustees of the John Voelcker Bird Book Fund.

House crow

Corvus splendens

Group: Passeriformes: Corvidae

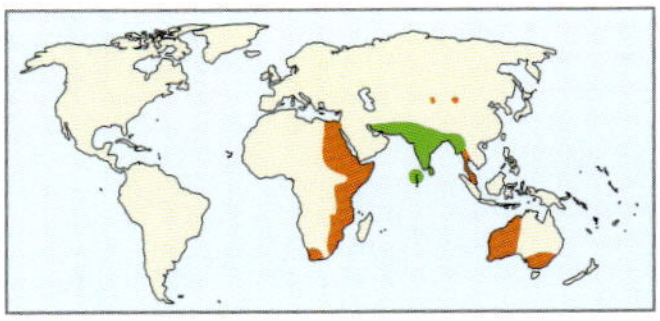

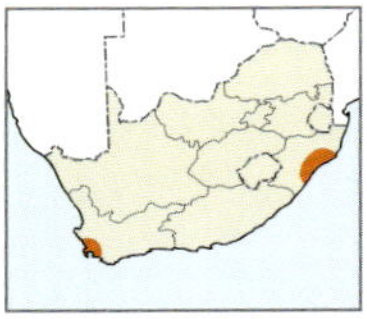

Habitat: Towns and cities.

Identification and biology: The smallest crow in South Africa. Glossy black, with grey or grey-brown nape, mantle and breast. Bill, legs and feet black. Gregarious, splits up into pairs when breeding. Roosts communally in large trees. Omnivorous, but eats mostly animal prey, especially eggs and nestlings of other birds. Also eats small mammals, fish, amphibians and invertebrates and steals food from other birds. Nests singly or in loose colonies. Builds a substantial nest, usually in a tree, rarely on buildings. Lays 3–5 eggs; in Durban, average of 1.3 fledglings/nest. **Introduction and spread:** Native to India and surrounds; widely introduced to tropical and subtropical mainland and island sites from eastern and southern Africa east to southern Australia. First recorded in South Africa in Durban between the late 1960s and early 1970s, possibly having spread from Zanzibar. Established in Cape Town in the early 1990s, perhaps from ship-assisted birds from Durban; founder population was fewer than 20 birds. Still confined to these two centres and Richards Bay. **Impacts:** Preys on eggs and nestlings of native birds, also eats small native animals. Mobs humans and pets. Potential air-strike hazard at Cape Town International Airport due to a nearby roost, which has now been removed. Occasionally destroys vegetable gardens in informal settlements. Vector for pathogens that cause cholera, typhoid, dysentery and salmonella poisoning. **Control:** By the late 1980s there were several thousand House crows in Durban and control measures (poisons) were introduced, reducing the population to about 150. Controls were then relaxed and the population rapidly returned to several thousand, prompting new eradication efforts. By late 2009 the population was less than 10 birds. Some failed efforts were made to control the founder population in Cape Town and large costs are now being incurred to control the resultant population, whose range has increased. Since 2007, c. 5 700 crows have been removed, but c. 8 000 birds remain. Plans target the species' eradication from Cape Town by 2015. Favoured control method is the avicide DRC 1339 (Starlicide).

Key reference: Allan, D.G. & Davies, G.B. 2005. Breeding biology of House Crows (*Corvus splendens*) in Durban, South Africa. *Ostrich*, 76:21-31.

C. Griffiths

Common starling

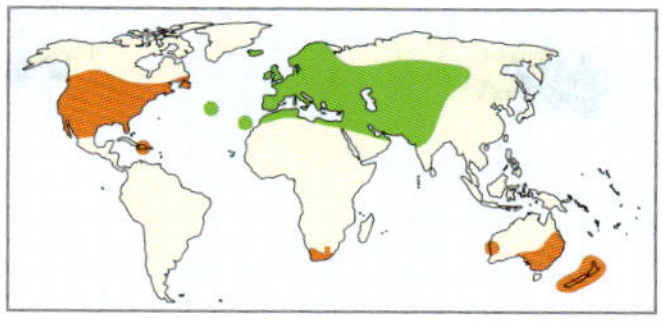

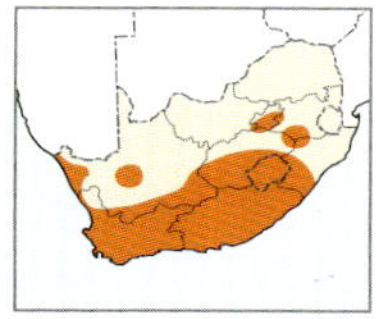

Sturnus vulgaris

Group: Passeriformes: Sturnidae

Habitat: Urban areas and agricultural lands. **Identification and biology:** Overall glossy black, with purple and green iridescence. Upperparts and underparts flecked buff and white, giving spotted appearance. Bill yellow, legs and feet red-brown. Sedentary and gregarious. Forages by day in small groups that coalesce at night to form large, communal roosts, often in reedbeds. Eats invertebrates, frogs, lizards, fruit, seeds, nectar and human discards. A hole nester, breeding singly or in loose colonies. Frequently nests in buildings, less often in natural cavities (sometimes evicting native species). Clutches contain up to six eggs. **Introduction and spread:** A highly successful invasive, native to temperate Europe and western Asia. Two introductions in New York at end of 19th century resulted in colonisation of most of North America and southern Canada. Within 25 years of arrival in California, one roost was estimated to contain five million birds. The South African founder population comprised 18 birds released by Cecil Rhodes at Groote Schuur in Cape Town in 1889. By 1910 they had reached Stellenbosch and by 1922 had crossed the adjacent mountains. By the mid-1990s they had spread north through the western lowlands to Oranjemund, Namibia, and eastwards to southern KwaZulu-Natal. Since then the range has expanded northwards to central and eastern Free State, and first colonists have reached Gauteng and Mpumalanga, resulting in increasing overlap with distribution of the closely related (alien) Common myna (p. 42), which utilises the same habitats and foods. **Impacts:** In south-eastern Australia, implicated in decreases in numbers of native species, especially parrots, probably due in part to competition for nest sites. A major pest of soft-fruit crops in many parts of its non-native range. Roof-nesting leads to fouling and introductions of ectoparasites, especially lice; implicated in the transmission of histoplasmosis. **Control:** Difficult to control. Over three years in the mid-1960s in Solamo County, California, more than nine million Common starlings were killed, but this did not eradicate the local population. No systematic control efforts in South Africa.

Key reference: Hockey, P.A.R., Dean, W.R.J. & Ryan, P.G. (eds). 2005. *Roberts birds of southern Africa*. 7th ed. Cape Town: Trustees of the John Voelcker Bird Book Fund.

C. Griffiths

Common myna

Acridotheres tristis

Group: Passeriformes: Sturnidae

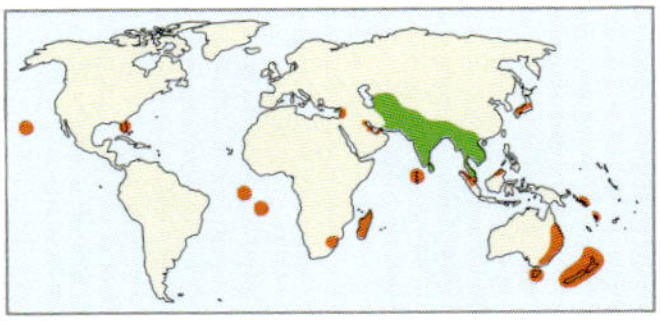

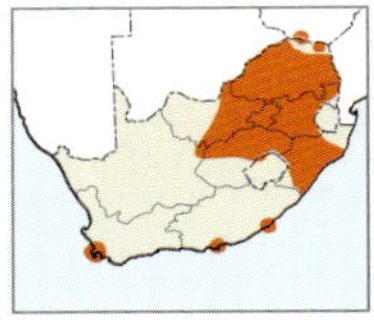

Habitat: Urban areas and farmland close to buildings. **Identification and biology:** Overall dark, blackish-brown colour contrasts with yellow bill, legs and feet, patch of bare yellow skin around eyes, conspicuous white wing patches (visible in flight) and white belly. Highly commensal with humans. Mostly gregarious, foraging in small groups by day, but roosting at night in large aggregations, often numbering thousands. Eats a mixture of plant and animal foods, including fruits, seeds, nectar and invertebrates. Also regularly eats human discards. Highly territorial when breeding, even chasing other bird species from the vicinity of nests. Usually nests on man-made structures: an untidy agglomeration of materials including cloth, twine and plastics. Occasionally nests in tree cavities and may oust native species to do so. Breeds throughout the year, laying up to six eggs in a clutch. **Introduction and spread:** Introduced from Asia to numerous localities worldwide, sometimes as an insect-control agent. A high proportion of introductions have been to islands, where mynas have been implicated in disease transfer, crop damage and predation of seabird eggs. The subspecies *tristoides* was introduced to Durban in 1888 and 1890, and the nominate subspecies was introduced to Johannesburg in 1938. Now widespread in the east of the country, extending west to Kimberley area. Population in Durban alone estimated in hundreds of thousands. Recent records from East London and Port Elizabeth may represent westward spread from KwaZulu-Natal, while occasional records and breeding attempts in Cape Town are probably of ship-assisted birds. **Impacts:** Probably minimal in natural and rural habitats. In urban areas may compete with native species for food and nest sites, and cause damage by fouling, especially at roosts and nest sites. **Control:** Large-scale control probably impossible, not least because of the species' close association with humans and urban landscapes. In Cape Town, occasional breeding pairs have been removed.

Key reference: Hockey, P.A.R., Dean, W.R.J. & Ryan, P.G. (eds). 2005. *Roberts birds of southern Africa*. 7th ed. Cape Town: Trustees of the John Voelcker Bird Book Fund.

P. Ryan

Mallard

Anas platyrhynchos

Group: Anseriformes: Anatidae

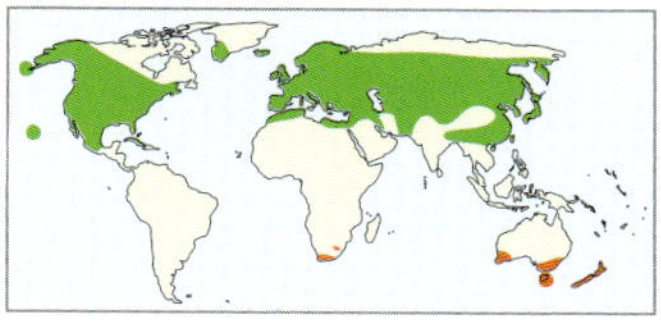

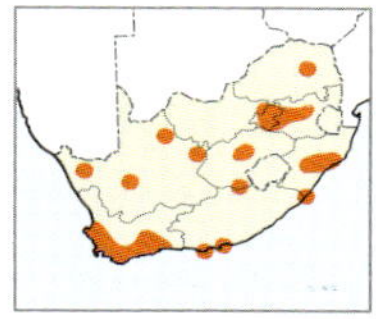

Habitat: Vegetated wetlands, including estuaries and lagoons. **Identification and biology:** Male's combination of yellow bill and bottle-green head and neck separated from chestnut breast by a narrow white band is diagnostic. Female's combination of grey bill and purple speculum shared only with African black duck (*Anas sparsa*), but latter much darker, with conspicuous white spotting above. A popular ornamental duck that regularly lives in close association with humans and is tolerant of high levels of human disturbance. Found singly or in pairs; sometimes with other waterfowl, especially Yellow-billed duck (*A. undulata*). Found in wide variety of wetlands, including man-made habitats, such as sewage plants, favouring wetlands that are well vegetated. Eats both vegetable (seeds, leaves, roots) and animal (molluscs, crustaceans, amphibians) material. Males aggressive during breeding season, mainly in guarding their mates against sneaky copulations by other males. Clutches large (European average 12.6; maximum 18). Nest built by female, who also undertakes all incubation (lasting about four weeks), covering the eggs with down when leaving nest. **Introduction and spread:** Native to most of northern hemisphere. Not deliberately introduced, but escapees have established feral, self-sustaining populations, primarily in the Western Cape and Gauteng, with regular escapees reported elsewhere. **Impacts:** Primary impact is hybridisation with native *Anas* ducks. Hybridises with Yellow-billed duck (regularly across its range) and Cape shoveler (*A. smithii*) in KwaZulu-Natal; suspected to hybridise with African black duck in KwaZulu-Natal. Extent of hybridisation difficult to quantify because female offspring of hybrid pairs not easily identified. Worldwide, hybrids recorded with *c.* 40 other waterfowl species. In New Zealand, may hybridise the native Pacific black duck (*A. superciliosa*) to extinction. Suggestions that it could hybridise South African Yellow-billed duck to extinction are probably overstated. **Control:** Localised control measures implemented in Western Cape, Gauteng, Mpumalanga and KwaZulu-Natal; a national control strategy is being developed. In Greater Cape Town >500 have been removed, mostly using the poison alpha-chlorolose, but >1 000 remain. Control measures are hampered by public resistance.

Key reference: Hockey, P.A.R., Dean, W.R.J. & Ryan, P.G. (eds). 2005. *Roberts birds of southern Africa*. 7th ed. Cape Town: Trustees of the John Voelcker Bird Book Fund.

C. Griffiths

House sparrow

Passer domesticus

Group: Passeriformes: Passeridae

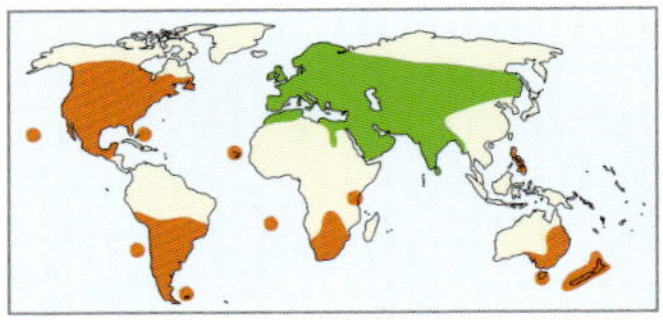

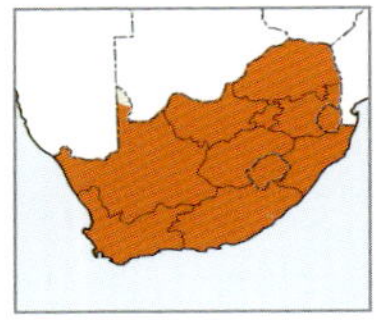

Habitat: Human settlements, from urban to rural. **Identification and biology:** Small, chunky and finch-like. Male has predominantly chestnut hind-neck, nape, mantle and upper wings, contrasting with grey crown and rump, white cheek patches and a bold, black bib. Underparts pale grey. Female paler, browner and lacks conspicuous markings. Distinguished from similar Great sparrow (*Passer motitensis*) by having a grey, not chestnut, rump. Gregarious outside breeding season. Eats seeds, flowers, nectar and invertebrates. Breeds year-round, mostly spring and early summer. Nests singly or in loose colonies: untidy balls of mixed materials, including grass, wool and feathers, usually with a side entrance. Favoured nesting sites are holes in buildings, but sometimes takes over nests of other species, especially swallows. Clutches contain up to six eggs, averaging 3–4. A single female can breed four times a season, producing as many as 18 young. **Introduction and spread:** Introduced from its native Europe, Asia and the Mediterranean to North and South America, southern Africa and Australasia, as well as to many associated islands. In South Africa there were two introductions to Durban (1893 and 1897) and one to East London (1927). For the first 50 years post-introduction the birds did not spread far, but by 1949 had reached Gauteng. These same founder populations have been responsible for spread as far north as the southern Democratic Republic of Congo. **Impacts:** Although common and widespread, the impacts on natural habitats are minimal, although, as elsewhere, they undoubtedly have some impact on agriculture, especially grain and fruit crops. Cause fouling on buildings, and their nests sometimes block gutters and drainpipes. Implicated in the transfer of poultry parasites, including cestodes and nematodes. There is some suggestion that numbers and range may be decreasing, especially around major urban centres; recent decreases have been reported elsewhere, noticeably in Great Britain. **Control:** No systematic control efforts in place in South Africa, and the species' distribution throughout the country, coupled with its association with humans, would probably render any such efforts ineffective.

P. Ryan

P. Ryan

Key reference: Hockey, P.A.R., Dean, W.R.J. & Ryan, P.G. (eds). 2005. *Roberts birds of southern Africa*. 7th ed. Cape Town: Trustees of the John Voelcker Bird Book Fund.

REPTILES

Class Reptilia

Only one exotic reptile species has thus far unequivocally been shown to have established viable wild populations in South Africa, but the potential for more species to do so is now greater than ever. South African pet owners currently keep at least 300 exotic reptile species in captivity, several of which have a high risk of establishment.

Certain snakes, such as Corn snakes (*Pantherophis guttatus*), are bred by the thousands each year and routinely traded in pet stores. Corn snakes are probably the most popular 'starter species' for pet owners and are regularly found roaming freely in city suburbs. Whether the species has established feral populations, however, remains open to speculation.

Other species, such as the Burmese python (*Python molurus bivittatus*), are common household pets in South Africa and have a proven ability to establish feral populations elsewhere in the world. Burmese pythons are now well established in large parts of Florida, USA, where they continue to spread and multiply despite concerted and expensive efforts at control. They have already had a negative effect on indigenous species in that area. Large parts of South Africa have a climate suitable for this species and it must be regarded as a high-risk species.

Legislated restrictions on the reptile trade in South Africa have probably had only limited impact on the potential for future establishment of feral populations. However, regulation has been effective in at least one case. The American red-eared slider (*Trachemys scripta elegans*), a small, freshwater terrapin, was a popular pet in South Africa in the 1970s. Sold as cute and colourful hatchlings, many of these terrapins soon outgrew their cages and their owners' attention, and were released into nearby freshwater bodies. Many feral individuals were observed in South African wetlands, and it was assumed that they had established several breeding populations, as has happened in Europe. Legislation brought trade in this species to a halt in South Africa, and this resulted in a reduction in the number of wild terrapins observed over subsequent years. No American red-eared sliders have been reported from the wild in South Africa in the past 20 years, even from wetlands where they were previously seen regularly. It can therefore now be assumed that the species no longer poses a conservation problem.

G. Alexander

Burmese pythons are common in the pet trade in South Africa.

Brahminy blind snake

Ramphotyphlops braminus

Group: Squamata: Typhlopidae

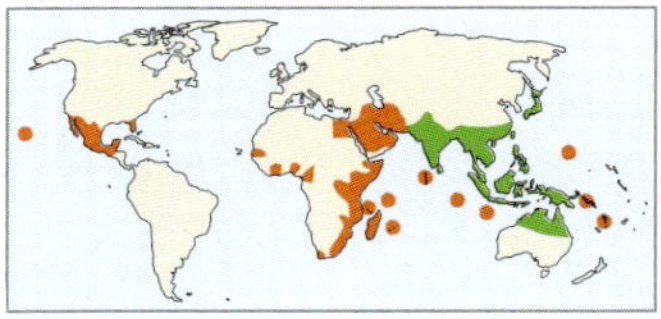

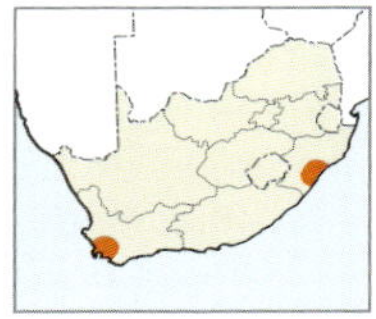

Habitat: Burrowing in soil.
Identification and biology: Easily confused with other blind snakes (Typhlopidae) and thread snakes (Leptotyphlopidae), but tends to be smaller than indigenous blind snakes, and slightly larger and lighter in coloration than most thread snakes. Body usually light to dark brown, but may be black or dark grey. Scales tight-fitting, overlapping, smooth and very glossy, occurring in 20 rows along entire length of the body. No broad belly scales, as in typical snakes, and maximum body length approximately 165 mm. Head indistinct, rounded and blunt. No hardened edge on the lip, which may limit this species to burrowing into friable soils. Eyes vestigial and visible as dark spots under the scales. Tail very short and has a small spine at the tip, used to anchor the snake in its burrow while burrowing. At first glance the two ends of the snake are indistinguishable, but the anterior is thinner and also identified by the snake's small, white forked tongue. Usually found under rocks in areas of soft, sandy soil or in rotting logs, where they may aggregate. Exposed snakes attempt to retreat down burrows. Thought to feed on ants and insect larvae and pupae, including those of ants, termites and beetles. Up to eight eggs laid per clutch. The species is parthenogenetic and all individuals are female. **Introduction and spread:** Thought to have originated in Southeast Asia, but populations now occur worldwide in tropical and even some temperate areas. Unintentionally dispersed during transport of garden soil and pot plants. First collected in Cape Town in the 1820s, Durban in 1985 and Worcester in 1997. Now relatively widespread and common in Cape Town and Durban, with a single record confirmed from Worcester. Does not appear to have established far from urban areas and may be limited by soil type. **Impacts:** Although very abundant in certain localised areas, distribution restricted and unlikely to have significant biological impacts. **Control:** Probably impossible due to its cryptic, burrowing habits and parthenogenetic reproduction.

Key reference: Broadley, D.G. 1983. *FitzSimons' snakes of southern Africa*. Johannesburg: Delta Books.

J. Marais

FISHES

Class Teleostei

Introductions of freshwater fishes into South Africa have occurred since the 18th century when Goldfish (p. 54) were reported in ornamental ponds in the Cape. Since then numerous fish species have been purposefully imported for recreational angling opportunities, control of mosquitoes and aquatic weeds, fish farming and the aquarium trade. Many of these fishes have found their way into rivers, lakes and dams, either through direct stocking for angling, or accidental escape from fish farms.

Most widespread are fishes that were imported for sport angling, such as Rainbow trout (p. 61), Largemouth bass (p. 59) and Common carp (p. 52). After importation these fishes were propagated in state-funded hatcheries, distributed throughout the country and intentionally stocked so widely that they now occur in parts of most major river systems. Unintentional introductions include escapees from fish farms. Examples are Silver carp (p. 53) and Vermiculated sailfin (p. 63). Until the Silver carp established a population in the Olifants-Limpopo River system and the Grass carp (p. 51) became so abundant in the Vaal Dam that anglers considered it a pest, it was thought that these fish were unable to adapt to South African conditions. Some indigenous fishes have also been translocated within South Africa, primarily for angling, but also through inadvertent introductions via inter-basin water transfer schemes, which divert river water and its associated fauna across river catchments. When an indigenous fish is translocated into an ecosystem where it did not occur historically, it is considered alien to that system.

Alien fishes have resulted in significant environmental impacts, including transfer of associated parasites, changes in aquatic invertebrate communities and extirpation of indigenous fishes by direct predation and competition. Extralimital introductions of indigenous fishes also include the threat of hybridisation, because they are often closely related to the local biota. Conversely, some alien fishes are valuable contributors to South Africa's economy because they are major target species for recreational anglers, contribute to food security in rural communities, and drive a growing aquaculture industry for ornamental and table fish. Once established in a river system, alien fishes are almost impossible to eradicate. Future management of alien fishes will need to focus on their exclusion from areas of biodiversity concern and on limiting their further spread.

Thus far no alien marine fish species have been recorded from southern African waters. However, the marine fish pet trade has increased considerably in recent years and it is important to remain vigilant to the threat, as elsewhere invasive marine fish species have become a serious problem. For example, Indian Ocean lionfish (*Pterois volitans*) are invading the Caribbean area and causing major ecological changes.

C. Griffiths

Rainbow trout crowding a hatchery.

Mosquitofish

Gambusia affinis

Group: Cyprinodontiformes: Poeciliidae

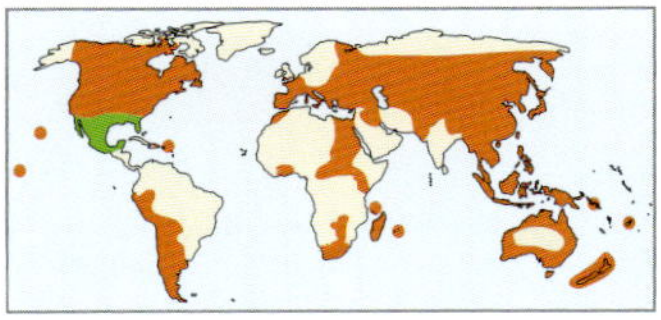

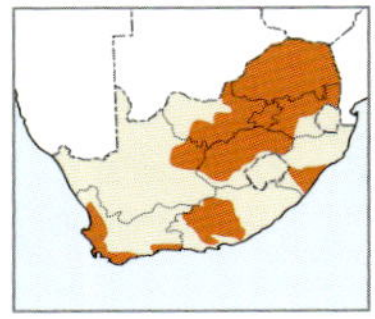

Habitat: Freshwater. Prefers vegetated margins of slower sections of rivers and streams, backwaters and vegetated habitats. Can tolerate brackish conditions and a wide range of temperatures (4.5–38°C). **Identification and biology:** A small species, with males attaining 35 mm and females 60 mm total length. Body slender in males, but stout in females, flat above, fins small, dorsal short, caudal rounded. Anal fin modified in males to form extended intromittent organ for internal fertilisation. Translucent light grey or brown, bluish tinges, eyes black with blue iris. Preys on mosquito larvae and other small invertebrates and has also been reported to feed on fish eggs and larvae. A live-bearer that releases fully developed fry after the internal development of the eggs. **Introduction and spread:** Native to southern USA and northern Mexico. Considered one of the most widely distributed fish species globally, Mosquitofish were deliberately introduced to South Africa in 1936 for mosquito control. The species has since spread widely and current locality records include many Cape drainages, from the Great Fish River to the Berg River, as well as parts of the Limpopo and Mvoti river systems. It is present in at least one estuarine system, the Wilderness lakes, and in a coastal lake, Lake Pleasant, in the Western Cape. Also probably more widespread than official records indicate. **Impacts:** Impacts have not been studied in South Africa, but it may alter aquatic ecosystems by reducing abundance of zooplankton and aquatic insect larvae. Its success as a mosquito control agent is disputed and Mosquitofish are now considered a pest in most countries. Documented impacts include competition and predation (on eggs, larvae and juveniles of native fishes, amphibians and invertebrates). This negatively affects the recruitment dynamics of indigenous fishes. **Control:** Once established, Mosquitofish are difficult to eliminate.

SAIAB, P. Skelton

Key references: De Moor, I.J. & Bruton, M.N. 1988. *Atlas of alien and translocated indigenous aquatic animals in Southern Africa*. South African National Scientific Programmes Report 144.

Skelton, P.H. 2001. *A complete guide to the freshwater fishes of Southern Africa*. Cape Town: Struik.

SAIAB, P. Skelton

Guppy

Poecilia reticulata

Group: Cyprinodontiformes: Poeciliidae

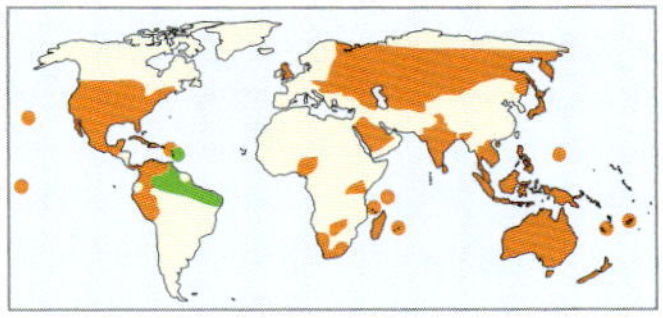

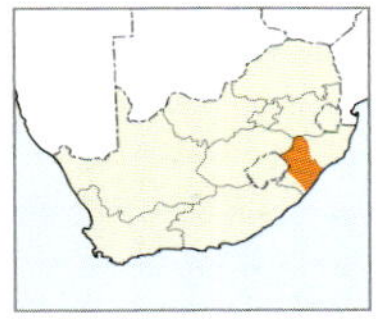

Habitat: Freshwater and brackish water habitats. Prefers slower-flowing, well-vegetated areas of rivers, streams, ponds and swamps. Optimum temperature range 22–24°C and unable to tolerate temperatures below 15°C. **Identification and biology:** Males smaller (30 mm total length) than females (60 mm total length). Males slender, caudal fin large and frequently elaborate, anal fin modified to form a gonopodium or intromittent organ for internal fertilisation of the female. Female deep-bodied, caudal fin rounded. Domesticated males brightly coloured and have fancy, extended finnage (shown, top image). Wild-type colours are less spectacular; a dark spot on body and at base of caudal fin is common. Livebearing. Feeds on small invertebrates. **Introduction and spread:** Native to South America and the Caribbean islands of Barbados and Trinidad. First introduced to the Western Cape from Barbados in 1912 for mosquito control, but that population did not survive the winter. Subsequent importations have all been through the pet trade. One of the most popular species in the aquarium trade and numerous different introductions are likely to have occurred as a result of releases by aquarists. Feral populations are limited to warmer regions of South Africa, mainly in small urban streams and ponds. **Impacts:** Impact in South Africa has not been assessed, but feral guppies are regarded as pests elsewhere. Impacts may be similar to those associated with Mosquitofish (p. 48). **Control:** Can be eliminated by poisoning where necessary, but prevention is far easier than control and the danger associated with release of ornamental fishes into the wild should be part of an educational process associated with the purchase of aquarium fishes.

Key references: De Moor, I.J. & Bruton, M.N. 1988. *Atlas of alien and translocated indigenous aquatic animals in Southern Africa*. South African National Scientific Programmes Report 144.

Skelton, P.H. 2001. *A complete guide to the freshwater fishes of Southern Africa*. Cape Town: Struik.

C. Griffiths

Per Harald Olsen, Wikimedia Commons

Swordtail

Xiphophorus helleri

Group: Cyprinodontiformes: Poeciliidae

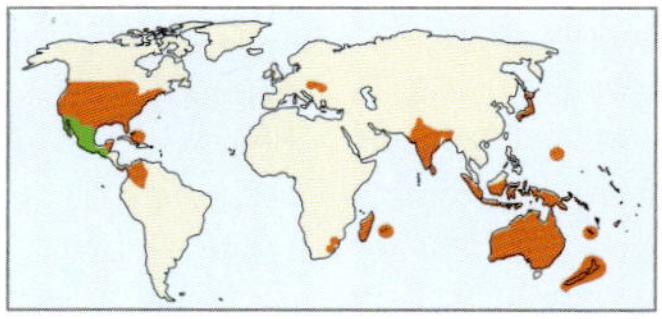

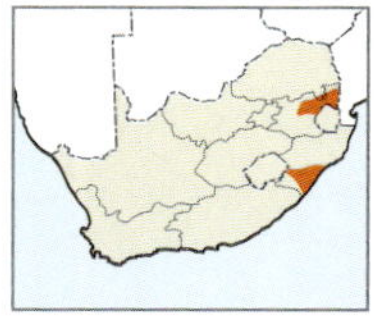

Habitat: Freshwater. Tolerant of a wide range of habitats, including flowing streams, swamps and ponds. Prefers vegetated areas. Ideal temperature range 22–28°C. **Identification and biology:** Deep, compressed body form. Dorsal fin large and sail-like, its origin in mid-body. In males, anal fin modified as a slender gonopodium, which is used for internal fertilisation of the female, and caudal fin has extended lower rays forming a pointed 'sword' from which the species derives its name. Head flat above, eyes large and lateral, mouth supero-terminal. Translucent brown above, silvery-white below, tricoloured orange-yellow-red band from snout to end of caudal peduncle. Males smaller (80 mm total length) than females (120 mm total length). Females less brightly coloured. Eats insects and invertebrates. A prolific live bearer that can produce up to 180 offspring per spawning. **Introduction and spread:** Native to Mexico from where this species was imported for the aquarium trade. A popular aquarium fish and was most likely introduced by aquarists releasing unwanted fish. Feral populations are restricted to more tropical regions in South Africa, where they are frequently recorded in urban streams and ponds. First feral specimen collected from the Crocodile River in 1974. **Impacts:** Not determined for South Africa, but internationally regarded as a pest species. Known to prey on the fry of other fish species and because of a similarity in diet its impacts are likely to be similar to those reported for Mosquitofish (p. 48). **Control:** In small ponds can be eliminated by poisoning, but in larger systems this is impractical. Prevention is far easier than control and the danger associated with release of ornamental fishes into the wild should be part of an educational process associated with the purchase of aquarium fishes.

SAIAB, R. Bills

Key references: De Moor, I.J. & Bruton, M.N. 1988. *Atlas of alien and translocated indigenous aquatic animals in Southern Africa.* South African National Scientific Programmes Report 144.

Skelton, P.H. 2001. *A complete guide to the freshwater fishes of Southern Africa*. Cape Town: Struik.

SAIAB, P. Skelton

Grass carp

Ctenopharyngodon idella

Group: Cypriniformes: Cyprinidae

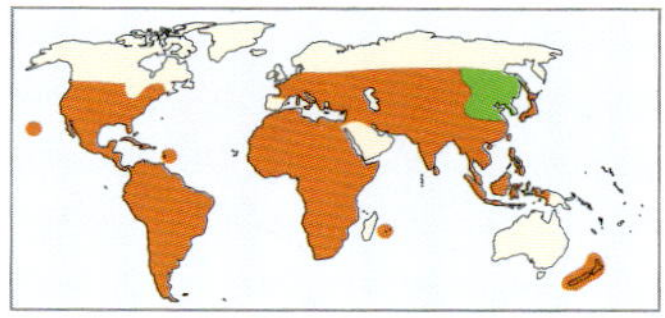

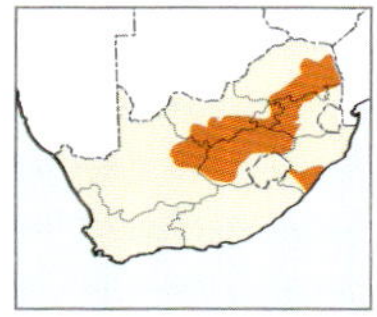

Habitat: Freshwater. Thrives in large, well-vegetated, standing or slow-flowing habitats. Can tolerate a wide range of temperatures from just above freezing to 35°C. **Identification and biology:** Sub-cylindrical shape with short-based median fins. The dorsal fin origin is mid-body above the pelvic fins. Caudal fin forked and scales large and conspicuous. Head relatively small, eyes far forward, mouth small and terminal. Colour generally silvery or deep olive along the back, the fin membranes being dark grey. A large fish, attaining a total length more than 1 m and 45 kg in weight. Feeds mainly on aquatic plants, but also takes insects and other invertebrates. Spawns on riverbeds with very strong currents. **Introduction and spread:** Native to eastern Asia, it was imported from Malaysia to Umgeni hatchery in KwaZulu-Natal in 1967 for aquaculture and aquatic vegetation control. Widespread stocking has subsequently occurred, especially in KwaZulu-Natal, for aquatic weed control. As a result of its specific spawning requirements it was thought that this fish was unable to breed in the wild. While this is the case in many smaller river systems, it has become well established in the Vaal, the Olifants-Limpopo, as well as lower reaches of the Limpopo in Mozambique. In the Vaal Dam, for example, it is so abundant that recreational anglers consider it a pest. **Impacts:** Used for aquatic weed control. Considered responsible for the introduction of many fish parasites, including Asian tapeworm (p. 203), a gut parasite that has been shown to infect indigenous cyprinids. An aggressive feeder on macrophytes, which may result in total denudation, habitat loss and increased turbidity. **Control:** None practical once species is established. Use for aquatic macrophyte control is currently restricted to stocking genetically modified (triploid) fish that are sterile and hence unable to breed.

Key references: De Moor, I.J. & Bruton, M.N. 1988. *Atlas of alien and translocated indigenous aquatic animals in Southern Africa*. South African National Scientific Programmes Report 144.

Skelton, P.H. 2001. *A complete guide to the freshwater fishes of Southern Africa*. Cape Town: Struik.

SAIAB, P. Skelton

Dezidor, Wikimedia Commons

Common carp

Cyprinus carpio

Group: Cypriniformes: Cyprinidae

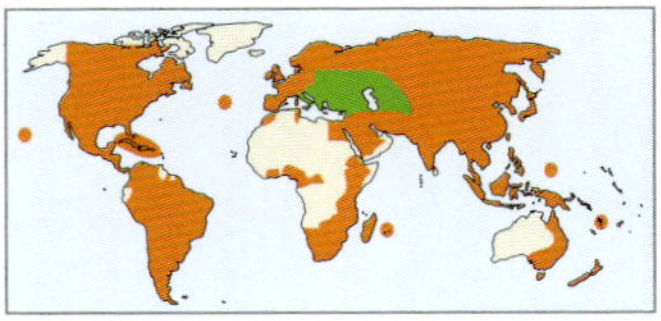

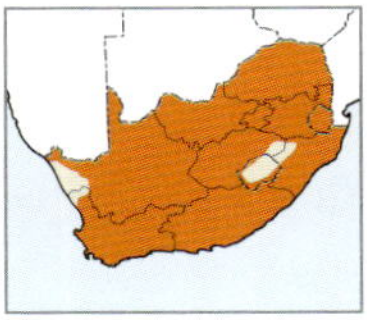

Habitat: Freshwater. Thrives in large, shallow, vegetated, standing water habitats, but is also found in soft-bottomed, slow-flowing reaches of larger rivers. **Identification and biology:** A large fish reaching over 1 m total length and 24 kg in mass, deep-bodied, ellipsoid-shaped. Scales variable, and full-scaled, mirror-scaled and naked forms occur. Dorsal fin extends along back. Anterior serrated spine on both dorsal and anal fin is distinctive for this species. Head deep, mouth terminal and protrusable, lips thin and with one pair of barbels. Generally brazen-gold in colour, with lighter ventral surfaces. The ornamental form, Koi carp, is available in a wide variety of often bright and varied colour versions and is a popular ornamental fish. Omnivorous, feeds by grubbing from bottom sediments. Spawns readily in shallow areas during summer, scattering eggs onto vegetation. **Introduction and spread:** Native to Europe and Asia, but domesticated forms have been transported by humans for hundreds of years. First introduced to South Africa from England as a food fish and for angling in 1859. Numerous introductions followed for angling and aquaculture. Now occurs in all major river systems and many farm dams. **Impacts:** Bottom feeding and sediment stirring by carp increase turbidity and nutrient cycling, which can lead to eutrophication. This results in increased costs of water treatment, biodiversity loss and effects on sight-feeding fishes. Used by subsistence, recreational and commercial fisheries. Koi carp are highly prized and a major component of the pet trade. Specialist Koi breeders often import breeding stock from Japan. Koi keeping in garden ponds is widespread in South Africa. Occasional feral Koi carp are encountered in natural habitats, where they readily interbreed with ordinary strains and colours revert to wild type within a few generations. **Control:** None practical on a river-basin scale. In small impoundments may be controlled by ichthyocides, such as Rotenone.

Key references: Balon, E.K. 2004. About the oldest domesticates among fishes. *Journal of Fish Biology*, 65: Suppl. A, 1-27.

De Moor, I.J. & Bruton, M.N. 1988. *Atlas of alien and translocated indigenous aquatic animals in Southern Africa*. South African National Scientific Programmes Report 144.

C. Griffiths

Silver carp

Hypophthalmichthys molitrix

Group: Cypriniformes: Cyprinidae

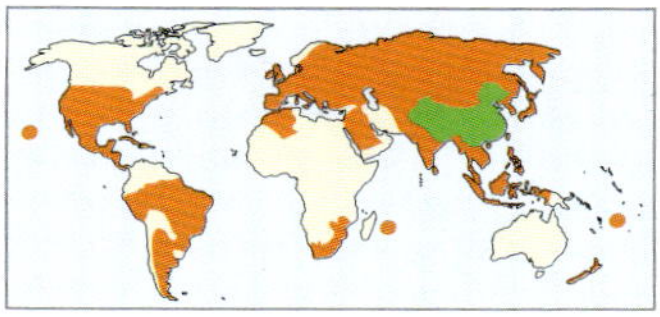

Habitat: Freshwater. Requires standing or slow-flowing conditions such as in lakes, dams or the backwaters of large rivers. **Identification and biology:** Body paddle- or club-shaped, scales minute. Dorsal fin short, with its origin behind pelvic fins. Caudal fin shallow and forked. Head large, eyes far forward and positioned low down below mouth level. Mouth supra-terminal projecting upwards. Generally silvery, brown or olive along back. A large fish that can attain a total length of up to 1 m and a weight of 50 kg. Filter-feeds on plankton and other small floating or free-swimming organisms and algae. Migrates upstream to breed. Spawning occurs in flowing water, with eggs and larvae floating downstream to floodplain zones. **Introduction and spread:** Native to China. First imported in 1975 from Germany to the Marble Hall experimental fish farm on the Olifants River. Following escape, Silver carp were reported in the Albasini Dam and the Kruger National Park in 1996. Floods in 2000 are thought to have spread fish downstream into lower Limpopo system. It is now considered well established in the Olifants and lower Limpopo rivers, with mass migrations having been observed in tributaries. **Impacts:** Not determined in South Africa. As a filter-feeding species it is likely to have effects on the food-web dynamics in aquatic ecosystems and may compete for zooplankton with juveniles of many native species. In the USA, documented effects include the reduction of native phytoplanktivorous fishes. Also harvested by artisanal fishers in Mozambique. Because of its jumping habits, considered dangerous to small craft handlers. **Control:** There are no practical control measures for this species, but further spread should be avoided.

Key references: De Moor, I.J. & Bruton, M.N. 1988. *Atlas of alien and translocated indigenous aquatic animals in Southern Africa*. South African National Scientific Programmes Report 144.

Skelton, P.H. 2001. *A complete guide to the freshwater fishes of Southern Africa*. Cape Town: Struik.

SAIAB, P. Skelton

Goldfish

Carassius auratus

Group: Cypiniformes: Cyprinidae

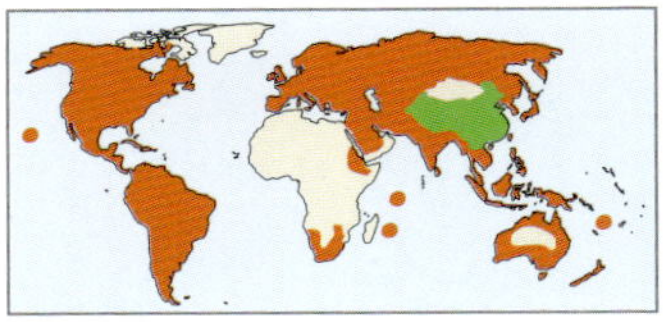

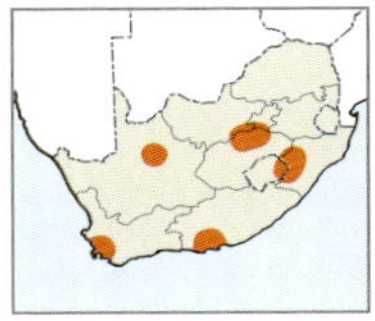

Habitat: Freshwater. Favours shallow, vegetated standing waters with soft sediments. **Identification and biology:** Total length 10–32 cm, body shape variable, wild type has a shallow-ellipsoid shape. Fins variable, dorsal and anal fins with anterior serrated spine. Dorsal fin extends along back, caudal fin forked, but frequently elaborately extended. Mouth sub-terminal, protrusable and has no barbels. Wild-type colour brazen-olive with lighter (off-white) ventral surfaces. Domesticated colours highly varied, but uniform orange-red common. Omnivorous, feeding on both plants and animals, mostly from the bottom. Spawns readily in summer at temperatures exceeding 20°C, laying eggs in vegetation. **Introduction and spread:** This eastern Asian species has been used for ornamental purposes in China for c. 1 000 years, and like the Common carp (p. 52), this species is domesticated. It was first imported, most likely by Dutch trading vessels from India, for ornamental purposes in 1726. There have been numerous subsequent importations. Goldfish are often introduced into the wild by aquarists when they outgrow aquaria, or accidentally during flooding of ornamental ponds. As a result feral populations are encountered in waters around many urban areas. However, there are few established populations in major river systems, despite numerous opportunities to invade. **Impacts:** An important species in the aquarium trade. Also associated with the spread of many protozoan, monogean, branchiuran and nematode parasites around the world. Because these parasites are often associated with other alien fishes the specific impact of Goldfish cannot be directly determined. **Control:** As this fish is a major species in the pet trade, numerous importations continue to occur. While it is unlikely that this species will become invasive throughout the country, the parasites for which it is a vector are likely to have considerable impact. It is therefore important that new importations are screened for parasites. As with other ornamental species, public education of the consequences of releasing unwanted pets into the wild is necessary.

Key references: Balon, E.K. 2004. About the oldest domesticates among fishes. *Journal of Fish Biology*, 65: Suppl. A, 1-27.

De Moor, I.J. & Bruton, M.N. 1988. *Atlas of alien and translocated indigenous aquatic animals in Southern Africa*. South African National Scientific Programmes Report 144.

C. Griffiths

Tench

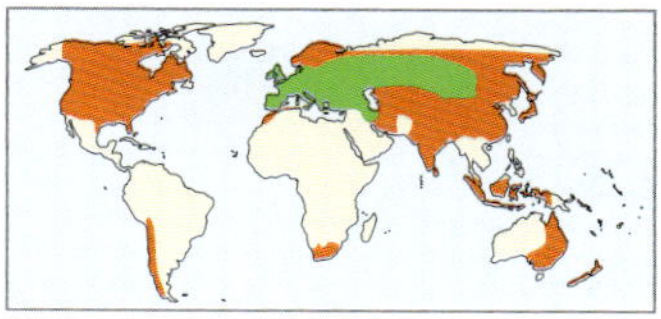

Tinca tinca

Group: Cypriniformes: Cyprinidae

Habitat: Freshwater. Deep, standing or slow-flowing vegetated waters in vleis, dams and large rivers. **Identification and biology:** Heavy-set, paddle-shaped body, minute scales. Fins large and rounded. Dorsal fin originates in mid-body and caudal fin emarginated. Head large and rounded. Eye small. Mouth large, terminal and protrusable, with one pair of short barbels. Colour dark olive-green, becoming lighter, sometimes golden, on the ventral surfaces, fins blackish. Mature ripe males have prominent thickened whitish leading edges to pelvic fins. Attains a total length of 64 cm and weights in excess of 5 kg. Juveniles feed on plankton, adults omnivorous bottom feeders that grub through soft sediment for insect larvae, worms, crustaceans and molluscs. Spawn in shallow water on vegetation. **Introduction and spread:** A Eurasian species introduced into the Western Cape in 1910 for angling. Subsequently stocked in numerous localities both for angling and as a fodder fish for predatory game fishes. Failed to establish in most locations and currently the only feral populations are in the Breede River and some isolated vleis and dams in the Western Cape. **Impacts:** Currently not assessed in South Africa. Feeding behaviour similar to that of Common carp (p. 52) and therefore likely to contribute to increased turbidity and nutrient cycling. Tench are also possible competitors with indigenous fish species. **Control:** While no practical control measures exist for large river systems such as the Breede River, it is unlikely that this species will spread further. Poisoning in small dams is an option, but unlikely to be economically justifiable given the limited spread of the fish.

Key references: De Moor, I.J. & Bruton, M.N. 1988. *Atlas of alien and translocated indigenous aquatic animals in Southern Africa*. South African National Scientific Programmes Report 144.

Skelton, P.H. 2001. *A complete guide to the freshwater fishes of Southern Africa*. Cape Town: Struik.

Piet Spaans, Wikimedia Commons

SAIAB, A. Chakona

Bluegill sunfish

Lepomis macrochirus

Group: Perciformes: Centrarchidae

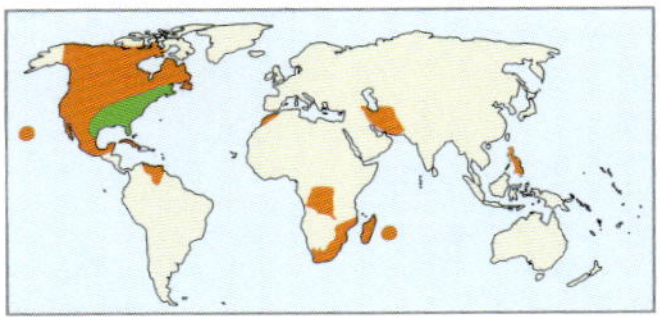

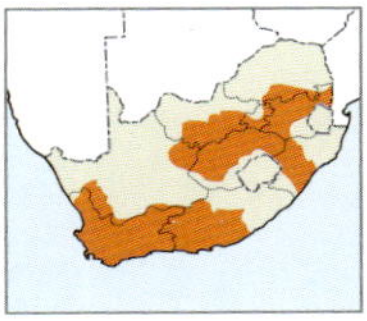

Habitat: Freshwater. Prefers slow-flowing or still waters such as dams and weirs with vegetation. **Identification and biology:** Compressed ovoid body shape, caudal peduncle distinct, dorsal fin extends along back with 9–10 spines and is soft-rayed towards its end; caudal fin emarginated, anal fin has three spines and soft rays. Head steep, eyes large, mouth terminal and inclined. Iridescent olive-grey, flanks silvery, darker vertical bars on body, distinct black upper corner to gill cover. Male breeding dress reddish below, blue chin. Preys on insects and invertebrates as well as small fishes and frogs. Territorial when breeding, males construct saucer-shaped nest and guard eggs and larvae. Maximum size about 35 cm total length and 1 kg weight in some localities, but more commonly attain weights of about 200 g. **Introduction and spread:** An angling species imported from its native USA in 1938 primarily as a forage fish for bass. Has been stocked widely both through formal stocking initiatives as well as illegally by anglers and has established populations in parts of most major South African river systems. **Impacts:** A prolific invader that dominates the fish fauna in many localities where it has been introduced. In dams it has a tendency to overpopulate, resulting in stunted populations dominated by small fish. As a result it exerts considerable predation pressure on invertebrates and on juveniles of indigenous fishes. In the Eastern Cape, indigenous cyprinid fish populations declined after stocking of Bluegill in the Krom and Kouga river systems. Has some subsistence angling value. **Control:** Internationally regarded as a pest species and recent legislation prohibits further movement of this fish. Once established, eradication is unlikely except in small water bodies.

Key references: De Moor, I.J. & Bruton, M.N. 1988. *Atlas of alien and translocated indigenous aquatic animals in Southern Africa.* South African National Scientific Programmes Report 144.

Skelton, P.H. 2001. *A complete guide to the freshwater fishes of Southern Africa.* Cape Town: Struik.

SAIAB, A. Chakona

SAIAB, P. Skelton

Smallmouth bass

Micropterus dolomieu

Group: Perciformes: Centrarchidae

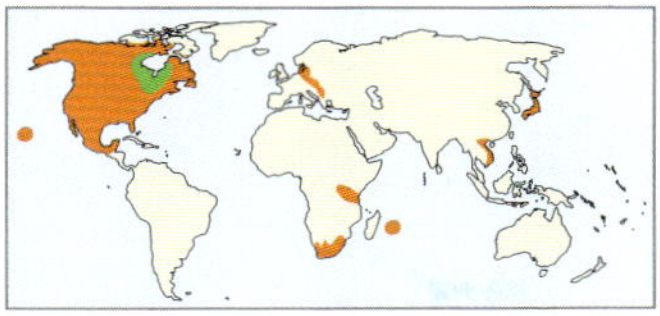

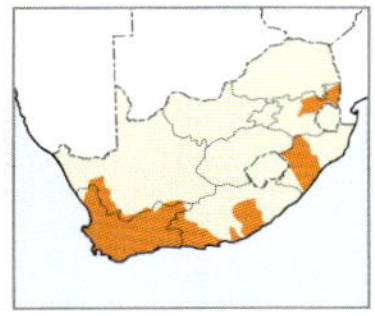

Habitat: Freshwater. Prefers rocky pools and deep runs in flowing streams and rivers. Also stocked in dams. **Identification and biology:** Body deep, caudal peduncle twice as long as deep, scales small, dorsal fin in two sections, intermembrane indented but not deeply cleft, 9 (+1) spines, caudal fin emarginated. Head large, eyes forward, mouth large, protrusable and terminal, extending to below orbits. Olive to coppery-yellow with deep olive-brown vertical bars and large spots, radial bars behind orbits. Juveniles with black-and-yellow-banded caudal fins. Predatory on insects and larger invertebrates when juvenile, becoming piscivorous and also taking crabs and other larger prey as adults. Territorial when breeding. Males guard eggs, larvae and fry. Attains 60 cm in total length and weights above 3.5 kg. **Introduction and spread:** An angling species imported from its native temperate USA in 1937 to fill the gap between high-altitude trout waters and slow-flowing, low-lying Largemouth bass zone. Subsequently spread by anglers into various river systems. Widespread in more temperate regions of South Africa, particularly in the Eastern and Western Cape. Also present in some dams in the Limpopo, Mvoti and Tugela catchments, but not well established in the rivers. **Impacts:** Predation on native fishes and invertebrates has been linked to ecosystem effects in the Western Cape, including changes in invertebrate community structure that resulted in differences in grazing and concomitant differences in algal biomass between invaded and non-invaded sections of the river. A major angling species in South Africa, with directed fishing tournaments, particularly in the Western Cape. **Control:** Once established, control is not feasible except in very small streams where ichthyocides may be used to eradicate populations. Further stocking and movement of these fish outside their current distribution ranges must be tightly controlled.

Key references: De Moor, I.J. & Bruton, M.N. 1988. *Atlas of alien and translocated indigenous aquatic animals in Southern Africa*. South African National Scientific Programmes Report 144.

Lowe, S.R., Woodford, D.J., Impson, D.N. & Day, J.A. 2008. The impact of invasive fish and invasive riparian plants on the invertebrate fauna of the Rondegat River, Cape Floristic Region, South Africa. *African Journal of Aquatic Science*, 33:51-62.

D. Woodford

Spotted bass

Micropterus punctulatus

Group: Perciformes: Centrarchidae

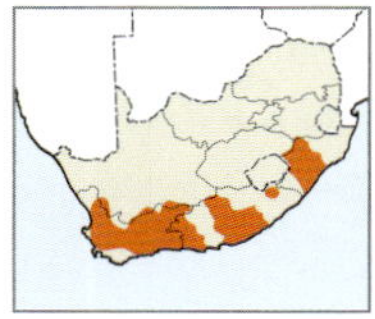

Habitat: Freshwater. Prefers sluggish-flowing water in rivers and streams, over soft bottoms. **Identification and biology:** Body slender, scales small, caudal peduncle about twice as long as deep. Caudal fin emarginated, dorsal fin in two sections, membrane indented between sections. Head large and pointed, eyes forward, mouth large, terminal and protrusable, reaching below orbits. Has a characteristic tooth patch on tongue. Olive above, light brown to off-white below, series of deep olive-brown bars and blotches form an irregular mid-lateral band. Differs from Largemouth bass (p. 59) in that hind margin of jaw does not extend past hind border of eye. Attains 60 cm total length. Juveniles feed on insects and invertebrates; adults on fishes, crabs and similar larger prey. Spawns in spring. Territorial males guard eggs, larvae and fry. **Introduction and spread:** Imported from its native central USA in 1939 for sport fishing. Intended for stocking in rivers that were too turbid and subject to flooding to suit Smallmouth bass (p. 57) and Largemouth bass. Although it failed to establish in many areas, small populations persist in dams and parts of many river systems, particularly in the Eastern and Western Cape. Most likely their current distributions are underestimated because they are often misidentified and stocked as Largemouth bass by anglers. **Impacts:** Little is known about the impact of this introduction, but likely impacts will be related to predation and similar to those of Smallmouth and Largemouth bass. A widely introduced angling species but not as well known as Largemouth or Smallmouth bass in southern Africa and often not recognised as a distinct species by anglers. **Control:** Not a recognised angling species and no further introductions should be permitted. Eradication is unlikely to be feasible.

Key references: De Moor, I.J. & Bruton, M.N. 1988. *Atlas of alien and translocated indigenous aquatic animals in Southern Africa.* South African National Scientific Programmes Report 144.

Skelton, P.H. 2001. *A complete guide to the fresh-water fishes of Southern Africa.* Cape Town: Struik.

SAIAB, P. Skelton

SAIAB, O. Weyl

Largemouth bass

Micropterus salmoides

Group: Perciformes: Centrarchidae

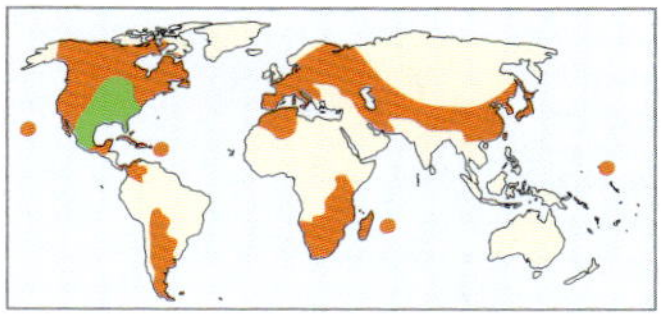

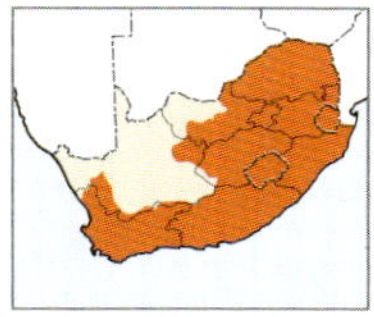

Habitat: Freshwater. Favours clear, slow-flowing or standing waters in rivers and dams, with vegetation and other cover. **Identification and biology:** Body oblong with long caudal peduncle, small scales, dorsal fin in two sections, 8–9 anterior spines, indented to soft-rayed posterior section, caudal fin emarginated. Head large, eyes forward above mouth, mouth very large, terminal, inclined to behind orbit. Olive-green, lighter on flanks and off-white below, broad, irregular deeper markings and spots along back and black band along mid-flanks to base of caudal fin, dark olive-black radial stripes behind eyes. Attains 70 cm total length and weights of over 6 kg. Juveniles insectivorous, becoming piscivorous and predatory on any larger aquatic and swimming terrestrial organisms as adults. Territorial when breeding, males defend a nest constructed on a firm surface, guarding the eggs, larvae and fry. **Introduction and spread:** A North American species imported in 1928 for angling. Initially spread by fisheries authorities and later by anglers, and probably the most widespread centrarchid in southern Africa. Common in slower sections of all larger rivers, in impoundments and in some coastal lakes and estuaries. The closely related Florida bass (*Micropterus floridanus*) was imported in the 1980s and has also been widely stocked by anglers. This latter species grows bigger than the Largemouth bass but to date has not been recorded locally as a separate species. **Impacts:** Negative impacts include predation on indigenous fishes, restructuring of invertebrate communities and fragmentation of native fish populations. Changes in invertebrate and indigenous fish community structure resulting from predation by Largemouth bass are well documented. Positive impacts are associated with the economic value of recreational angling for this very popular game fish. The fishery includes both casual anglers and anglers organised under the South African Bass Anglers' Association (SABAA), who compete at local and international level. **Control:** Once the species is established, control is not feasible or economically viable. Efforts should be made to educate the public on current legislature controlling movement of this species.

Key references: De Moor, I.J. & Bruton, M.N. 1988. *Atlas of alien and translocated indigenous aquatic animals in Southern Africa*. South African National Scientific Programmes Report 144.

Skelton, P.H. 2001. *A complete guide to the freshwater fishes of Southern Africa*. Cape Town: Struik.

SAIAB, P. Skelton

Perch

Perca fluviatilis

Group: Perciformes: Perciidae

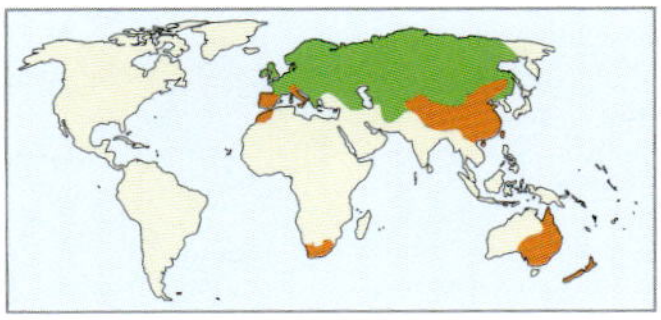

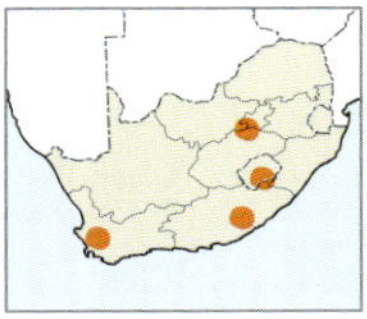

Habitat: Freshwater. Favours slow-flowing parts of rivers and still-water habitats in lakes and dams. Can tolerate brackish water environments. **Identification and biology:** Body deep, scales small, caudal peduncle twice as long as deep, dorsal fin large and in two sections, deeply cleft between sections, 13–17 (+1–2) spines, caudal fin shallowly forked. Head large and pointed, eyes forward, mouth large, terminal and protrusable, gill covers have a spine. Silvery to light olive, black vertical bars on body, black hind corner to spined dorsal, fins yellowish-orange. Maximum total length 60 cm. Predatory, with juveniles feeding on zooplankton and adults preying on both invertebrates and fish. Not territorial. Spawns in spring and lays eggs in long ribbons on vegetation, rocks and other submerged structures. Does not protect eggs or larvae. **Introduction and spread:** Native to Europe and Asia, introduced from England for angling purposes in 1915. It was distributed from the Jonkershoek hatchery in the Western Cape to suitable waters from 1927. Despite stocking numerous localities in the Western Cape it failed to establish in most systems. Current locality records include a few small farm dams in the Western Cape and the status of this fish in South Africa is not clear. **Impacts:** Impacts would be related to predation on indigenous fauna. However, appears not to have established significant feral populations in South Africa and is not currently considered an invasive threat. **Control:** It is unlikely that this fish will become invasive throughout the country, but further spread should be avoided. Little is known about the current distribution, and new locality records should be reported to the South African Institute for Aquatic Biodiversity and relevant nature conservation authorities.

Key references: De Moor, I.J. & Bruton, M.N. 1988. *Atlas of alien and translocated indigenous aquatic animals in Southern Africa.* South African National Scientific Programmes Report 144.

Skelton, P.H. 2001. *A complete guide to the freshwater fishes of Southern Africa.* Cape Town: Struik.

Piet Spaans, Wikimedia Commons

Rainbow trout

Onchorhynchus mykiss

Group: Salmoniformes: Salmonidae

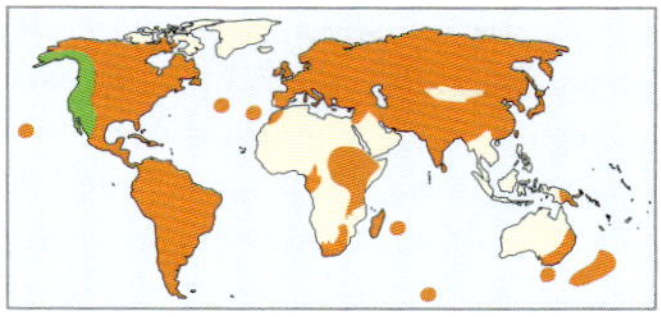

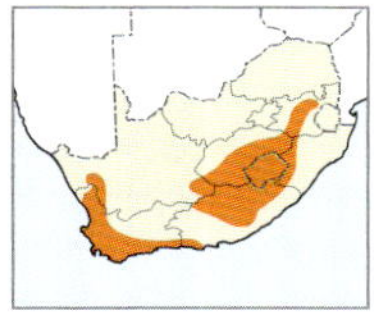

Habitat: Freshwater. Requires cold (<20°C), unpolluted, well-oxygenated water. Populations are therefore restricted to pools in cool, clear, flowing mountain streams and rivers or deep runs over rocks and rapids. Also thrives in open waters of cool upland dams. **Identification and biology:** Oblong, streamlined, spindle-shaped body, can be heavy-set when large and well fed. Scales tiny, soft-rayed dorsal fin in mid-body, small lobate adipose fin, caudal fin emarginated. Head obtusely pointed, eyes smallish and far forward, mouth large and terminal, teeth large. Silvery, olive-brown to bluish above, iridescent pinkish lateral band, silvery-white ventral surfaces. Males darken considerably. In South Africa has been recorded at total lengths exceeding 70 cm and weights in excess of 6 kg in dams. In rivers fish are smaller, seldom exceeding 50 cm total length and 1.5 kg in weight. Feeds on drifting insects and invertebrates, as well as small fishes, crabs and frogs. Breeds in winter into early spring, spawning over gravel beds. **Introduction and spread:** A North American fish introduced via England for sport fishing in 1897. Stocked widely and introductions continue to support a growing recreational fishery. Self-sustaining populations are present in many high-altitude rivers and streams in South Africa. In other regions populations are maintained by continuous stocking from hatcheries for sport fisheries. **Impacts:** A valued game fish and cultured food fish with several large hatcheries producing stock with considerable economic benefits for South Africa. Has been linked to the decline and extirpation of several small indigenous fishes, amphibians and invertebrates. **Control:** In many areas populations are sustained by continuous stocking. Practical control measures include regulation of stocking practices.

Key references: Cambray, J.A. 2003. The global impact of alien trout species – a review; with references to their impact in South Africa. *African Journal of Aquatic Sciences*, 28:61-67.

De Moor, I.J. & Bruton, M.N. 1988. *Atlas of alien and translocated indigenous aquatic animals in Southern Africa*. South African National Scientific Programmes Report 144.

C. Griffiths

Brown trout

Salmo trutta

Group: Salmoniformes: Salmonidae

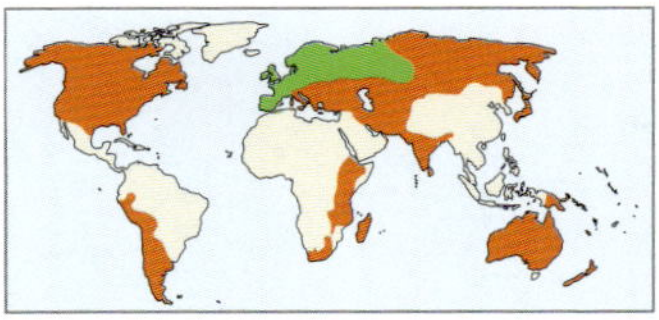

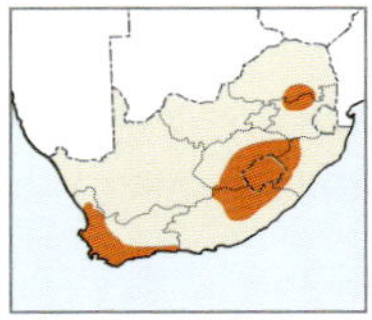

Habitat: Freshwater. Cool mountain streams and reservoirs with maximum temperatures <17°C. **Identification and biology:** Oblong, streamlined, spindle-shaped body, very small scales. Dorsal fin soft-rayed, positioned slightly ahead of pelvics, lobate adipose fin, caudal fin truncate. Head pointed, eyes smallish, mouth terminal, jaws large and hooked in mature males. Silvery, olive-brown to yellowish, lighter below, large reddish-brown spots over dorsal half. Attains 75 cm total length and weights of over 6 kg in dams. Smaller in rivers. Feeds in current or at current shearwater at the head of pools and runs, on drifting insects and invertebrates, as well as fishes, frogs and crabs. Breeds in autumn or early winter over gravel beds, males establishing territories. **Introduction and spread:** Native to Europe and western Asia, imported into southern Africa in 1892 for sport angling. Bred and stocked from hatcheries with some wild populations self-sustaining in mountain regions of most major river systems in South Africa, but because of their sensitivity to low water quality, not as widespread as Rainbow trout (p. 61). **Impacts:** Major sport angling species. Has been linked to the decline of some indigenous fish species, most notably the Maloti minnow (*Pseudobarbus quathlambae*) in the Umkomazana River in KwaZulu-Natal. Community-altering predation effects on indigenous invertebrate, fish and amphibian populations have also been reported. Brown trout are a popular angling species. **Control:** Control of use in angling by restricting stocking to areas of low conservation concern.

Eric Engbretson for U.S. Fish and Wildlife Service, Wikimedia Commons

Key references: Cambray, J.A. 2003. The global impact of alien trout species – a review; with references to their impact in South Africa. *African Journal of Aquatic Sciences*, 28:61-67.

Skelton, P.H. 2001. *A complete guide to the fresh-water fishes of Southern Africa*. Cape Town: Struik.

Vermiculated sailfin

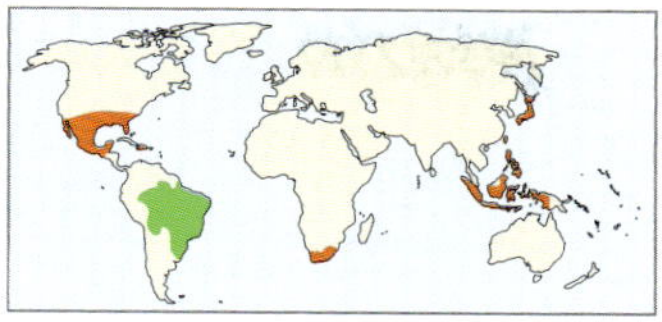

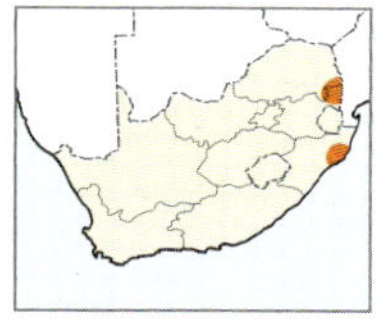

Pterygoplichthys disjunctivus

Group: Siluriformes: Locariidae

Habitat: Freshwater. Favours slow-flowing or still water. In its natural range inhabits floodplain lakes and swamps. Tolerates poorly oxygenated waters by air breathing using a specialised stomach that can function as an accessory respiratory organ. **Identification and biology:** A medium-sized, armoured catfish that can attain a total length of 70 cm. Body covered with flexible bony plates. Dark in colour with the dark-grey abdomen covered in white spots irregularly joined to form a vermiculate pattern. A thick-toothed spine present at origin of dorsal and pectoral fins. Head large, eyes small, and mouth ventral with rasp-like teeth. Feeds primarily on detritus, benthic insects and attached algae. Breeding males excavate burrows into which females lay eggs. Male exhibits parental care of young. **Introduction and spread:** Native to the Amazon River in Bolivia and Brazil. A major species in the pet trade, introduced into South African rivers by escaping from fish farms, ornamental fish ponds or through release by aquarists. The first record was in 2000 from Lake Empangeni. Subsequently collected from the Mhlatuze River and the Mposa/Nseleni confluence. Currently considered established in rivers in the Richards Bay area. Has also been recorded from dams in the Crocodile River, Mpumalanga. **Impacts:** Not evaluated in South Africa. Elsewhere, reported impacts include siltation and shoreline instability resulting from burrowing and potential displacement of native fishes. Impacts seriously on gill-net fisheries where these occur. **Control:** None practical once species is established. Releases of this and other aquarium fishes should be discouraged through an educational campaign.

Key reference: Alfaro, M.E.M., Cudmore, B., Orr, R., Fisher, S.J.P., Balderas, S.C., León, N., Courtenay, W.R., Osorio, P.K., Mandrak, N., Torres, P.A., Damián, M.A., Michoacán, J., Gallardo, C.E., Sanguinés, A.G., Greene, G., Lee, D., Orbe-Mendoza, A., Martínez, C.R. & Arana, O.S. 2009. *Trinational Risk Assessment Guidelines for Aquatic Alien Invasive Species: Commission for Environmental Cooperation Project Report QA07.64.* Montreal, Canada.

SAIAB, O. Weyl

Vaal-Orange smallmouth yellowfish

Labeobarbus aeneus

Group: Cypriniformes: Cyprinidae

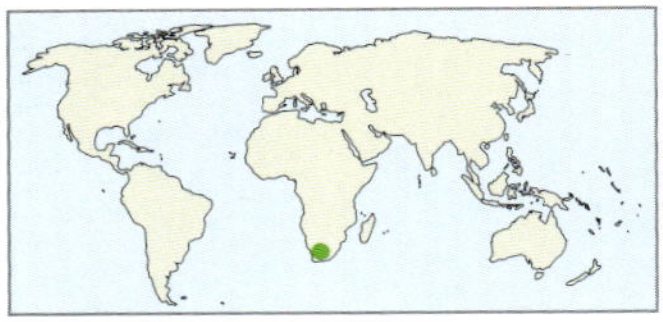

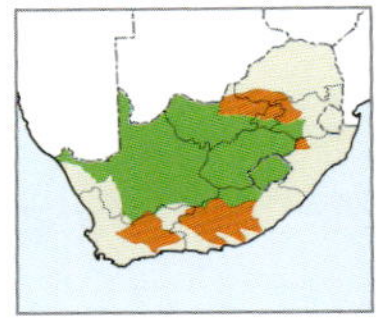

Habitat: Freshwater. Occurs in a wide range of riverine habitats from flowing runs and riffles to deep pools and also large standing waters of impoundments. **Identification and biology:** Body stout and blade-shaped, moderately deep as fish grows, dorsal fin positioned behind origin of pelvic fins, primary ray large and spinous, anterior anal rays extend in adults, caudal fin forked. Head rounded, mouth sub-terminal with variable lips from thin and sharp to full and fleshy ('rubberlips'), two pairs of barbels. Silvery with yellow tinge when juvenile, brazen-olive above with silvery flanks and creamy ventral surfaces when adult, anal fin with orange infusion. Large fish attain more than 50 cm total length and 7.8 kg mass. Breeds in summer after rains, spawning fish assemble in nuptial schools over gravel beds, eggs are adhesive, larvae drift downstream to nursery reaches. Omnivorous, feeding on invertebrates, algae and organic sediments. **Introduction and spread:** Endemic to the west-flowing drainage of the Vaal-Orange River system in southern Africa. A popular angling species and as a result of active stocking, mainly during the 1950s, together with inter-basin water transfer schemes, its distribution range has expanded significantly. Locality records include the east-flowing Gouritz, Sundays, Great Fish, Kariega, Kei and Tugela river systems, and north-flowing Limpopo River.

SAIAB, O. Weyl

SAIAB, R. Bills

Impacts: Not assessed, but there is potential for hybridisation with other *Labeobarbus* species. Able to penetrate small streams that are important conservation areas for indigenous fishes, particularly small cyprinids on which it may prey. May compete with other indigenous fishes for food. It is also a recreational angling target and therefore has some positive economic impacts. Used for subsistence by fishers in rural communities. **Control:** None possible once species is established. Stocking outside its current distribution range in South Africa should be prohibited.

Key references: Impson, N.D, Bills, I.R. & Wolhuter, L. 2008. *Technical report on the state of yellowfish in South Africa 2007.* WRC Report No. KV 212/08.

Skelton, P.H. 2001. *A complete guide to the freshwater fishes of Southern Africa.* Cape Town: Struik.

Orange River mudfish

Labeo capensis

Group: Cypriniformes: Cyprinidae

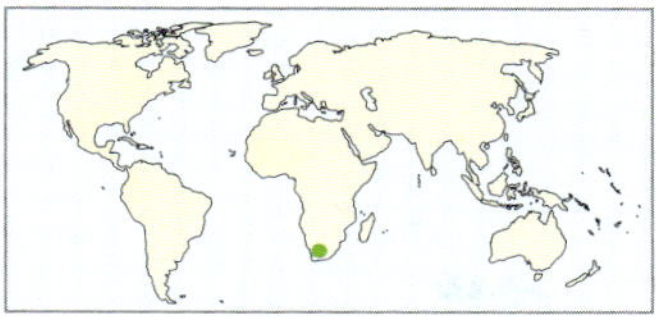

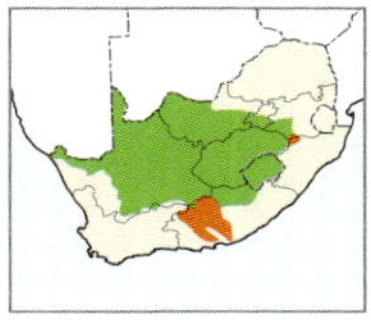

Habitat: Freshwater. Prefers runs and riffles of large rivers, grazing off rocky and other firm surfaces. Also abundant in reservoirs. **Identification and biology:** Body heavy-set and blade-like, nape rises in adults, scales moderately small, dorsal fin origin in advance of pelvic fins, caudal fin forked. Head obtuse, eyes small, snout large, mouth inferior, lips fleshy and papillose, two pairs of short barbels. Attains a total length of 60 cm and weights of 4 kg. Feeds on detritus, diatomaceous floc and algae. Migrates upstream, spawning instream on gravel spawning sites after rains. Also breeds in dams. **Introduction and spread:** This Orange-Vaal River endemic was accidentally introduced into the Great Fish and Sundays rivers following an inter-basin transfer in the late 1970s. Also transferred to the Tugela River system by a Vaal-Tugela inter-basin transfer. **Impacts:** Hybridises with the the indigenous Moggel (*Labeo umbratus*) in the Darlington Dam on the Sundays River system. The driving mechanisms and population-level consequences of such hybridisation are not well understood but at the very least they will result in wasted reproductive effort (if hybrids are sterile), or at worst result in complete introgression and loss of the Moggel populations in invaded systems. Similar hybridisation with indigenous Labeos in the Tugela system is possible. While this fish is targeted by recreational and subsistence anglers in the Orange-Vaal system, it is not considered an important angling species elsewhere and therefore introductions have little positive economic impact. **Control:** None feasible after introduction, particularly when this is facilitated through continuous new introductions via inter-basin water transfers. Direct stocking outside its current distribution range in South Africa should be prohibited.

Key references: De Moor, I.J. & Bruton, M.N. 1988. *Atlas of alien and translocated indigenous aquatic animals in Southern Africa*. South African National Scientific Programmes Report 144.

Skelton, P.H. 2001. *A complete guide to the freshwater fishes of Southern Africa*. Cape Town: Struik.

SAIAB, P. Skelton

Banded tilapia

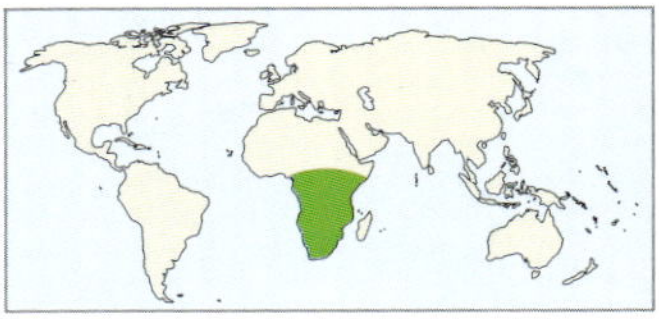

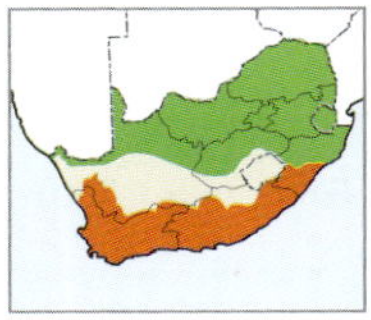

Tilapia sparrmanii

Group: Perciformes: Cichlidae

Habitat: Freshwater. Prefers shallow, quiet or standing water with vegetation in rivers and vleis. **Identification and biology:** Ovoid body shape, dorsal fin with 13–15 spines, caudal fin truncate. Head convex, eyes moderately large, mouth short, teeth fine. Colour variable, often deep olive with dark body bars, tilapia spot on anterior base of the soft dorsal fin. Lower lip blue, pelvic fins sooty, median fins with red edges. A small species, seldom exceeding 20 cm in total length and 300 g in weight. Omnivorous, feeding on epiphytic and benthic algae, macrophytes, invertebrates, small crustaceans and fish fry. Breeds in summer and breeding pairs guard eggs and larvae in a saucer-shaped nest excavated into the substratum. **Introduction and spread:** Widespread in central and southern Africa. In South Africa the natural distribution is limited to the Vaal-Orange and warmer east-flowing rivers. The initial translocation is reported to have occurred in 1941 when fish were transported from the Vaal River system to the Jonkershoek hatchery in the Western Cape. After successful breeding it was widely distributed as a fodder fish for bass from the 1950s. Initially promoted as an alternative to Bluegill sunfish (p. 56), this fish has since been spread by illegal introductions by anglers. Highly invasive and tolerant of a wide range of temperatures (4–32°C) and low oxygen concentrations. It has been observed to jump over barriers as high as 1 m and can therefore penetrate small streams. As a result, populations have established in almost all major river systems in the country. **Impacts:** Competition with indigenous fishes has resulted in the decline of a number of threatened fishes. Also a vector for at least 14 species of fish parasite. **Control:** Restrict further introductions.

Key references: De Moor, I.J. & Bruton, M.N. 1988. *Atlas of alien and translocated indigenous aquatic animals in Southern Africa*. South African National Scientific Programmes Report 144.

Skelton, P.H. 2001. *A complete guide to the fresh-water fishes of Southern Africa*. Cape Town: Struik.

C. Griffiths

Redbreast tilapia

Tilapia rendalli

Group: Perciformes: Cichlidae

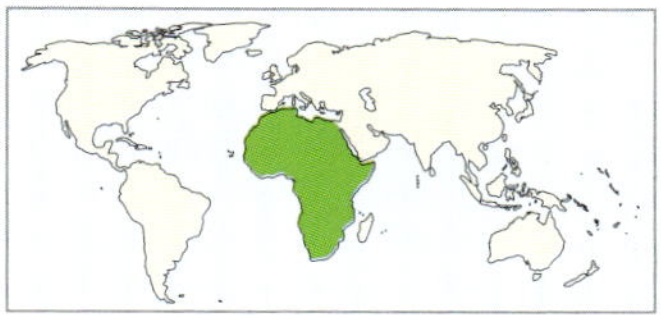

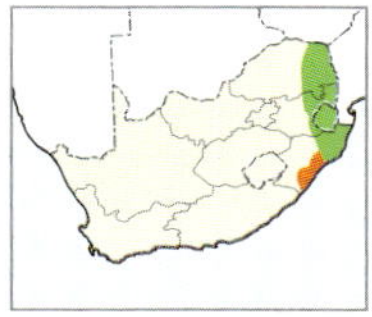

Habitat: Freshwater. Prefers quiet, deep waters in larger rivers, lakes, dams and lagoons with submerged or floating plants. Can tolerate brackish water and temperatures as low as 11°C. **Identification and biology:** Body deep and ovoid, dorsal fin with 14–16 spines, caudal fin truncate. Head profile steep, slightly convex, eyes moderately large, mouth short, prominent and 'beak like', teeth bicuspid. Brown with olive back and vertical bars, cream ventral surfaces which on chin, throat and chest turn pink to deep blood-red in adults, dorsal fin with tilapia spot, caudal fin bicoloured, upper half lighter from yellow to red, lower half darker, sooty olive. Attains about 40 cm total length and a maximum weight of 2 kg. Feeds on plant leaves and algae, as well as invertebrates. Breeding pairs excavate pit-like nest with tunnel chambers in shallow sandy areas, guard eggs and brood. **Introduction and spread:** Native to, and widespread through Africa; in South Africa it occurs naturally in warmer regions down to Lake Sibaya. Initially bred in the Umgeni hatchery in KwaZulu-Natal and subsequently distributed in the 1950s for aquatic weed control and as a forage species for bass. Has since established extralimital populations in many of the warmer river systems in KwaZulu-Natal. **Impacts:** Rapidly denudes dams of aquatic vegetation. This has resulted in total denudation, habitat loss and increased turbidity of dams where this fish was stocked. It is also a popular table fish and angling species. Elsewhere in Africa an important small-scale aquaculture species. **Control:** None feasible once established in a river system. Further spread of this species should be controlled by prohibiting stocking outside its natural distribution range.

Key references: De Moor, I.J. & Bruton, M.N. 1988. *Atlas of alien and translocated indigenous aquatic animals in Southern Africa*. South African National Scientific Programmes Report 144.

Skelton, P.H. 2001. *A complete guide to the freshwater fishes of Southern Africa*. Cape Town: Struik.

SAIAB, P. Skelton

Mozambique tilapia

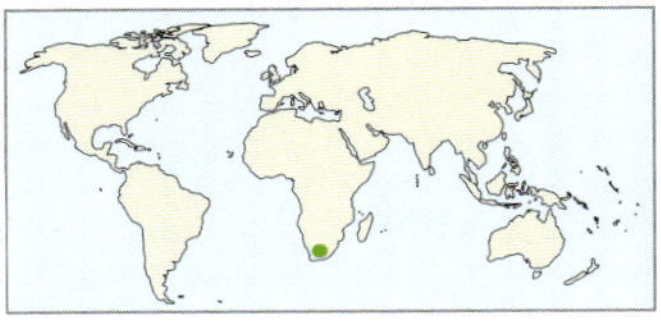

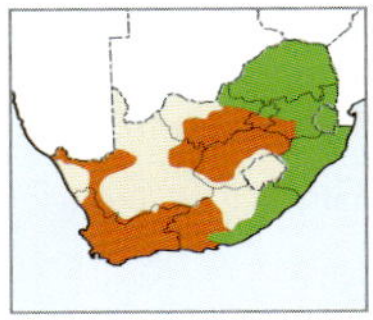

Oreochromis mossambicus

Group: Perciformes: Cichlidae

Habitat: Freshwater and brackish water to above 35 parts per thousand. Thrives in shallow standing waters with vegetation, including rivers, lakes and coastal lagoons and estuaries. **Identification and biology:** Deep-bodied, dorsal fin with 15–17 spines, caudal fin truncate. Head profile concave, especially in breeding males, eyes moderately small, mouth fairly large, protruding in large breeding males, teeth fine in broad bands. Silvery when smaller, deep grey with lighter lower flanks and ventral surfaces, dorsal and caudal edges red. Males turn black with white chin and throat in breeding dress. Attains 40 cm total length. Feeds mainly on bottom sediments, detritus and algae. A mouthbrooder, males form large saucer-shaped nest in shallow sandy areas. Intolerant of cold water below about 12°C. Able to survive at very high salinities. **Introduction and spread:** Native to the warmer east-flowing rivers from the Zambezi in the north down to the Bushman's River in the Eastern Cape. Has been translocated widely for both angling and as a fodder fish for bass. Extralimital populations are now present in freshwater regions of all major river systems in the country, but limited inland by cold winter temperatures. In addition, many temperate estuaries have also been invaded. **Impacts:** Documented impacts include competition with indigenous species. Has also been an important vector for the introduction of various fish diseases. A popular table fish and angling species used by recreational, subsistence and commercial fishers. Also cultured in the freshwater food fish industry. **Control:** None feasible, except in small isolated water bodies.

Key references: Canonico, G.C., Arthington, A., McCrary, J.K. & Thieme, M.L. 2005. The effects of introduced tilapias on native biodiversity. *Aquatic Conservation: Marine and Freshwater Ecosystems*, 15:463-483.

Skelton, P.H. 2001. *A complete guide to the freshwater fishes of Southern Africa*. Cape Town: Struik.

SAIAB, P. Skelton

Nile tilapia

Oreochromis niloticus

Group: Perciformes: Cichlidae

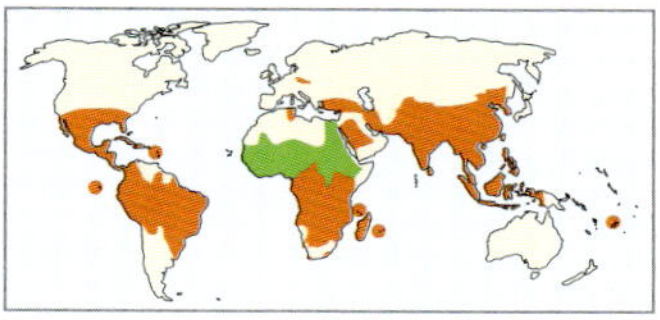

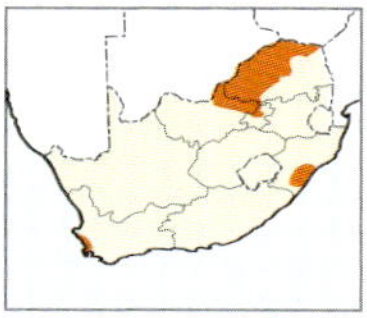

Habitat: Freshwater, tolerant of brackish water. Favours shallow vegetated standing waters of lake shores and large rivers. Intolerant of temperatures below 14°C. **Identification and biology:** Deep-bodied, dorsal fin with 16–18 spines, caudal fin truncate. Maximum total body length 28 cm. Head profile straight, eyes moderately large, mouth terminal and short. Silvery with olive-brown and grey-to-black body bars and spots, bluish tinges on cheeks and head, caudal fin vertically striped in black and white. Breeding males have red flush on head and body. Males construct large saucer-shaped nests in sandy substrates. Females brood eggs within the mouth cavity. Primarily feeds on phytoplankton, but will also take invertebrates and even small fish. **Introduction and spread:** Native to central and north Africa. One of the 10 most introduced species in the world. In South Africa, introduced as fodder fish for bass in 1959 and 1978. Fish were released into farm dams in the Stellenbosch area, but this introduction was unsuccessful. Widely spread in other southern African countries for aquaculture purposes in the early 1980s. As a result of escape from aquaculture facilities and through intentional introduction by anglers, has become established in the Incomati and Limpopo river systems. **Impacts:** Most commonly used tilapia for fisheries enhancements and aquaculture worldwide. Ecological effects of Nile tilapia invasions include decreased abundance and extinction of native species resulting from habitat and trophic overlaps and competition for spawning sites, habitat destruction and water-quality changes, and hybridisation with other *Oreochromis* species. **Control:** This aggressive invader cannot be controlled once established. Further introductions should be prevented by enforcing current legislature banning stocking of this species. Aquaculture facilities using Nile tilapia must be escape-proof and based on recirculating systems.

Key references: Canonico, G.C., Arthington, A., McCrary, J.K. & Thieme, M.L. 2005. The effects of introduced tilapias on native biodiversity. *Aquatic Conservation: Marine and Freshwater Ecosystems*, 15:463-483.

Skelton, P.H. 2001. *A complete guide to the fresh-water fishes of Southern Africa*. Cape Town: Struik.

SAIAB, O. Weyl

African sharptooth catfish

Clarias gariepinus

Group: Siluriformes: Clariidae

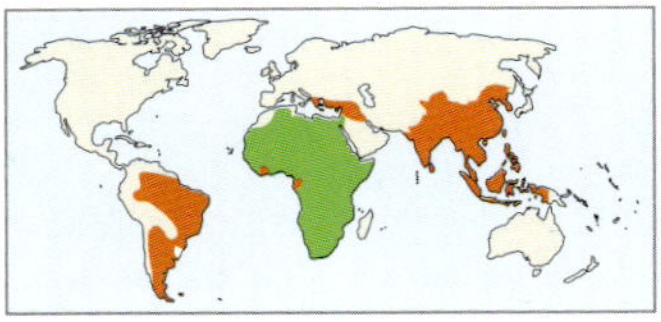

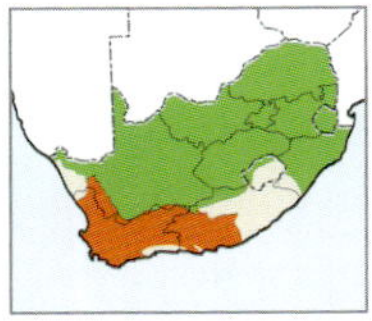

Habitat: Freshwater. Rivers, ponds, floodplains, reservoirs and lakes. Tolerant of poor water quality, temperatures of 8–35°C and salinities up to 10 parts per thousand. Favours slow-flowing parts of large rivers and lakes. **Identification and biology:** A large fish, capable of attaining over 30 kg and 1.5 m total body length. Long-bodied and scaleless, with extended soft-rayed dorsal and anal fins, spines only in pectoral fins, caudal fin rounded. Head large, depressed, heavily bony-armoured; eyes small, lateral and forward on head; mouth terminal and wide with four pairs of long simple barbels. Special air-breathing apparatus in chambers above gills. Colour varies from dark mottling above with creamish ventral surfaces to plain brown with creamish ventral surfaces. Extremely resilient and versatile; invariably the last fish species surviving in drying out pools of savannah rivers. Can air-breathe and will emerge at night or in overcast wet conditions to move overland if necessary. Omnivorous, with food items ranging from plankton to fishes, frogs, reptiles and nestling birds. Migrates to flooded vegetated marginal spawning sites after rain; eggs are laid on vegetation. **Introduction and spread:** Widespread with an almost pan-African distribution extending into Turkey. In South Africa native to the warmer east-flowing drainages and to the Orange-Vaal River system. As a result of illegal stocking, escape from hatcheries and through inter-basin water transfers, extralimital populations now common in all major drainages of the Eastern and Western Cape. **Impacts:** While the impact of this large omnivore has not been directly assessed in South Africa, predation effects and changes in vertebrate and invertebrate community structure are likely. Also likely to compete for resources with indigenous species. Also a valued food source for humans and other vertebrates including crocodiles and fish-eating birds and mammals. **Control:** None feasible once this species is introduced. Could be targeted by directed fishing to control population size in some reservoirs.

Key references: De Moor, I.J. & Bruton, M.N. 1988. *Atlas of alien and translocated indigenous aquatic animals in Southern Africa*. South African National Scientific Programmes Report 144.

Skelton, P.H. 2001. *A complete guide to the freshwater fishes of Southern Africa*. Cape Town: Struik

P. Gourley

SEA SQUIRTS Ascidiacea

Despite their sedentary lifestyle and simple appearance, sea squirts (also termed ascidians or tunicates) are close relatives of vertebrates, as evidenced by the presence of a notochord during their tadpole-like larval stage. Adults are sessile and can be large and solitary or tiny and colonial. All draw water in through an incurrent siphon, filter it through a sieve-like branchial basket and then expel the filtered water through an excurrent siphon. Sea squirts are exclusively marine and can reach considerable abundance in low intertidal and sub-tidal habitats, forming large aggregations that structure benthic communities and filter large amounts of plankton and organic debris. When introduced away from their native range, normally via shipping, they can spread rapidly.

Light-bulb sea squirt

Clavelina lepadiformis

Group: Clavelinidae

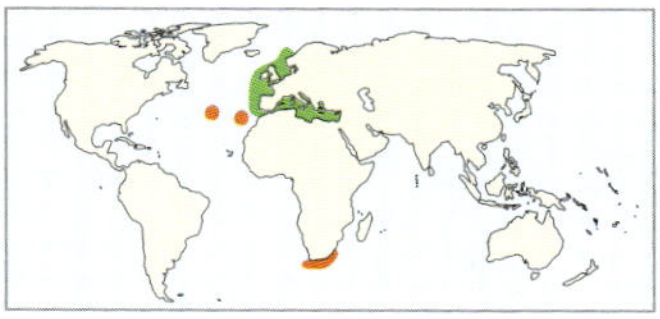

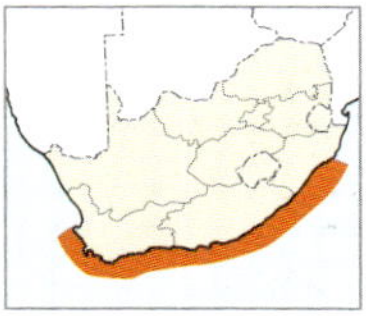

Habitat: On man-made structures in harbours, particularly wharfs and ropes. **Identification and biology:** A strikingly beautiful and unmistakable colonial ascidian. Body elongated and enclosed in a completely transparent outer coating (test) with a bright white stripe up the side and around the siphons. Sieve-like pharynx clearly visible below siphons and bright orange visceral organs located in lower portion of test. Individuals typically 3 cm long and colonies 5–10 cm in diameter. **Introduction and spread:** A recent introduction from the north-east Atlantic Ocean, first recorded in South Africa only in 2001, but already common in harbours and estuaries from Durban to Saldanha Bay. Not recorded on the open coast, but can be very abundant in sheltered harbours. **Impacts:** Since it does not appear to have invaded the open coast, unlikely to have significant impacts, although can be a significant fouling species in harbours. **Control:** No specific control methods known, but can be scraped from structures or killed by removing them from the water.

C. Griffiths

Key references: Monniot, C., Monniot, F., Griffiths, C.L. & Schleyer, M. 2001. South African ascidians. *Annals of the South African Museum*, 108:1-141.

Turon, X., Tarjuelo, I., Duran, S. & Pascual, M. 2003. Characterising invasion processes with genetic data: an Atlantic clade of *Clavelina lepadiformis* (Ascidiacea) introduced into Mediterranean harbours. *Hydrobiologia*, 503:29-35.

Jelly crust tunicate

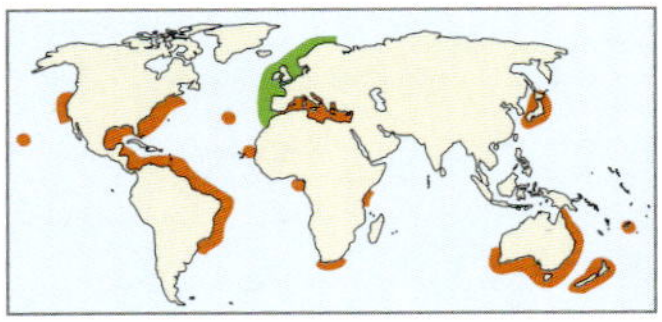

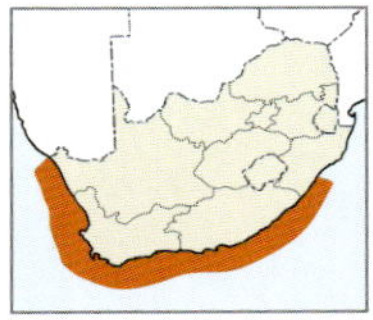

Diplosoma listerianum

Group: Didemnidae

Habitat: On rocky shores and artificial structures. **Identification and biology:** Forms thin, fragile, pale yellow to dark grey jelly-like sheets, within which pale, rounded and microscopic zooids are embedded. Breaks up easily and therefore often overlooked, or misidentified as a sponge. The zooids have independent incurrent siphons, through which water is sucked and filtered through the sieve-like pharynx. Filtered water discharged through larger, transparent and inconspicuous excurrent siphons dispersed over colony surface. Colonies grow over all types of substrata, and can reach diameters of 50 cm. **Introduction and spread:** Accidentally introduced from European waters prior to 1949, probably as a fouling organism. Now widespread in harbours and on sheltered shores from Alexander Bay to Durban. **Impacts:** Common and significant fouling species, especially of other ascidians. Impacts probably minimal because of fragile structure. **Control:** None known or proposed.

C. Griffiths

Key references: Millar, R.H. 1955. On a collection of ascidians from South Africa. *Proceedings of the Zoological Society of London*, 125:169-221.

Monniot, C., Monniot, F., Griffiths, C.L. & Schleyer, M. 2001. South African ascidians. *Annals of the South African Museum*, 108:1-141.

Vase tunicate

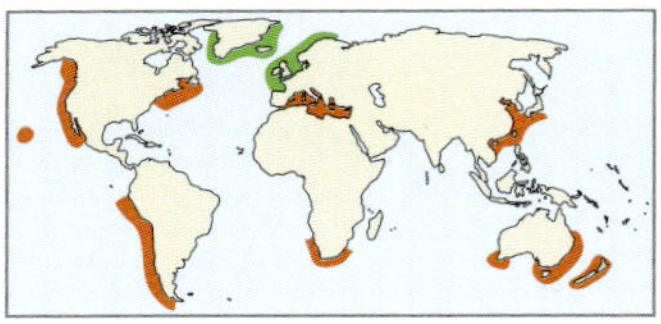

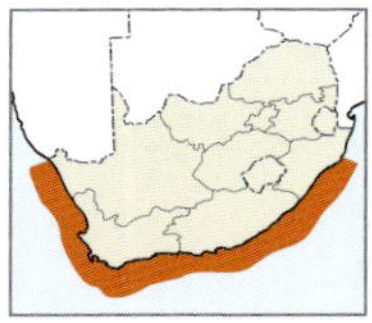

Ciona intestinalis

Group: Cionidae

Habitat: On floating pontoons, ship hulls, submerged ropes and wharfs, preferring shaded surfaces. **Identification and biology:** Tall, cylindrical, yellowish solitary ascidian with a soft, floppy, transparent test, revealing internal organs. Reaches 15 cm in length. Forms large aggregates on submerged structures. **Introduction and spread:** Accidentally introduced from the North Atlantic before 1955, now abundant in harbours and lagoons from Saldanha Bay to Durban. **Impacts:** Important pest, quickly coating marine structures. Dense growths smother and kill mussels on aquaculture facilities, especially on mussel ropes. **Control:** Mechanical removal tedious and needs frequent repetition; treating culture ropes with acetic acid, brine solution, hydrated lime and pressurised water effective; biological control using rock crabs feasible.

C. Griffiths

Key references: Carver, C.E., Chisholm, A. & Mallet, A.L. 2003. Strategies to mitigate the impact of *Ciona 516 intestinalis* (L.) biofouling on shellfish production. *Journal of Shellfish Research*, 22:621-631.

Robinson, T.B., Griffiths, C.L., McQuaid, C.D. & Rius, M. 2005. Marine alien species of South Africa – status and impacts. *African Journal of Marine Science*, 27:297-306.

Crevice ascidian

Ascidia sydneiensis

Group: Ascidiidae

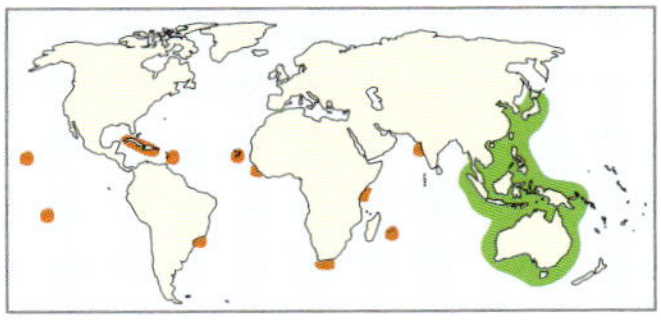

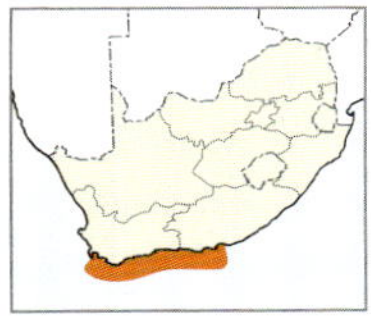

Habitat: Harbours and marinas, rocky shores, usually in crevices or under rocks. **Identification and biology:** Large (up to 15 cm in length), solitary, sometimes semi-transparent but mostly covered by other fouling species. Lies on its side among other ascidian species forming large clumps, or in narrow cracks and crevices, often covered in gravel and pieces of shell. Siphons elongated and ridged, with lobed openings. Tunic breaks easily when collected. **Introduction and spread:** Probably introduced as a fouling organism from the north-eastern Pacific on ships and first recorded from False Bay in 1932. Now common in harbours and under boulders on rocky shores from False Bay to Bushman's River. **Impacts:** Common and significant fouling species in harbours. **Control:** None attempted and would be difficult, given the cryptic habitat.

C. Griffiths

Key references: Millar, R.H. 1955. On a collection of ascidians from South Africa. *Proceedings of the Zoological Society of London*, 125:169-221.

Monniot, C., Monniot, F., Griffiths, C.L. & Schleyer, M. 2001. South African ascidians. *Annals of the South African Museum*, 108:1-141.

Dirty sea squirt

Ascidiella aspersa

Group: Ascidiidae

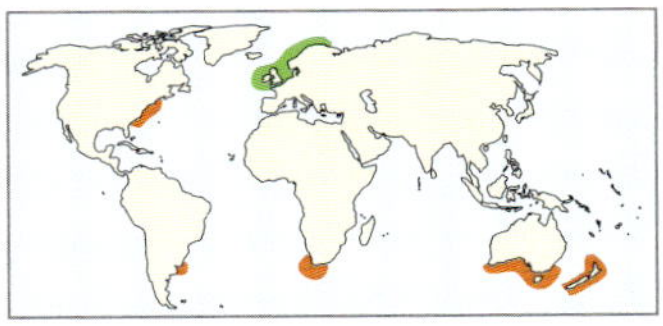

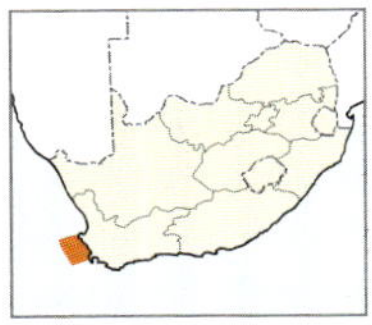

Habitat: Attached to harbour ropes and floating pontoons. **Identification and biology:** Up to 10 cm long, solitary, rounded and elongated. Incurrent siphon located at the top of the body, excurrent siphon about one-third down the side of the body. Tunic grey and semi-transparent, sometimes papulate and pale red around the siphons, usually covered with other sessile life forms (epibionts) that grow on it and/or detritus. Tunic stronger than Crevice ascidian (above) but still fragile. Individuals can form aggregates with others of the same species or other fouling species. **Introduction and spread:** A European species occurring from Saldanha Bay to Table Bay. **Impacts:** Not known, but together with Vase tunicate (p. 72) and Leathery sea squirt (p. 74) form important clumps of fouling in Table Bay harbour. **Control:** Likely the same methods suggested for Vase tunicate.

C. Griffiths

Key references: Locke, A. 2009. A screening procedure for potential tunicate invaders of Atlantic Canada. *Aquatic Invasions*, 4:71-79.

Monniot, C., Monniot, F., Griffiths, C.L. & Schleyer, M. 2001. South African ascidians. *Annals of the South African Museum*, 108:1-141.

Pleated sea squirt

Styela plicata

Group: Styelidae

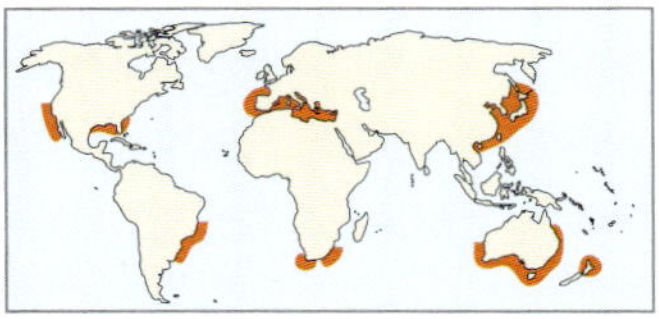

Habitat: Harbours and marinas. **Identification and biology:** Squat, off-white, solitary ascidian, with four pairs of brown stripes on each siphon. Often covered by fouling species such as algae and other ascidians. Body height 5–10 cm, occasionally reaching 15 cm. **Introduction and spread:** Introduced through shipping supposedly from Asia and now common in harbours from Richard's Bay to Mossel Bay. In 2010 a single specimen was spotted in Saldanha Bay. **Impacts:** Can out-compete other native ascidians by affecting some of their early life-history stages. Found on mussel ropes in South Africa in low abundances and forms large aggregations in harbours where it can be an important fouling organism. **Control:** No known method.

M. Rius

Key references: Barros, R.C., Rocha, R.M. & Pie, M.R. 2009. Human-mediated global dispersion of *Styela plicata* (Tunicata, Ascidiacea). *Aquatic Invasions*, 4:45-57.

Rius, M., Turon, X. & Marshall, D.J. 2009. Non-lethal effects of an invasive species in the marine environment – the importance of early life-history stages. *Oecologia*, 159:873-882.

Leathery sea squirt

Cnemidocarpa humilis

Group: Styelidae

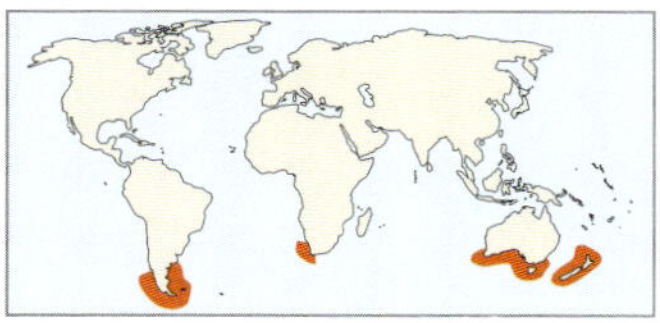

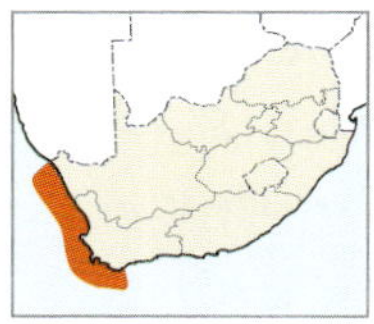

Habitat: On wharfs, harbour ropes, etc. in harbours. **Identification and biology:** A solitary species with leathery tunic normally covered by epibionts. The tunic is extremely tough, especially when stressed. The siphons are fairly prominent and elongated with wide openings. Typically reaches a maximum length of about 5 cm. **Introduction and spread:** Accidentally introduced via shipping and now common in harbours and marinas in various parts of the world. It is a common species in New Zealand, Australia and the southern part of South America. Recorded from outside harbours, including those close to Simon's Town. **Impacts:** Can be an important fouling species on marine structures, especially on floating pontoons and harbour ropes. **Control:** Unknown.

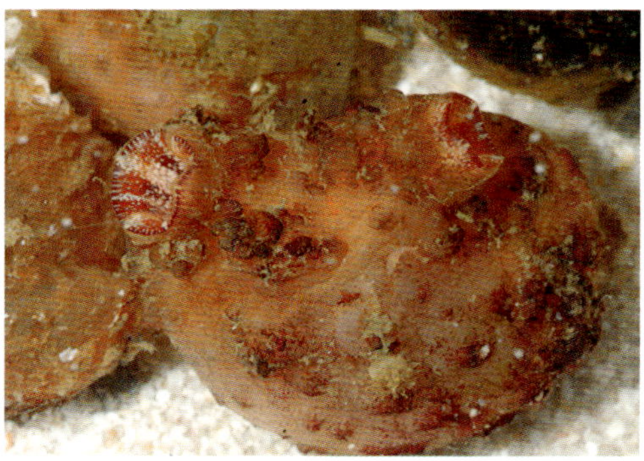

C. Griffiths

Key references: Monniot, C., Monniot, F., Griffiths, C.L. & Schleyer, M. 2001. South African ascidians. *Annals of the South African Museum*, 108:1-141.

Primo, C. & Vázquez, E. 2008. Zoogeography of the southern New Zealand, Tasmanian and southern African ascidian fauna. *New Zealand Journal of Marine and Freshwater Research*, 42:233-256.

Star sea squirt

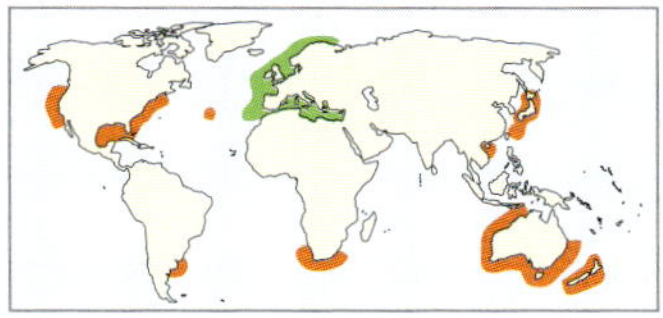

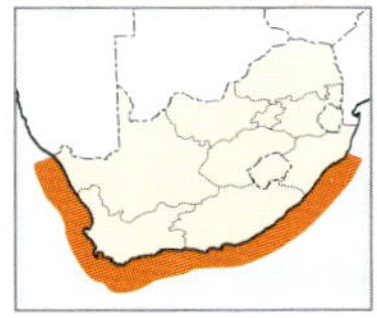

Botryllus schlosseri

Group: Styelidae

Habitat: On rocky shores, artificial structures and seaweeds. **Identification and biology:** Colonial, forming firm, gelatinous sheets on rocks, seaweeds and other hard structures. Colony diameter typically 5–10 cm; individual zooids do not exceed 2–3 mm in diameter. Background colour usually chocolate, zooids white and arranged in distinct star-shaped groups of 8–12 individuals. However, this species presents several colour morphs, so the most distinctive characteristic is the shape of the zooids. Each zooid has an independent incurrent siphon; filtered water is ejected through a common exhalant opening in the centre of each circular group. **Introduction and spread:** Considered a European species, it was first recorded in South Africa in 1946, now occurring in harbours and sheltered lagoons from Alexander Bay to Durban. **Impacts:** An encrusting filter-feeder that can coat man-made structures creating a nuisance, but unlikely to have significant impacts. **Control:** No control methods known.

C. Griffiths

Key references: López-Legentil, S., Turon, X. & Planes, S. 2006. Genetic structure of the star sea squirt, *Botryllus schlosseri,* introduced in southern European harbours. *Molecular Ecology*, 15:3957-3967.

Monniot, C., Monniot, F., Griffiths, C.L. & Schleyer, M. 2001. South African ascidians. *Annals of the South African Museum*, 108:1-141.

Blunt-spined microcosmus

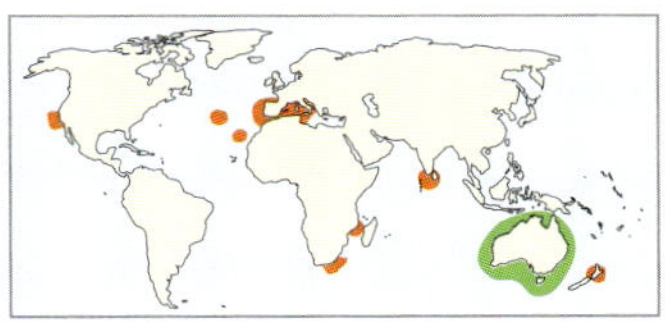

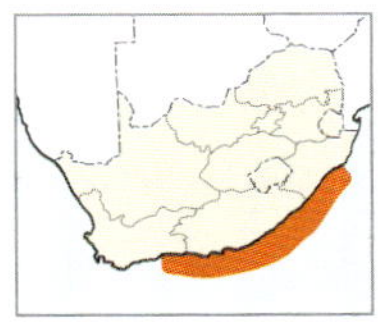

Microcosmus squamiger

Group: Pyuridae

Habitat: Aggregates on wharfs, ropes, etc. in harbours. **Identification and biology:** Solitary, with irregularly oval, red-brown tunic typically covered with other fouling organisms, making it hard to spot. Attaches to substratum by its bottom and side, the closely spaced short siphons ending in striped red-and-orange bands. It usually forms large aggregates, mostly with other individuals of the same species, but also with other species of ascidians and mussels. Body usually no longer than 5 cm. **Introduction and spread:** Accidentally introduced from Australia via shipping, now common in harbours and marinas from Richard's Bay to Mossel Bay. **Impacts:** Can be an important fouling species on marine structures. On the west coast of North America found fouling oyster farms, and in the Mediterranean found outside harbours, covering rocky reefs and excluding native species. **Control:** Unknown.

C. Griffiths

Key references: Rius, M., Pascual, M. & Turon, X. 2008. Phylogeography of the widespread marine invader *Microcosmus squamiger* (Ascidiacea) reveals high genetic diversity of introduced populations and non-independent colonizations. *Diversity and Distributions*, 14:818-828.

Rius, M., Pineda, M.C. & Turon, X. 2009. Population dynamics and life cycle of the introduced ascidian *Microcosmus squamiger* in the Mediterranean Sea. *Biological Invasions*, 11:2181-2194.

ECHINODERMS Phylum Echinodermata

This group comprises the starfish, sea urchins, brittle stars, feather stars and sea cucumbers. All are characterised by possessing spicules or spines in the skin, after which the phylum is named (echinos = a hedgehog, derma = skin). The spicules are developed to very differing degrees in the various groups, being most obvious in sea urchins, which have a hard shell and conspicuous spines, and least so in sea cucumbers, where they are reduced to minute spicules in a soft, flexible skin. The body form of echinoderms is highly unusual in that most are five-sided, with no obvious head, sense organs, or indeed front and back ends. Movement is achieved by numerous hydraulically operated tube-feet, another unique feature characteristic of this group. Sexes are separate and reproduction usually involves release of eggs and sperm into the water, followed by fertilisation and development of widely dispersing planktonic larvae. A few species brood their eggs within the body and give birth to fully developed young.

Little six-arm brittle star

Ophiactis savignyi

Group: Ophiuroidea: Ophiactidae

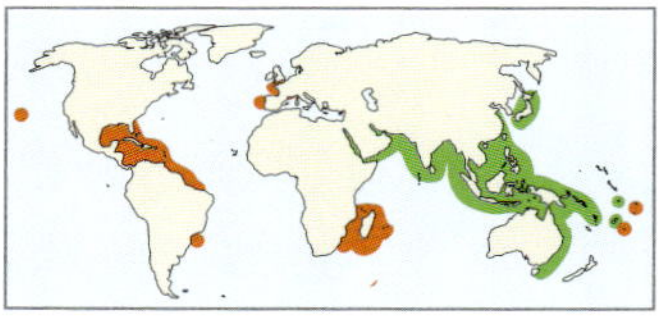

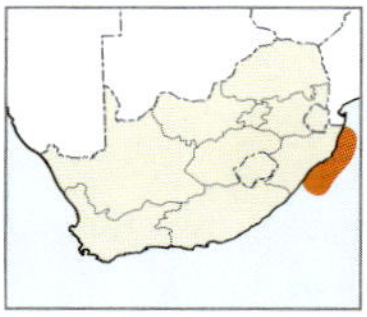

Habitat: Marine, on tropical rocky shores and reefs, under rocks, or among sponges and other fauna. **Identification and biology:** Tiny brittle star with arms less than 20 mm long. Disc variegated light and dark, arms moderately spiny and banded with green or yellow, alternating with white. Unusual in that most specimens have six, not the normal five, arms. Capable of reproducing asexually by division, leading to three-armed individuals, and to colonies of genetically identical animals. Most populations dominated by males, suggesting that males undergo fission more than females.

Introduction and spread: Widely introduced around the world via shipping and reputed to be the commonest and most widespread brittle star. Introduced to Durban harbour prior to 1950, almost certainly via shipping from Asia. **Impacts:** Unlikely to have significant impacts due to its small size and infaunal, detritus-feeding habits. **Control:** No control needed or proposed.

Key reference: Clark, A.M. & Courtman-Stock, J. 1976. *The Echinoderms of Southern Africa*. London: British Museum (Natural History).

J. Olbers

Black sea urchin

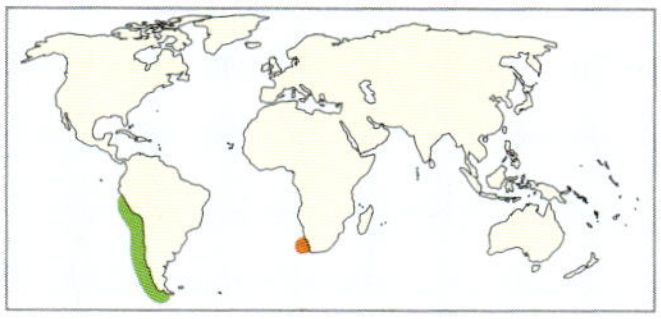

Tetrapygus niger

Group: Echinoidea: Arbaciidae

Habitat: Marine, rocky shores, feeding on seaweeds, especially kelps. **Identification and biology:** A black urchin with equal black spines. Shell (test) of dead urchin reddish-purple, somewhat flattened and oval, diameter up to 60 mm. **Introduction and spread:** Introduced from Chile to oyster farms in Alexander Bay on the west coast, almost certainly with oyster spat, and has now established a breeding population. Has not yet spread to the open coast. **Impacts:** Could pose a significant threat if it becomes established on the open shore, since known to impact kelp beds in native Chile. **Control:** Given their present restricted range could be eliminated chemically, or by manual collection, but will be difficult to control should they escape into the open ocean.

Key reference: Haupt, T.M, Griffiths, C.L., Robinson, T.B. & Tonin, A.F.G. 2010. Oysters as vectors of marine aliens, with notes on four introduced species associated with oyster farming in South Africa. *African Journal of Zoology*, 45:52-62.

C. Griffiths

C. Griffiths

AUTHOR: C. JANION

SPRINGTAILS Class Collembola

Springtails are primitive, wingless invertebrates (Apterygotes) and among the most abundant and widespread organisms in the world, being especially common in soil. They are best recognised by their unique forked jumping organ, or furca, folded under the abdomen, and the 'ventral tube', an abdominal organ used for water balance. There are more than 7 000 species. Although they resemble insects, some scientists argue that they are more closely related to crustaceans.

Clover springtail

Sminthurus viridis

Group: Symphypleona: Sminthuridae

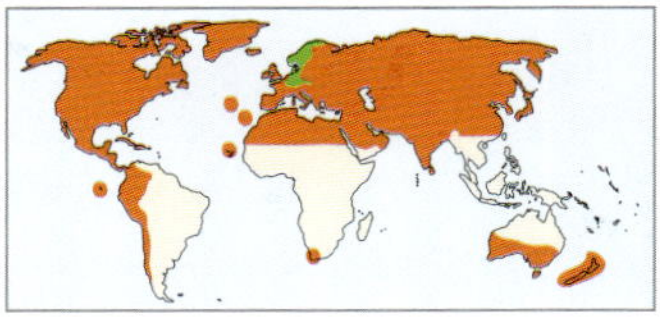

Habitat: Terrestrial, plant-feeding. Occurs in a wide variety of habitats, including grasslands and field crops, such as clover. **Identification and biology:** Typical spherical body shape, 2–4 mm long, usually green or yellowish-brown; depending on environmental conditions can produce several generations per year. Eggs laid in soil. **Introduction and spread:** Origin likely Europe, now widely spread in temperate parts of the world, including Australia, where it is a pest of legumes and cereals and is known as the 'lucerne flea'. Probably introduced to South Africa from Australia. **Impacts:** Can cause damage to leguminous crops such as alfalfa (lucerne). Green leaves of crops are frequently eaten, and seedling death may occur with heavy infestations. **Control:** Systemic and contact insecticides have been used. The predatory Pasture snout mite (*Bdellodes lapidaria)* was released in Caledon in 1963 as a biocontrol agent.

Key reference: Wallace, M.M.H. 1974. Present and probable world distribution of *Sminthurus viridis* and prospects for its biological control. *Pedobiologia*, 14:238-243.

C. Janion

Marsh springtail

Isotomurus palustris

Group: Entomobryomorpha
Isotomidae

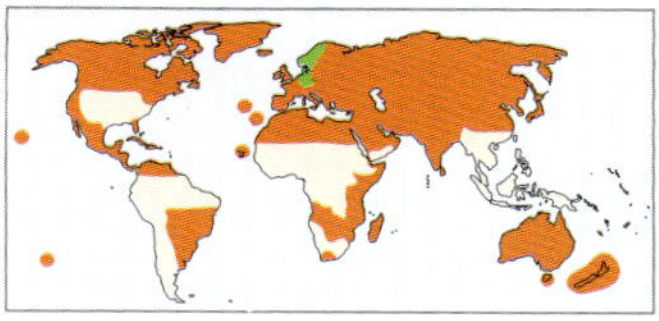

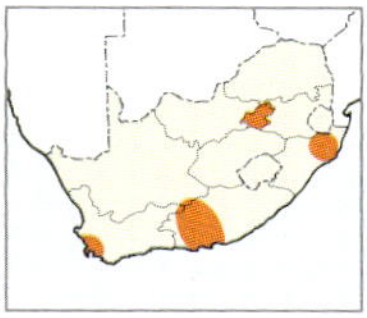

Habitat: Terrestrial, in wet habitats. **Identification and biology:** Up to 2.5 mm in length, with a yellow-brown colour pattern and darker central dorsal stripe. Long, thin hairs on the dorsal side of abdomen on the second to fourth abdominal segments are characteristic of the genus. **Introduction and spread:** A northern European species that has spread globally and is now one of the most frequently recorded species, with a cosmopolitan distribution in temperate areas. Found in damp habitats associated with standing water. **Impacts:** Causes plant damage when occurring in large numbers, but effects on plants usually secondary as feeds mostly on decaying plant material. **Control:** None recorded, but have been used in toxicity tests.

C. Janion

Key reference: Fjellberg, A. 2007. The Collembola of Fennoscandia and Denmark, Part II. Entomobryomorpha and Symphypleona. *Fauna Entomologica Scandinavica* Vol. 43. Leiden: Brill Academic Publishers.

Mushroom springtail

Hypogastrura manubrialis

Group: Poduromorpha: Hypogastruridae

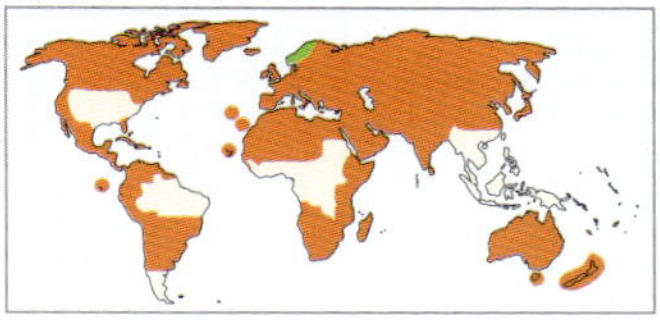

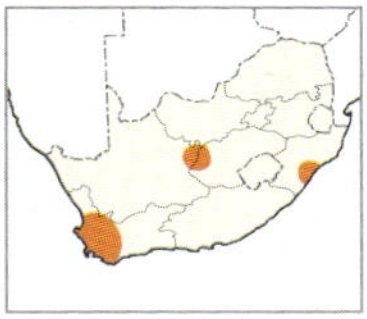

Habitat: Terrestrial, usually in upper soil and litter layer. **Identification and biology:** Up to 1.5 mm in length, colour grey to reddish-blue. **Introduction and spread:** Originally from northern Europe, now cosmopolitan in distribution, including scattered localities in South Africa. Usually found in disturbed areas or near human dwellings. **Impacts:** Considered a pest of mushrooms as outbreaks have occurred on mushroom farms in Europe. However, this is no longer a problem as soil sterilisation is now standard procedure among commercial mushroom growers. Found in manure in glasshouses in Australia, but no damage recorded. **Control:** Measures against outbreaks include disinfection during preparation of mushroom beds and use of sterile manure. Known to have a high resistance to insecticides.

C. Janion

Key reference: Fjellberg, A. 1998. The Collembola of Fennoscandia and Denmark, Part I. Poduromorpha. *Fauna Entomologica Scandinavica* Vol. 35. Leiden: Brill Academic Publishers.

Moss springtail

Neanura muscorum

Group: Poduromorpha: Neanuridae

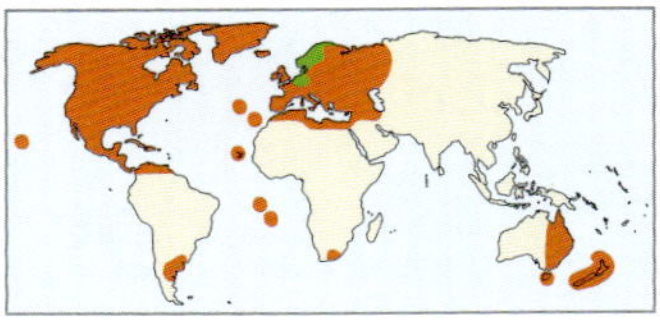

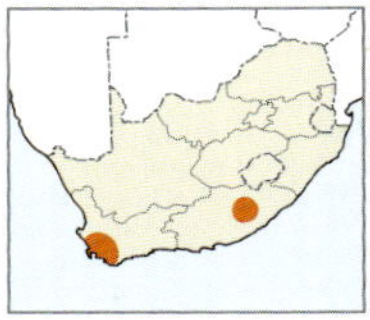

Habitat: Terrestrial, in a wide variety of habitats from coastal to alpine, usually in litter and under rotting wood. **Identification and biology:** Size up to 3.5 mm, colour bluish-grey to lilac. Body has a distinctive warty appearance, hair long and spiny and three eyes on each side of head. Aromatic substances secreted may play a role as defence against predators. Mostly parthenogenetic, the females laying unfertilised eggs that develop into viable offspring, but sexual forms are found. Like all Collembola, they moult throughout their adult life. **Introduction and spread:** A northern European species now found in most temperate regions of the world. Localised in the Eastern and Western Cape provinces of South Africa. **Impacts:** None known, and given their small size and benign feeding habits unlikely to be harmful. **Control:** None recorded.

C. Janion

Key reference: Coates, T.J. 1986. The Collembola of South Africa. 1. The genus *Neanura*. *Journal of the Entomological Society of Southern Africa*, 31:185-195.

C. Janion

SILVERFISH

Class Insecta, Order Thysanura

Silverfish, or 'fishmoths', are a small group of primitive, wingless insects with soft, elongated, tapering bodies ending in three long, bristly 'tails'. The body is coated in tiny overlapping scales that rub off easily, giving it a metallic appearance and 'greasy' feel. Most silverfish are nocturnal and have poorly developed eyes, relying on their long antennae and body hairs to feel their way around. Eggs are laid in cracks and crevices and develop into miniature adults that live for approximately three years. Silverfish are known for their ability to absorb water vapour from the atmosphere and can survive without drinking. Three introduced species occur in the region, but they are difficult for the non-specialist to distinguish, so only one is detailed here.

•Grey silverfish

Ctenolepisma longicaudata

Group: Lepismatidae

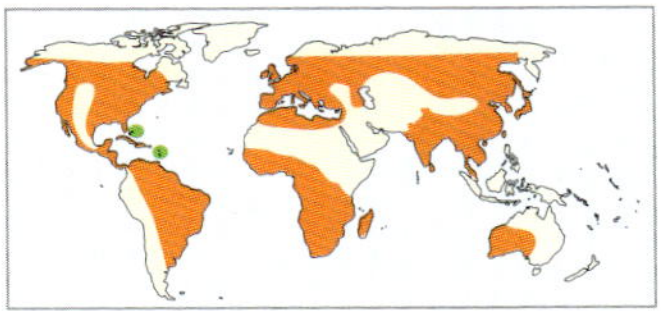

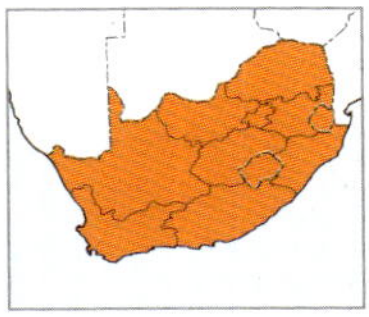

Habitat: In buildings, particularly among books and stored papers, also in kitchens and behind loose wallpaper. **Identification and biology:** Adults typically 15 mm long, excluding tails, and covered in dense coating of scales and hairs, silver-grey above and off-white below, antennae long, three long tails held splayed. Abundant in buildings, feeding on starch, glue, paper, dry foods, wool and the like. Often encountered when sorting old books and papers, or when cleaning out drawers and cupboards. Characteristic gnawed damage to binding of books or stored documents often more evident than the cryptic insects themselves. Sometimes found trapped in smooth washbasins, bowls or light fittings, where they can remain alive for extended periods. **Introduction and spread:** Widely dispersed globally from their native Caribbean. Introduced to South Africa from Europe before 1900 and now abundant in buildings throughout the region. **Impacts:** Can cause considerable economic damage in libraries and to valuable books, documents and photographs, particularly attracted to the glued binding of books. **Control:** Valuable libraries and archives are best protected by annual fumigation. Silverfish in the home can be controlled by sprinkling Borax powder in cupboards or bookshelves, or with commercially available insecticide baits.

Key reference: Theron, J.G. 1963. The domestic fish moths of South Africa (Thysanura: Lepismatidae). *South African Journal of Agricultural Science*, 6:125-130.

C. Griffiths, M. Picker

COCKROACHES Class Insecta, Order Blattodea

An ancient, unmodified group of relatively large, oval and flattened insects with biting mouthparts and running legs. The forewings are leathery and the hindwings thin and transparent. In some species only males are winged and in others both sexes are wingless. Most are nocturnal and feed on plant material, although a wide range of foodstuffs is accepted. Eggs are produced in a characteristic large, purse-shaped egg case that is sometimes seen protruding from the abdomen of females. Nymphs resemble small, wingless adults. The few domestic and pest species are best known, but numerous wild and endemic species also occur in natural habitats.

American cockroach/ Common cockroach

Periplaneta americana

Group: Blattidae

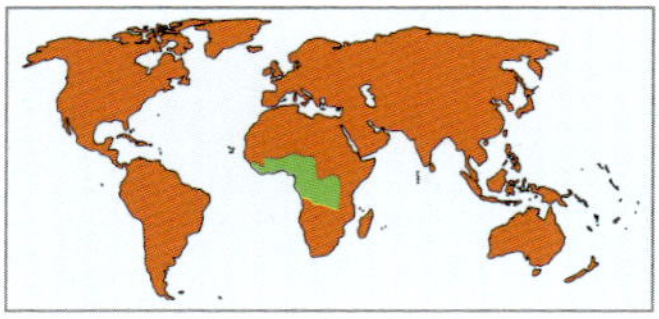

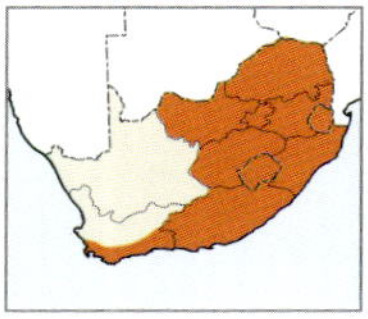

Habitat: The most common household cockroach in the region, abundant in houses, especially kitchens and bathrooms, in drains and sewers, indeed wherever humans live. Most abundant near the coast and in warm, humid situations. **Identification and biology:** Familiar large, red-brown cockroach with long, spiny legs; attains over 40 mm body length. Wings are fully formed in both sexes and adults fly readily. Occur in mixed groups of nymphs and adults. Infected sites have a characteristic smell. Females produce purse-like egg cases containing a double row of cylindrical pockets, each containing a single, elongated egg. The young resemble tiny, wingless adults and moult six or seven times before attaining sexual maturity at 6–12 months of age, depending on temperature and food availability. Adults live about a year. **Introduction and spread:** Native to West Africa, and probably secondarily introduced from Europe by early settlers and now abundant in human habitations in all but the driest regions of South Africa. Capable of flying, but more often passively dispersed by 'hitch-hiking' with goods. **Impacts:** An important and greatly disliked pest species that consumes food materials and spreads bacteria by walking over contaminated materials. Also causes allergic reactions. **Control:** Best controlled by a combination of improved sanitation, elimination of suitable habitat and use of one or more of a wide variety of commercially available cockroach traps or insecticides.

C. Griffiths

Key reference: Skaife, S.H. 1979. *African insect life*. Cape Town: Struik Publishers.

German cockroach

Blattella germanica

Group: Blattellidae

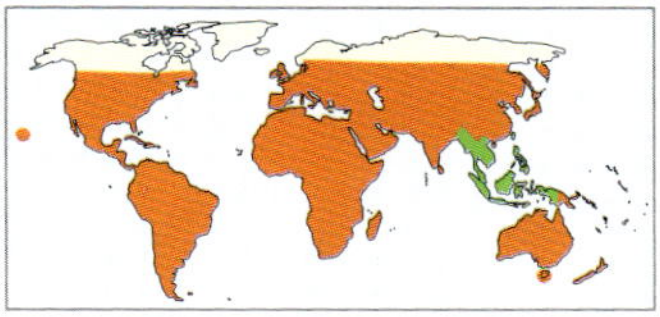

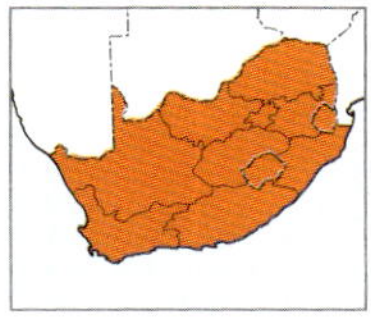

Habitat: In human dwellings, preferring warm and more humid locations, such as kitchens and drains. The most common cockroach at inland locations in South Africa and in much of the world. **Identification and biology:** A relatively small species, with adult body length 11–15 mm. Yellowish-brown with two parallel dark stripes extending down thorax. Both sexes are fully winged, but are unable to sustain prolonged flight. Live in mixed groups of adults and nymphs, feeding on almost any organic material available. Females use a sex pheromone to attract males and continue to carry the egg case projecting from the abdomen as the eggs develop (shown below). Fast growing, reaching maturity in 12 weeks. Lifespan only 6–9 months. **Introduction and spread:** Originated from Asia, but almost certainly an early secondary introduction to South Africa from Europe and now found in human habitations throughout the region, although the American cockroach dominates near the coast. **Impacts:** One of the most unpopular of insects. Spread pathogenic organisms, both on their body surfaces and in their faeces, transmitting not only bacterial diseases, but also the eggs of parasites, and hence posing significant threats to human health. Also emit a powerful allergen that causes allergic reactions, especially in patients with asthma. **Control:** More difficult to control than the American cockroach because its smaller size allows it to invade smaller crevices and because of its faster rate of reproduction. Best managed using an integrated approach, in which access to shelter is minimised through sealing of apertures and good food hygiene is practised to limit access to food and water. A wide range of commercial baits, traps and insecticides are also available for control of existing infestations.

Key references: Picker, M., Griffiths, C.L. & Weaving, A. 2002. *Field guide to insects of South Africa*. Cape Town: Struik Publishers.

Skaife, S.H. 1979. *African insect life*. Cape Town: Struik Publishers.

C. Griffiths

AUTHOR: C. GRIFFITHS

TERMITES Class Insecta, Order Isoptera

Termites (sometimes inappropriately termed 'white ants') are social insects that live in large colonies made up of distinct forms or 'castes'. These comprise winged reproductives ('kings and queens', also termed 'flying-ants'), soldiers that can be recognised by their enlarged heads and jaws, and workers that are pale and eyeless and are the most numerous caste. Most termite species build large, distinctive nests or termitaria from soil, but some live in galleries excavated in wood. Termites play an enormously important ecological role as consumers of live or dead plant material, but may also cause major economic damage to wooden structures or crops.

West Indian drywood termite/House termite

Cryptotermes brevis

Group: Kalotermitidae

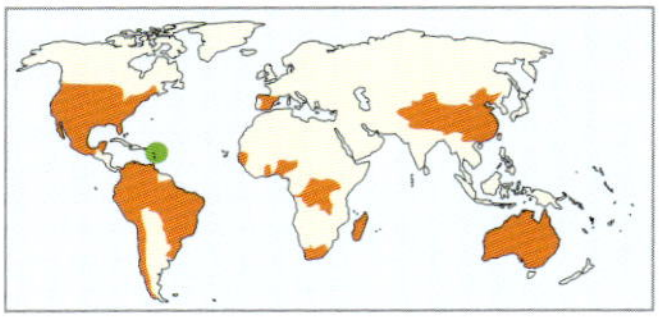

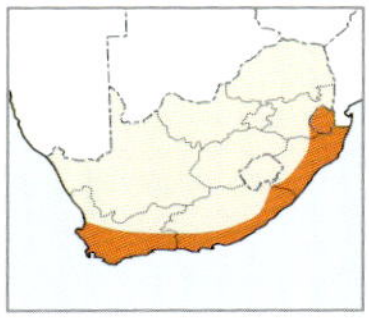

Habitat: Dead wood of all types, including both structural timber and furniture and the dead branches of forest trees. Prefers dry habitats (as the name implies). Hollows out cells in wood connected by narrow channels and seldom emerge from hiding. **Identification and biology:** Soldiers easily recognised by a combination of small body size (5–8 mm long) and unusual flattened, rounded head with undulating front margin and short, broad jaws. Form small colonies of only a few hundred individuals, living in galleries excavated in dry wood. Unusually, there is no true worker caste; the nymphs of the reproductive caste act as workers. As in other wood-eating termites, digestion is assisted by a rich fauna of protists that live inside the gut. **Introduction and spread:** Perhaps the most widely distributed of all termites, originating from the West Indies, but transported around the world with timber. Thought to have been introduced to South Africa in about 1918, probably in the wood of crates or other wooden products. **Impacts:** One of the most destructive termites in the world, causing severe damage to wooden structures of all types. **Control:** Best achieved by fumigation, since there can be many small colonies present within the same building. Individual items of infected furniture can be treated by injecting insecticide into the galleries, or by heating the entire piece to at least 60°C for 30 minutes.

Key references: Skaife, S.H. 1979. *African insect life*. Cape Town: Struik.

Uys, V.M. 2002. *A guide to the termites of Southern Africa*. Plant Protection Research Institute Handbook 15. Pretoria: Agricultural Research Council.

C. Griffiths, M. Picker

Formosan subterranean termite

Coptotermes formosanus

Group: Rhinotermitidae

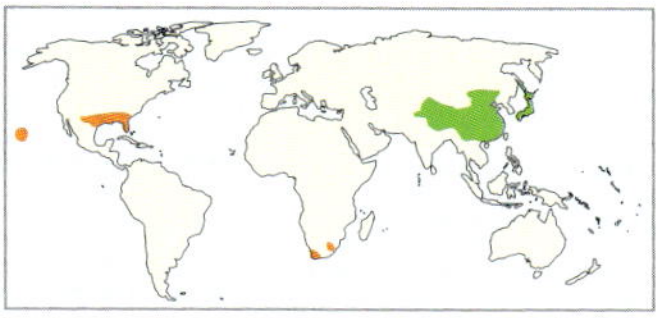

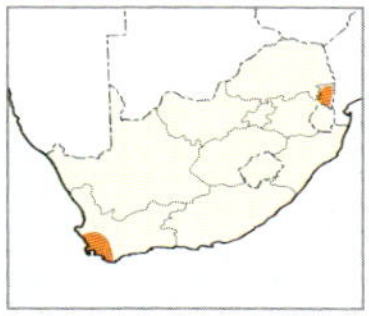

Habitat: Forms large subterranean colonies and can feed either on living plants, including ornamental trees, or utility poles and other structural timber, also paper and cardboard. Colonies usually need to maintain contact with the ground in order to obtain necessary moisture. **Identification and biology:** Soldiers 12–15 mm long, yellow-brown with oval head, best recognised by well-developed tube-like structure (fontanelle) on front margin of head, above the curved jaws. This emits a milky-white fluid when the soldier is provoked. Form very large colonies comprising millions of individuals and have a foraging range up to 100 m. As with other termites, spread by releasing up to 70 000 winged reproductives, which fly a short distance before losing their wings and seeking small crevices or hollows in which to establish a new colony. **Introduction and spread:** Native to Asia and accidentally introduced to Simon's Town in 1924, where responsible for considerable damage to seasoned timber, both within and removed from buildings; also reported from Komatipoort. Widely introduced internationally. **Impacts:** Can be very destructive to both living trees and structural timber. Feeding generally follows grain of the wood, but eventually large hollow spaces are created that can be filled with a characteristic nest-like material called 'carton' made of termite excrement and macerated wood, mixed with saliva and soil. A colony can consume 400 g of wood a day and destroy wooden structures within a few months. Nicknamed 'super termite' in the USA, where it is one of the most destructive of all pest species, costing over US$1 billion annually in either damage or preventive and remedial treatment. **Control:** Very difficult to control because of its subterranean habits. In infected areas wooden structures in contact with the soil should be treated with creosote or other chemicals. Chemical barriers can also be introduced into soil to prevent termite invasions, while a variety of baits can be used to eliminate existing infestations

Key references: Harris, W.V. 1966. On the genus *Coptotermes* in Africa (Isoptera: Rhinotermitidae). *Proceedings of the Royal Entomological Society of London* Series B, 35:161-171.

Uys, V.M. 2002. *A guide to the termites of southern Africa*. Plant Protection Research Institute Handbook 15. Pretoria: Agricultural Research Council.

Scott Bauer (USDA), Wikimedia Commons

C. Griffiths

EARWIGS Class Insecta, Order Dermaptera

Earwigs are a fairly small and uniform order of insects with about 2 200 species worldwide and 47 in South Africa. They are long and slender, typically brown and easily identified by the conspicuous forceps at the end of the abdomen. These are not venomous, but can inflict a painful pinch and are used to fold and unfold the intricate wings. Females care for the eggs and first instar nymphs. Earwigs typically hide under stones, bark, leaf axils and other narrow spaces.

Ring-legged earwig

Euborellia annulipes

Group: Anisolabididae

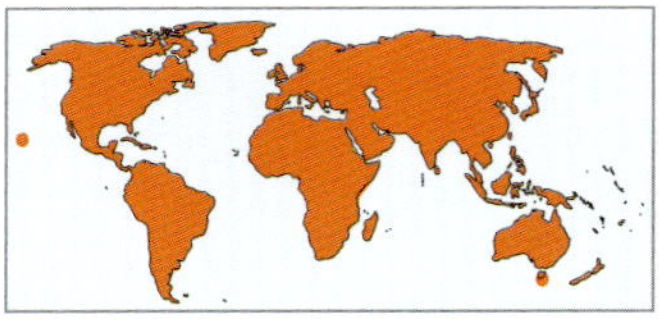

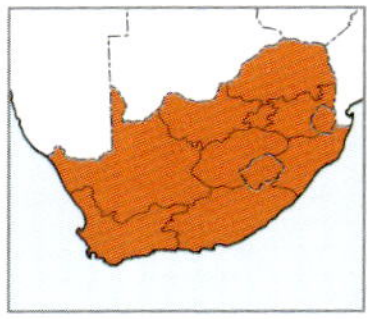

Habitat: Compost heaps in gardens, litter in pine plantations. **Identification and biology:** Body 10–15 mm long, completely wingless, smooth and shiny brown, legs light brown to yellow, often with dark ring on femur, forceps short and stout, triangular at base, tips typically crossing. Omnivorous, with preference for carnivory. **Introduction and spread:** Other species in the same genus extremely widely distributed, with many species on all continents, as well as single species on several continents, so natural range unclear. *Euborellia* species prefer warmer climates, and were originally confined to southern regions of Europe. First recorded in South Africa in 1912. **Impacts:** No negative impacts reported. **Control:** No control necessary, however, trials undertaken to use a closely related species as a biological control for other pests.

Key reference: Situmorang, J. & Gabriel, B.P. 1988. Biology of two species of predatory earwigs *Nala lividipes* (Dufour) (Dermaptera: Labiduridae) and *Euborellia* (*Euborellia*) *annulata* (Fabricius) (Dermaptera: Carcinophoridae). *Philippine Entomologist*, 7:215-238.

C. Griffiths, M. Picker

Sand earwig/Striped earwig

Labidura riparia

Group: Labiduridae

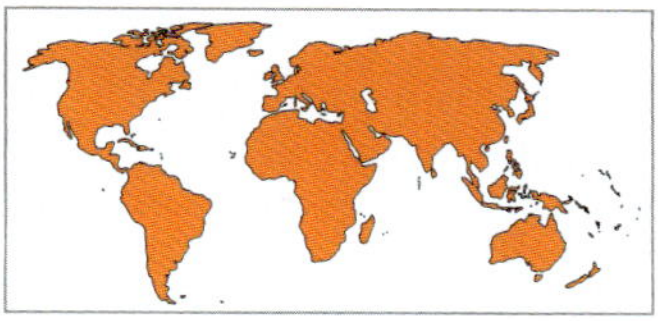

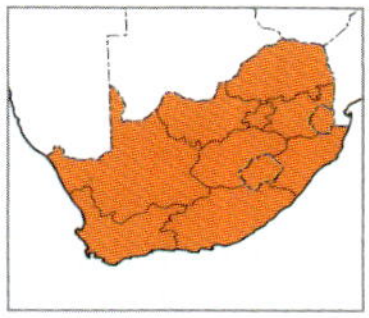

Habitat: Sandy soil, often along rivers and beaches, where it shelters under stones and debris. **Identification and biology:** Large species, up to 30 mm body length including slightly curved forceps; males and females similar. Body yellowish or light brown, dark with reddish stripes on pronotum and forewings. Forewings well developed, hindwings sometimes reduced and hardly visible. Fly well, at least some populations. Forceps used in defence, but also to attack and catch prey, typically other arthropods. They have some supporting function during copulation, but male never pinches female. Strictly carnivorous. **Introduction and spread:** Centre of origin unclear, occurs worldwide (first description of this species was from Siberian populations), distribution probably partially by passive transport via logs and plants floating on sea currents and partially by human introduction. First recorded in South Africa in 1863. **Impacts:** No negative impacts reported. **Control:** Occasionally experimented with as a biocontrol agent.

Key reference: Situmorang, J. & Gabriel, B.P. 1988. Biology of *Labidura riparia* (Pallas) and its predatory capacity on the Asian corn borer. *Philippine Entomologist*, 7:195-214.

C. Griffiths, M. Picker

C. Griffiths

Lesser earwig

Labia minor

Group: Spongiphoridae

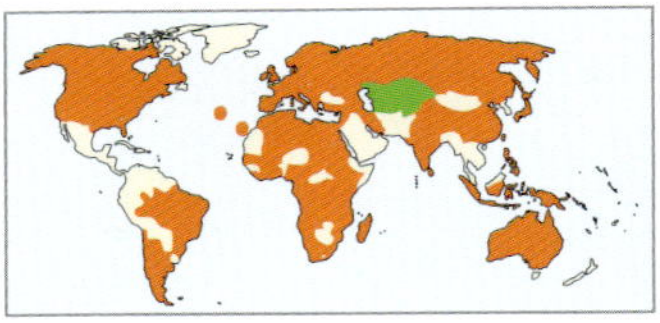

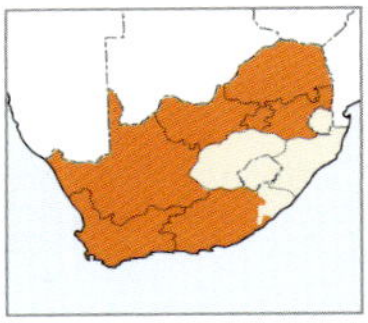

Habitat: Dung heaps used for nesting, hibernation and hunting grounds.
Identification and biology: Length 5–6 mm, always fully winged and regularly seen flying. Male forceps gently curved, with small teeth on inner margin, separated at base; those of female somewhat shorter, straight and close at base. Colour uniform brown to dark brown, legs brighter, head somewhat darker. Body hairy/pubescent. Carnivorous, may also feed on decaying plant matter. **Introduction and spread:** Widely distributed worldwide, with other species in the same genus in Asia. Often associated with dung heaps. Probably spread by agriculture and domestication of animals over the last few thousand years. First recorded in South Africa in 1900. **Impacts:** No negative impacts reported. **Control:** No control necessary.

Whitney Cranshaw, Colorado State University, Bugwood.org, Wikimedia Commons

Key reference: Mourir, H. 1986. Notes on the life history of *Labia minor* (L.) (Dermaptera), a potential predator of housefly eggs and larvae (Diptera, *Musca domestica* L.). *Entomologiske Meddelelser*, 53:143-148.

Black field earwig

Nala lividipes

Group: Labiduridae

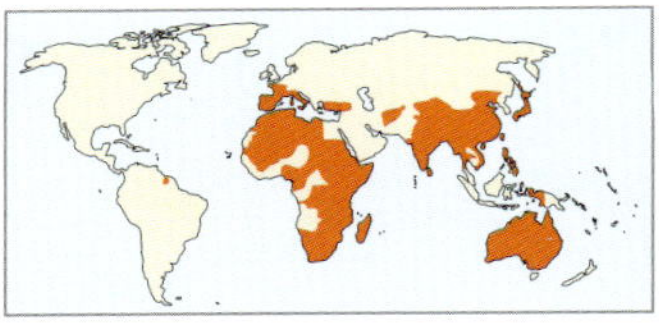

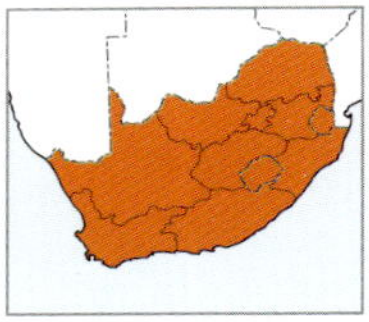

Habitat: Rotting wood and other damp places in subtropical regions.
Identification and biology: Body length 7–10 mm including forceps. Uniformly dark brown, with forewings and wing tips lighter brown. Entire body covered in fine hairs. Male forceps short, gently curved and widely separate at base, in female almost straight and contiguous at base. Omnivorous. **Introduction and spread:** Widely distributed in the Old World and Australia, few records from the Americas. Other species in the same genus occur in Africa and Asia; apparently the natural range is warm (but not hot) tropics. In artificial conditions fed on germinating seeds of wheat, sorghum, maize and sunflower; other sources claim essentially carnivorous. First recorded in South Africa in 1900. **Impacts:** Some crop damage reported from Australia. **Control:** Pesticides, but occasionally experimented with as a biocontrol agent.

Donald Hobern from Canberra, Australia, Wikimedia Commons

Key reference: Simpson, G.B. 1993. Effects of temperature on the development, longevity and fecundity of *Nala lividipes* (Dufour) (Dermaptera: Labiduridae). *Australian Journal of Entomology*, 32:265-272.

AUTHOR: C. GRIFFITHS

STICK INSECTS Class Insecta, Order Phasmatodea

Large but cryptic insects with elongated, slender bodies and very long, thin walking legs. Wings may be present or absent, but are usually only fully developed in males, which are generally smaller, thinner and less numerous that females. Stick insects are slow-moving, nocturnal herbivores and their main defence mechanism is their uncanny ability to mimic twigs. Reproduction is remarkably casual – the female simply drops eggs to the ground one at a time as she continues to feed. The eggs are large, look exactly like seeds and hatch into nymphs that resemble tiny adults. Parthenogenesis, where unfertilised eggs develop and hatch in the normal way, is common and in some species males are rare, or even unknown.

•Indian stick insect/ Laboratory stick insect

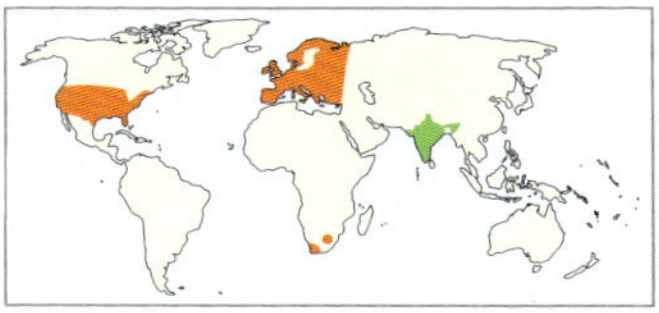

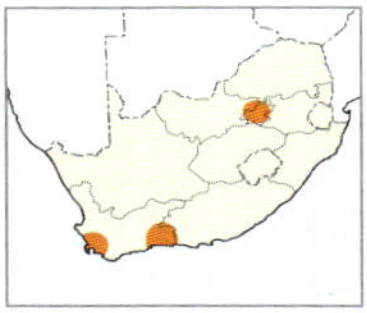

Carausius morosus

Group: Phasmatidae

Habitat: Common in both suburban gardens and in natural vegetation, particularly in and around the Cape Peninsula. **Identification and biology:** Females reach about 80 mm in body length, are wingless and relatively fat for a stick insect. Thorax may be smooth or bear small knobs, often arranged in a line along the sides. Colour generally uniform yellow, brown or green, but always with a characteristic bright red patch at base of inner surface of front legs. Males rare; thinner, shorter (50–60 mm) and brown, but have not been found in South Africa, where the population appears to be entirely parthenogenetic. Eggs (shown below left) are large, round and brown with a yellow knob, closely resembling seeds. Eggs hatch in 4–6 months and nymphs take 4–6 months to attain maturity, adults living a further six months. Feed mainly on ivy and privet, but will accept a wide range of other plants. **Introduction and spread:** Originally from India, but widely dispersed by research and pet trade. Exact mode and date of introduction unknown, but thought to have been introduced for laboratory use, or as pets, and to have subsequently spread into the wild, perhaps through careless disposal of eggs. **Impacts:** Reported to reach pest proportions in gardens in the USA. **Control:** No specific control methods known, but can easily be manually collected at night.

Key reference: Picker, M., Griffiths, C. & Weaving, A. 2002. *Field guide to insects of South Africa*. Cape Town: Struik.

C. Griffiths, M. Picker

C. Griffiths, M. Picker

WEB-SPINNERS Class Insecta, Order Embioptera

Embiids are a little-known order of insects consisting of only about 200 described species. They have long, cylindrical, soft-skinned bodies and short legs, but their most distinctive feature is the greatly swollen tarsal ('hand') segments of the first pair of legs. These are equipped with silk glands used to spin the silken tunnels and galleries within which embiids typically live. Females are wingless, but males have two similar pairs of elongated wings. Most species live in small colonies under bark or stones, in leaf litter and in other dark, damp places, where they feed on bark, lichen, moss and dead plant material. Nymphs resemble miniature adults, but are paler.

Saunders embiid

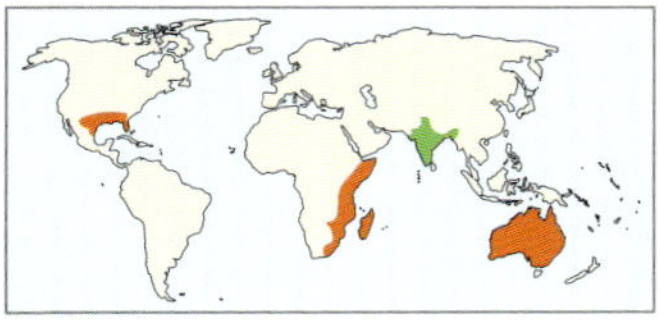

Oligotoma saundersii

Group: Oligotomidae

Habitat: Within silken webs spun around cracks in the bark of large shade trees. **Identification and biology:** Adults elongated and cylindrical, about 10 mm in length and dark reddish-brown to chocolate in colour, with short legs. Females are wingless, but males have two pairs of similarly long, flexible wings that extend to the end of the abdomen. Live in small family groups within a network of webbing radiating outwards from a cavity or crack in the bark of large trees. The silken galleries are often camouflaged with finely chewed fragments of wood and lichen and by faeces of the occupants. Feed on lichens and mosses gathered from the bark surface. Females seldom emerge from the galleries, but males fly between colonies and are sometimes attracted to lights, where they are often mistaken for small flying ants (the reproductive caste of termites). **Introduction and spread:** Introduced from their native India to Durban, almost certainly with timber, and now common in gardens and on street trees. Possibly more widespread, but have been poorly sampled because of their cryptic habits. **Impacts:** Given their specialised and benign habits, unlikely to have any significant ecological or economic impacts. Elsewhere forms webbing on export grapes. **Control:** No specific control methods known or required, although could be eliminated by insecticide sprays.

Key reference: Picker, M., Griffiths, C. & Weaving, A. 2002. *Field guide to insects of South Africa.* Cape Town: Struik.

C. Griffiths, M. Picker

BOOKLICE

Class Insecta, Order Psocoptera

Booklice or psocids are a diverse group of over 3 000 species of small, cryptic, soft-bodied insects with large, round heads usually bearing long, slender antennae and large eyes. Some superficially resemble true lice, but as the two groups are only distantly related, and as psocids are not parasitic, the common name 'booklice' is misleading. Most species have delicate, transparent wings that show reduced venation and are held in a characteristic tilted, roof-like posture over the body. They feed on algae, fungi, lichens, plant debris or stored food products. The introduced species are unusual in that they are wingless and have relatively short antennae and small eyes.

Domestic booklouse

Liposcelis bostrychophila

Group: Liposcelidae

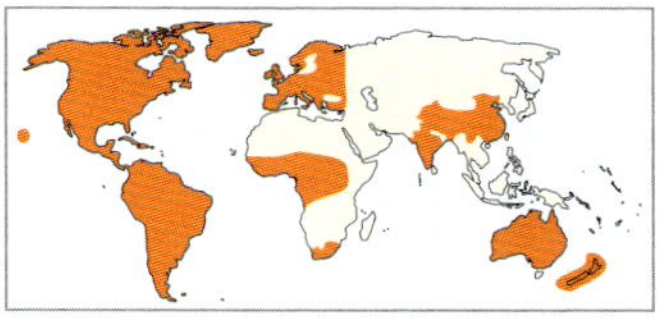

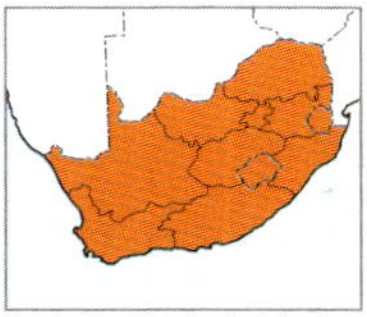

Habitat: In buildings, usually associated with either paper or stored food products. Prefers warm, damp and undisturbed habitats. **Identification and biology:** Tiny (1–2 mm long), brown, wingless insect with flattened body and enlarged, rounded head with biting mouthparts and small eyes. The basal segments (femora) of the third pair of legs are distinctively enlarged, but are not used for jumping. Usually found scuttling around on mouldy paper and on stored food. Can reproduce parthenogenetically (no males required). Females lay large, fertile, dust-covered eggs singly or in small groups. These hatch into small colourless nymphs that pass through four nymphal stages before reaching sexual maturity after about one month. Lifespan is typically about one year. **Introduction and spread:** Thought to have origins in tropical Africa, but subsequently introduced worldwide along with paper and food products. Found in dwellings throughout South Africa. **Impacts:** Feeds on paste and glue of books, stored grain and dried meat products, insect collections and the like, also on microscopic moulds growing on these products. Causes damage to documents and spoilage and loss of value of food materials. May also elicit allergic reaction in sensitive persons. **Control:** Best restricted by simply reducing the availability of food sources and by lowering humidity levels below 50%, either by increasing air circulation, or by use of a dehumidifier. Also susceptible to a wide range of commercially available insecticides. Borax or sodium borate scattered in infected areas can control minor infestations, while severely infected areas can be fumigated.

C. Griffiths, M. Picker

Key reference: Turner, B.D. 1994. *Liposcelis bostrychophila* (Psocoptera: Liposcelididae), a stored food pest in the UK. *International Journal of Pest Management*, 40:179-190.

BUGS

Class Insecta, Order Hemiptera

Members of the order Hemiptera, which includes the suborders **Heteroptera, Sternorrhyncha** and **Auchenorrhyncha,** have piercing, sucking mouthparts. They feed on plant cells and sap, seeds, insects and occasionally vertebrate blood. The majority of alien bug species in South Africa (well over 200 species) belong to the Sternorrhyncha, comprising mostly aphids and scale insects. Their economic impact on local farming and on international trade (through their quarantine importance) is considerable. Use of insecticides for controlling Sternorrhyncha pests also has significant environmental effects. Their impact on native vegetation seems largely unquantified, but there are various injurious exotic species that do not occur here yet, but may pose a future threat.

Eucalyptus thaumastocorid

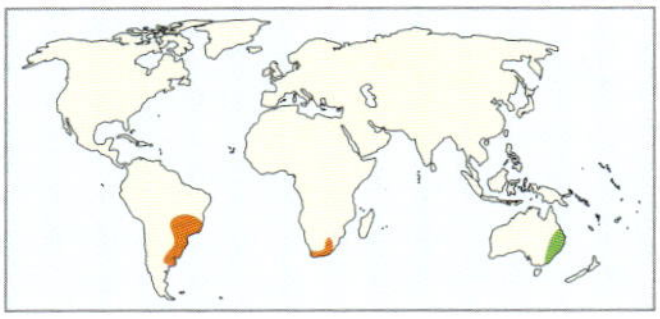

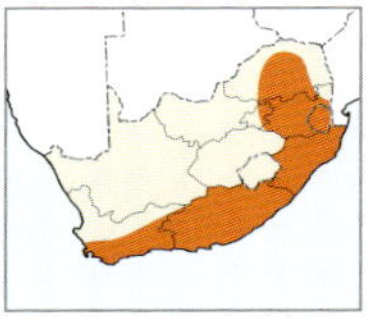

Thaumastocoris peregrinus

Group: Heteroptera: Thaumastocoridae

Habitat: Eucalyptus trees, where they occur in plantations, or in clumps in rural areas. **Identification and biology:** Small (body length 3 mm) flat insects, somewhat elongated; light brown with characteristic dark bulging eyes. Lays eggs in black capsules on leaves, usually in a cluster. Short life cycle of about 35 days, with several generations per year. Feeds on numerous *Eucalyptus* species. **Introduction and spread:** Native to Australia. A recent introduction to South Africa – the earliest record is from 2003. Now also in South America. **Impacts:** Feeding causes tree foliage to change to a yellow-brown colour, and leaves to drop off. Dieback, stunted growth and even death of trees occurs. Poses a threat to the timber industry. Tree debilitation also reduces nectar production, which in turn affects beekeeping. **Control:** Insecticides, and possible natural enemies such as parasitoids, are being investigated.

Key reference: Jacobs, D.J. & Neser, S. 2005. *Thaumastocoris australicus* Kirkaldy (Heteroptera: Thaumastocoridae): a new insect arrival in South Africa, damaging to eucalyptus trees. *South African Journal of Science*, 101:233-236.

C. Griffiths, M. Picker

Green peach aphid

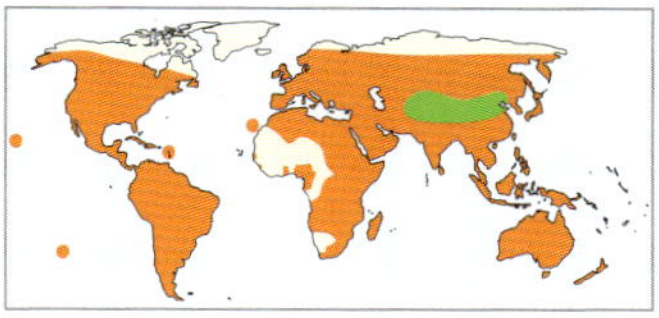

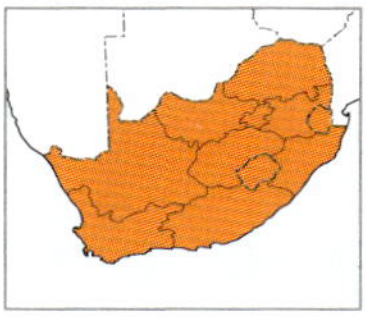

Myzus persicae

Group: Sternorrhyncha: Aphididae

Habitat: Common in agricultural and urban areas. **Identification and biology:** Small (body length 1–2 mm), colour very variable, ranging from whitish-green to mid-green, or reddish. Wingless and winged forms occur together, the latter with a dorsal black patch on its abdomen, resembling an 'H' turned sideways. Reproduction mainly parthenogenetic (occurs without mating), females bear live young, resulting in rapid build-up of huge populations. Feeds on plants in many families, forming colonies on growing shoots and under leaves. **Introduction and spread:** Probably native to Asia, now worldwide. Date of arrival in South Africa not known, but probably introduced (or spread) here long ago. **Impacts:** A serious pest of crops such as potato. Transmits many plant virus diseases. Often causes distortion of young leaves and shoots. Known to feed on some indigenous plants. **Control:** Chemical control often used; cultivation methods such as removing weedy host plants around potato fields are strongly promoted in agriculture.

D. Visser

Key reference: Blackman, R.L. & Eastop, V.F. 2000. *Aphids on the world's crops: an identification and information guide*. 2nd ed. Chichester: John Wiley & Sons.

Woolly apple aphid

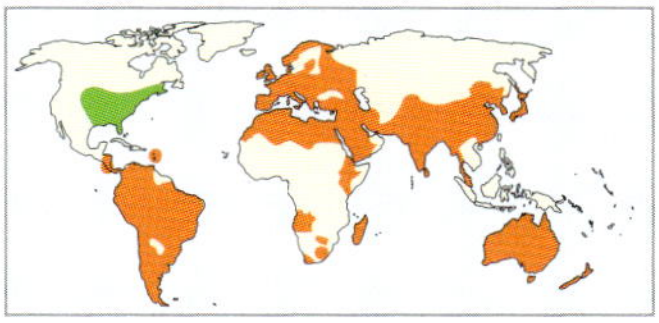

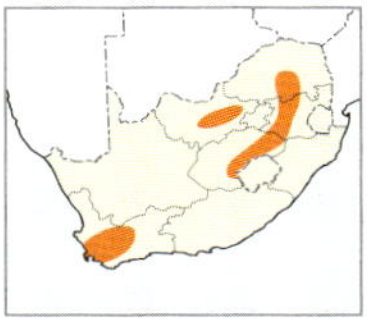

Eriosoma lanigerum

Group: Sternorrhyncha: Aphididae

Habitat: Occurs in deciduous fruit orchards and gardens. **Identification and biology:** Small (body length 2–3 mm), purple or brown, covered with woolly white wax, body fluid bright red. Feeds mainly on apple trees, also occurs on some other woody plants in the rose family such as pear. Forms colonies on the roots of its host plant in winter, and the aerial parts in summer; reproduces parthenogenetically. **Introduction and spread:** Possibly native to North America, now occurs worldwide wherever apples are grown. First recorded in South Africa in 1895. **Impacts:** An important pest of apple trees. Causes galls on roots and branches and may kill young growth. **Control:** The Woolly apple aphid wasp parasitoid (*Aphelinus mali*) kills apple aphids and has an impact on the aphids' aerial populations. Chemical control is used in apple orchards.

Pollinator at en.wikipedia, Wikimedia Commons

Key reference: Blackman, R.L. & Eastop, V.F. 2000. *Aphids on the world's crops: an identification and information guide*. 2nd ed. Chichester: John Wiley & Sons.

Cotton aphid/Melon aphid

Aphis gossypii

Group: Sternorrhyncha: Aphididae

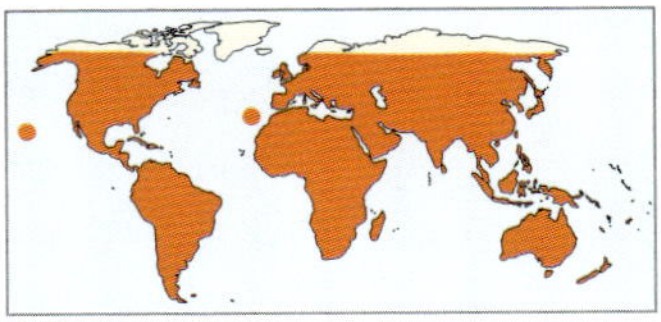

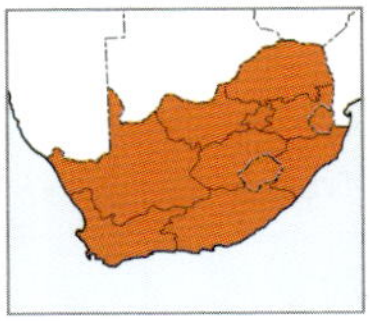

Habitat: Common on a wide range of plants including potato, cucumber, citrus, cotton and diverse garden plants. In the veld, occurs on Karee trees (*Searsia* species). **Identification and biology:** Small (body length 1.5 mm), pale yellow-green, dark green or blackish-grey, depending on the host plant. Winged or wingless, with long thin legs. Head, legs and tubes arising from abdomen all very dark. Extremely polyphagous, feeding on a great number of plant species in many families. Reproduces mainly parthenogenetically. **Introduction and spread:** Cosmopolitan, geographical origin not known. Time of arrival in South Africa unknown, but was probably introduced (or spread to here) long ago. **Impacts:** Feeds on buds, damaging growth tip, and underside of leaves. Forms outbreaks when attended by ants. An important pest of cotton and cucurbits (cucumber family). Capable of transmitting numerous plant virus diseases. **Control:** Chemical control is available for cases of serious outbreaks.

C. Griffiths, M. Picker

Key reference: Blackman, R.L. & Eastop, V.F. 2000. *Aphids on the world's crops: an identification and information guide*. 2nd ed. Chichester: John Wiley & Sons.

Russian wheat aphid

Diuraphis noxia

Group: Sternorrhyncha: Aphididae

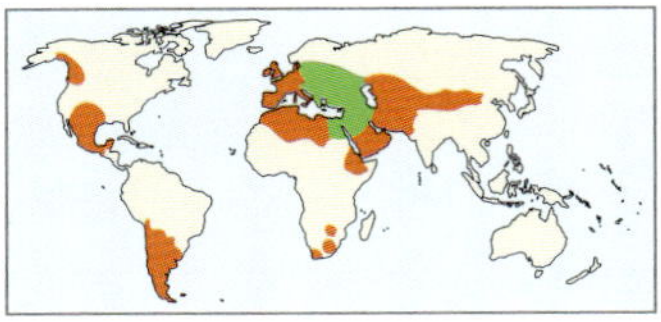

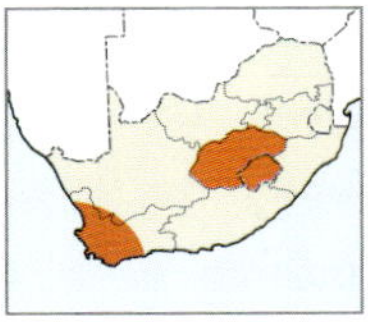

Habitat: In wheat fields.
Identification and biology: Small (body length about 2 mm), spindle-shaped, colour pale green. Both winged and wingless forms occur. Reproduces mainly parthenogenetically. Feeds mainly on wheat and barley, forming colonies on the leaves and ears. **Introduction and spread:** Originated in southern Russian and Middle Eastern parts of the Palaearctic region, and is now widespread. Arrived in South Africa in 1978.
Impacts: A serious pest of wheat. Causes streaks on leaves, which roll up at the edges into tubes, and ears become bent. **Control:** Chemical spraying is used. Cultivation practices such as timing of planting and clearing of auxiliary host plants around fields are important. Resistant wheat cultivars are being investigated.

Peggy Greb, Wikimedia Commons

Key reference: Walters, S.S., du Toit, F., Eulitz, E.G., Carnegie, A.J.M., van Rensburg, G.D.J. & Millar, I.M. 1989. Aphids. In: Myburgh, A.C. (ed.). Crop pests in southern Africa. Vol. 4. *Field crops and pastures*. Bulletin 416. Pretoria: Department of Agriculture and Water Supply, pp. 78-86.

Black pine aphid

Cinara cronartii

Group: Sternorrhyncha: Aphididae

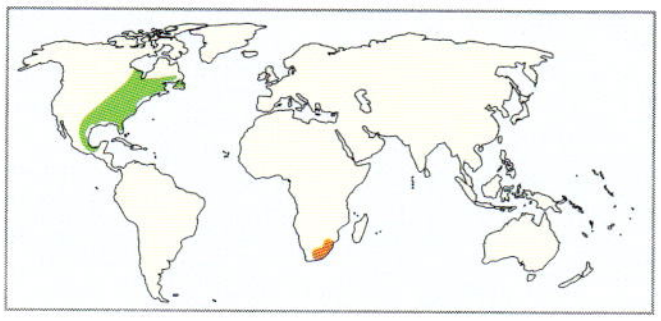

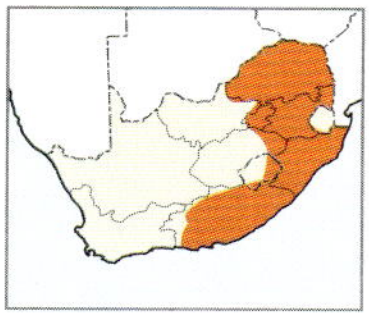

Habitat: In pine tree plantations. **Identification and biology:** Medium-sized aphid (body length 2–4 mm), stout, colour dark brown, legs predominantly dark. Both winged and wingless forms occur. Feeds on pine trees, forming colonies on the trunks and branches. In South Africa, occurs on the aerial parts of pine trees in winter and survives on the roots in summer. So far, known to reproduce only parthenogenetically (genus shown). **Introduction and spread:** Indigenous to the eastern USA; accidentally introduced to South Africa, where it was first observed in 1974. **Impacts:** A pest in pine plantations. Severe infestations are reported to cause tree dieback, and result in poor wood quality and pulp yield. **Control:** Chemical spraying is uneconomical. A parasitoid wasp, *Pauesia cinaravora*, was introduced into South Africa in 1983 for biocontrol of this aphid, and has been a successful control measure.

Frank Mikley, Wikimedia Commons

Key reference: Kfir, R., van Rensburg, N.J. & Kirsten, F. 2003. Biological control of the black pine aphid, *Cinara cronartii* (Homoptera: Aphididae), in South Africa. *African Entomology*, 11:117-121.

Cypress aphid

Cinara cupressivora

Group: Sternorrhyncha: Aphididae

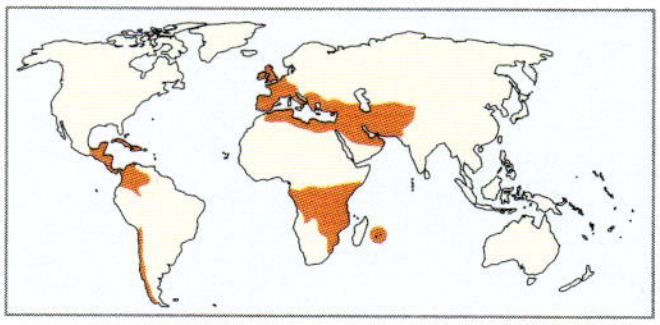

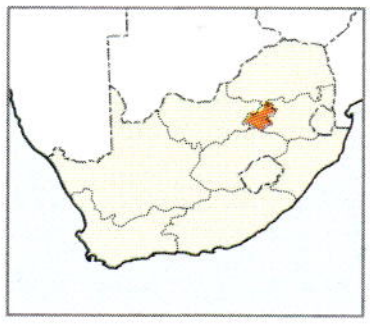

Habitat: Various species of trees in the cypress family, in gardens. **Identification and biology:** Medium-sized aphid (body length 2–4 mm), stout, orange-brown with grey bands across back. Both winged and wingless forms occur. Forms colonies on smaller twigs in foliated branches. Apparently becomes abundant on cypress trees in South Africa only during the winter months. Reproduces mainly parthenogenetically. **Introduction and spread:** Natural range uncertain, possibly eastern Greece. Now in North America, Europe, parts of Southwest Asia and South America. In Africa, first discovered in Malawi in 1986, spreading into neighbouring countries, and detected in South Africa in 1993. **Impacts:** An important pest of cypress and juniper trees, causing severe damage through dieback. Recorded from Mountain cedar (*Widdringtonia nodiflora*), an indigenous conifer in South Africa. **Control:** Chemical control feasible, especially with contact insecticides on small trees. A wasp parasitoid, *Pauesia juniperorum*, was introduced into Malawi for biological control of this aphid and became established, but has not yet been introduced into South Africa.

D. Ward

Key reference: Blackman, R.L. & Eastop, V.F. 1994. *Aphids on the world's trees: an identification and information guide*. Wallington: CAB International.

Red scale

Aonidiella aurantii

Group: Sternorrhyncha: Diaspididae

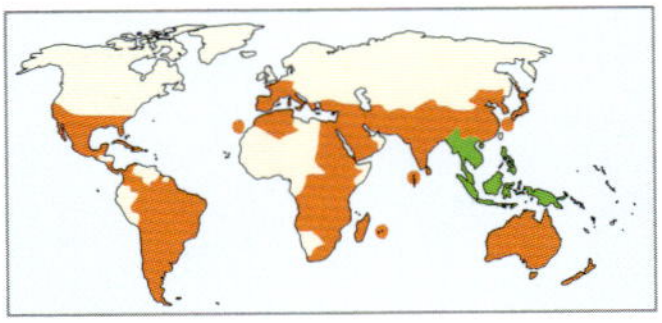

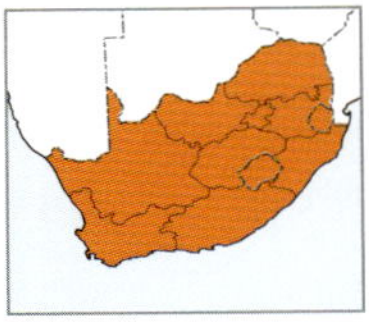

Habitat: Leaves, stems and fruit of many garden and agricultural plants**. Identification and biology:** Small (1.5 mm diameter). Adult females reside under round, translucent and flattened scales which they secrete, and which take on the reddish-brown colour of the insect. Each scale has a nipple-like central protuberance. **Introduction and spread:** Probably native to Southeast Asia, now cosmopolitan. Time of arrival in South Africa unknown, but probably introduced (or spread) here long ago. **Impacts:** Pest of citrus, banana, granadilla and 200 other host plants. Also occurs on at least 12 species of indigenous tree. Reduces host vigour, the apparently toxic saliva causing little yellow spots. Found on virtually every citrus tree, restricting trade of affected fruit. **Control:** Reliance on natural biological regulation (through wasp parasitoids and lady beetle predators) more successful than insecticides. Insecticide resistance is a problem; mineral oils are a useful alternative.

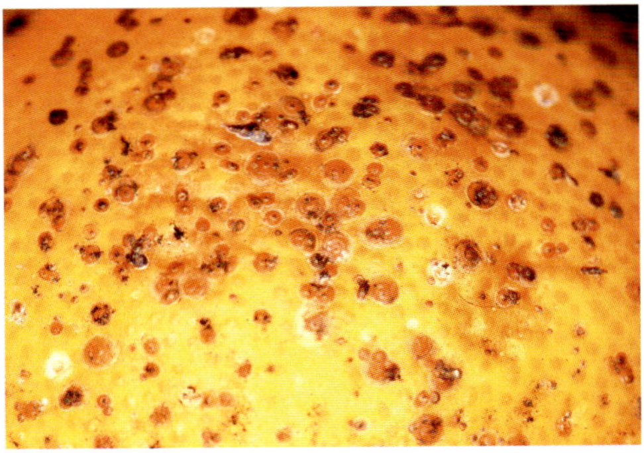

C. Griffiths

Key reference: Bedford, E.C.G., van den Berg, M.A. & de Villiers, E.A. (eds). 1998. *Citrus pests in the Republic of South Africa*. 2nd ed. Nelspruit: ARC-Institute for Tropical and Subtropical Crops.

Pernicious scale

Diaspidiotus perniciosus

Group: Sternorrhyncha: Diaspididae

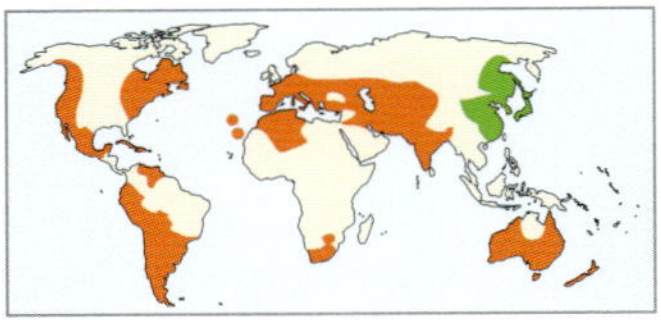

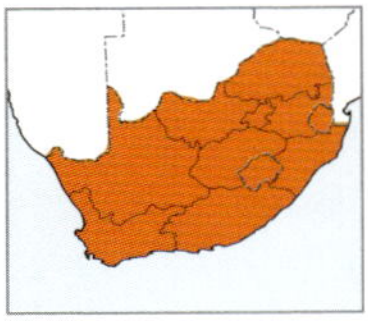

Habitat: Deciduous fruit orchards, gardens. **Identification and biology:** Adult females small (about 1 mm long), flattish, yellow, pear-shaped insects with no appendages. They live under circular, flattened covers that are light brown or grey, with a central pale nipple-like area. Male scale covers are similar, but smaller and elongated oval in shape. Polyphagous, feeding on a great many plant species, and recorded from numerous fruit trees. **Introduction and spread:** Probably originated in the Far East, around northern China; now widespread. First reported in South Africa in 1911. **Impacts:** An important pest of deciduous fruit trees, such as apples, and many ornamental plants. Causes leaf drop, branch dieback and even death of trees. On fruit, blemishing spots form around feeding individuals. An important quarantine pest, as no infested fruit or plant material may be imported or exported. **Control:** Insecticidal spraying is necessary. Biocontrol, using parasitoid wasps, still has to be investigated further in South Africa.

I. Millar

Key reference: Annecke, D.P. & Moran, V.C. 1982. *Insects and mites of cultivated plants in South Africa*. Durban: Butterworths.

Soft brown scale

Coccus hesperidum

Group: Sternorrhyncha: Coccidae

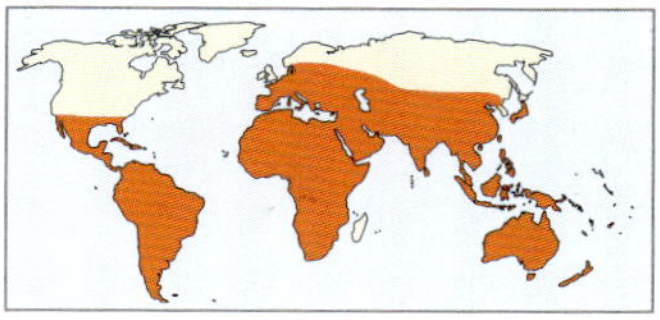

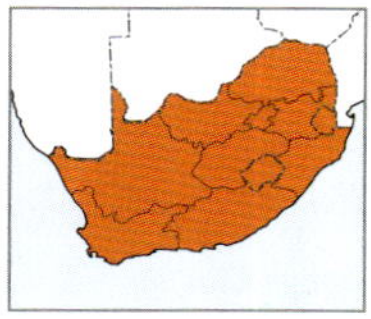

Habitat: Indigenous and exotic plants. Feeds on a wide variety of hosts such as citrus and other subtropical fruits, garden trees, shrubs and even ferns. Inconspicuous when numbers low, their presence revealed by ant trails to their honeydew secretions. **Identification and biology:** Females medium-sized (4.5 mm diameter), oval, light brown and flattened when young, turning darker brown and becoming domed, with radiating pattern of stripes, when older. Parasitised individuals turn black. **Introduction and spread:** Origin possibly Afro-Ethiopian, now a ubiquitous cosmopolitan species. First record in Europe in 1829. **Impacts:** Important pest of citrus; also occurs on other fruit trees and numerous ornamental and garden plants. Secretes sticky honeydew that coats plant, and on which sooty mould grows. **Control:** Ants (which tend and protect these insects from predators and parasitoids) should be controlled. Insecticides (if necessary) should be applied judiciously so as not to disrupt the impact of predators and parasitoids.

C. Griffiths, M. Picker

Key reference: Bedford, E.C.G., van den Berg, M.A. & de Villiers, E.A. (eds). 1998. *Citrus pests in the Republic of South Africa*. 2nd ed. Nelspruit: ARC-Institute for Tropical and Subtropical Crops.

Grapevine mealybug

Planococcus ficus

Group: Sternorrhyncha: Pseudococcidae

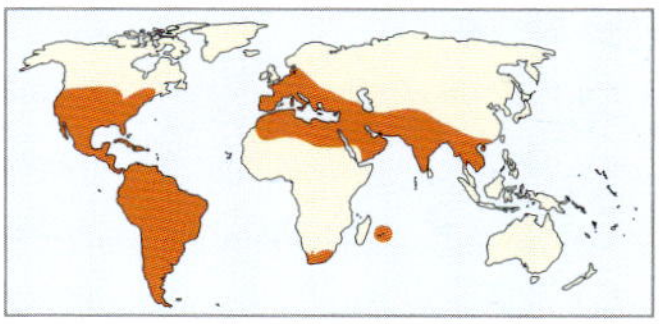

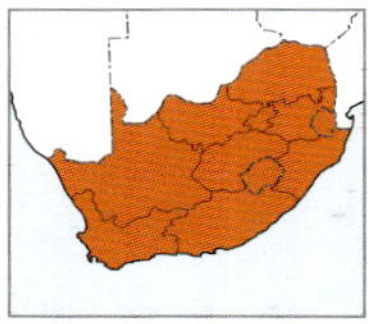

Habitat: In vineyards. **Identification and biology:** Females oval, medium-sized (body length 4 mm), segmented and pink with a thin white powdery wax layer, and waxy fringe around body. In South Africa, feeds only on grape plants and cultivated figs. **Introduction and spread:** Origin unknown, now cosmopolitan wherever grapes are grown. May be confused with Citrus mealybug (*Planococcus citri*), but usually has 17, not 18, marginal waxy filaments on each side of the body. Moves from stems to growing shoots in spring, soon infesting shoots and fruit. **Impacts:** Infests bunches of grapes, impacting negatively on export grapes, but not a serious problem for wine grapes. **Control:** Consumed by a number of indigenous lady beetles, a small introduced parasitic wasp (*Leptomastix dactylopii*), and indigenous lacewings (*Chrysopa* species). A range of pesticides are used in chemical control.

C. Griffiths

Key reference: Walton, V.M. & Pringle, K.L. 2004. Vine mealybug, *Planococcus ficus* (Signoret) (Hemiptera: Pseudococcidae), a key pest in South African vineyards. A review. *South African Journal of Enology and Viticulture*, 25:54-62.

Australian bug/Fluted scale

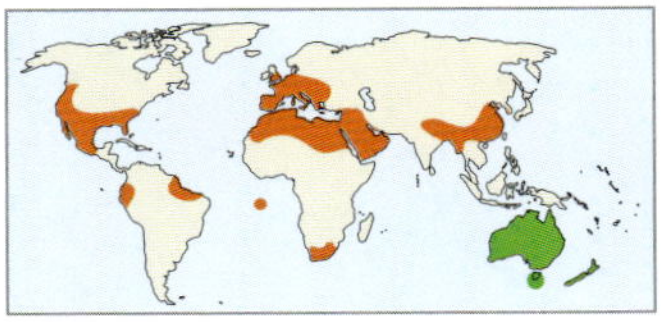

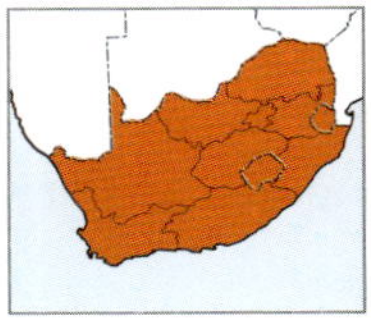

Icerya purchasi

Group: Sternorrhyncha: Monophlebidae

Habitat: On many species of cultivated plants grown in orchards, nurseries and gardens. **Identification and biology:** Unmistakeable, fairly large scale (body length 8 mm), with characteristic large fluted white egg mass carried on reddish-brown body, and filled with up to 1 000 red eggs. Hermaphroditic females may (rarely) produce small winged males. Nymphs are reddish, and occur mostly along the midrib on the undersides of leaves. They move from leaves to stems while maturing. Highly polyphagous. **Introduction and spread:** Originally Australian, recorded in South Africa in 1873. Now a cosmopolitan pest of woody plants. **Impacts:** When not under biological control it damages a range of trees, especially citrus and various ornamentals, causing dieback and premature leaf and bud fall. Excretes honeydew, which fouls the plant foliage. **Control:** Successfully controlled in South Africa (since 1892) and elsewhere by introduced lady beetle, the Vedalia beetle (p. 114). Insecticidal control is also feasible.

C. Griffiths

Key reference: Annecke, D.P. & Moran, V.C. 1982. *Insects and mites of cultivated plants in South Africa*. Durban: Butterworths.

Prickly pear cochineal

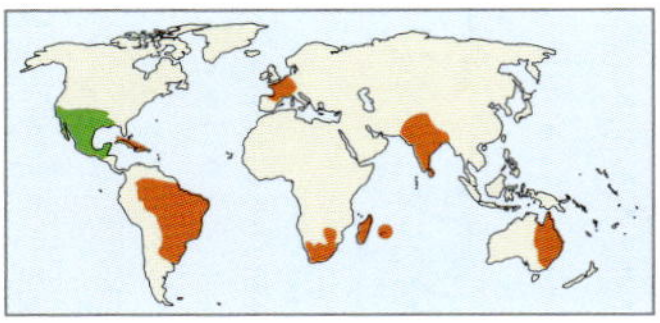

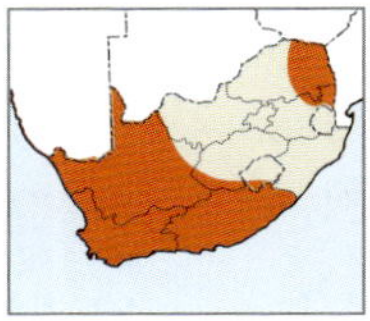

Dactylopius opuntiae

Group: Sternorrhyncha: Dactylopiidae

Habitat: On prickly pear plants (*Opuntia* species). **Identification and biology:** Female small (2–3 mm long), soft-bodied and covered in white waxy filaments, and attached to base of cactus spines. Body filled with red carminic acid. Sucks sap from cactus pads, causing them to yellow and rot. Eggs hatch into 'crawlers' that are dispersed by wind. Small, winged males rarely seen. **Introduction and spread:** Introduced to South Africa in 1937 from North America. Spread globally to control Prickly pear (*Opuntia ficus-indica*) infestations. **Impacts:** Effectively controls invasive *Opuntia* infestations but also impacts crop production of cultivated spineless Prickly pear, especially in drier parts. **Control:** A useful biocontrol agent, but is predated by two lady beetles; Cassava mealybug predator (*Exochomus flaviventris*) and Mealybug destroyer (*Cryptolaemus montrouzieri*). Can be controlled by application of pesticides.

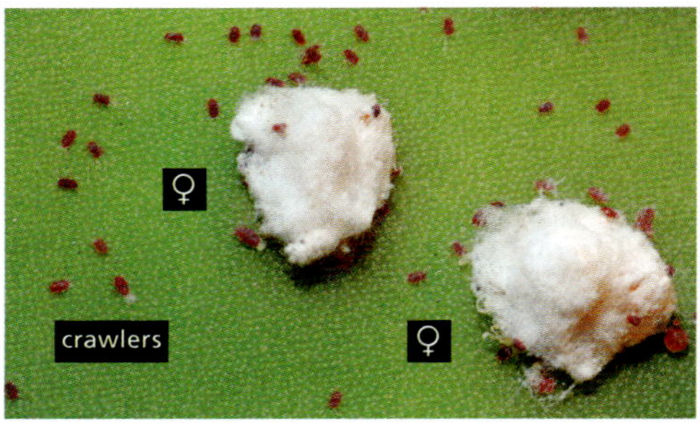

C. Griffiths

Key reference: Hoffmann, J.H., Moran, V.C. & Zeller, D.A. 1998. Long-term population studies and the development of an integrated management programme for control of *Opuntia stricta* in Kruger National Park, South Africa. *Journal of Applied Ecology*, 35:156-160.

Blue gum psyllid

Ctenarytaina eucalypti

Group: Sternorrhyncha: Psyllidae

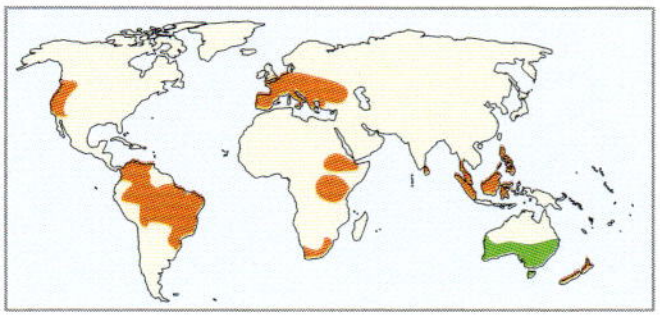

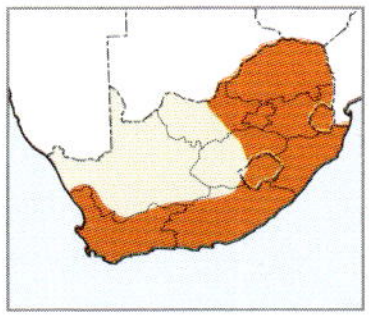

Habitat: Eucalyptus trees; in plantations, or where they occur in clumps in rural areas. **Identification and biology:** Small (adults about 1.5 mm long), dark brown, resembling miniature cicadas. Nymphs are flattened, oval, and are sessile on the leaves. Feeds on *Eucalyptus* species. **Introduction and spread:** Native to Australia. Now occurs in Europe, California, and has been present in South Africa since at least 1923. **Impacts:** A pest of eucalyptus trees in certain countries, but apparently not serious in South Africa. Feeding causes stunting, distorted growth and defoliation. **Control:** Biocontrol has been successfully achieved in California using the encyrtid wasp parasitoid, *Psyllaephagus pilosus*.

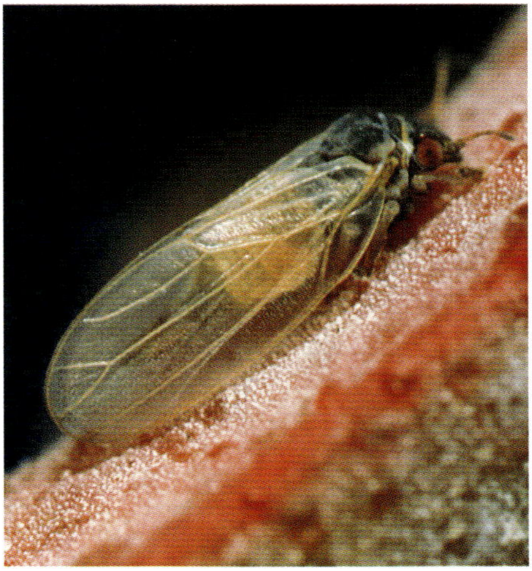

Jack K. Clark/AgstockUSA/Science Photo Library

Key reference: Hodkinson, I.D. 1999. Biocontrol of eucalyptus psyllids *Ctenarytaina eucalypti* by the Australian parasitoid *Psyllaephagus pilosus*: a review of current programmes and their success. *Biocontrol News and Information*, 20:129-134.

Eucalyptus shoot psyllid

Blastopsylla occidentalis

Group: Sternorrhyncha: Psyllidae

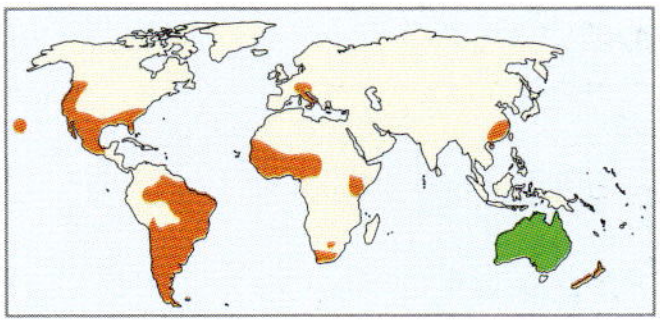

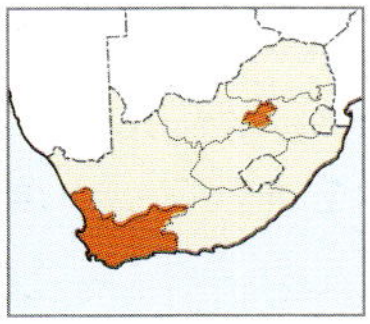

Habitat: Eucalyptus trees; in plantations, or where they occur in clumps in rural areas. **Identification and biology:** Small (1.5 mm long), yellowish with dark transverse bands across abdomen and dark linear patches on thorax. Adults resemble miniature cicadas. Feeds on various *Eucalyptus* species. Nymphs secrete a waxy substance, which appears as a fuzzy white ball above the nymphs. Early nymphal development occurs on young leaves. **Introduction and spread:** Native to Australia, now widespread (North and South America, East and West Africa, New Zealand). Present in South Africa from about 2004. **Impacts:** A pest of a range of species. Causes leaf drop from young shoots; flower buds abort. Resultant lack of nectar is detrimental to beekeeping. Higher numbers may occur in urban areas. **Control:** In Brazil the encyrtid wasp parasitoid, *Psyllaephagus pilosus*, was found to be a successful control agent in plantations of Tasmanian bluegum (*Eucalyptus globulus*).

C. Griffiths, M. Picker

Key reference: Neser, S. 2007. *Blastopsylla occidentalis*: another new Eucalyptus pest in South Africa. *Plant Protection News* No. 72, pp 2-3. [Newsletter of the ARC-Plant Protection Research Institute, Pretoria.]

Woolly whitefly

Aleurothrixus floccosus

Group: Sternorrhyncha: Aleyrodidae

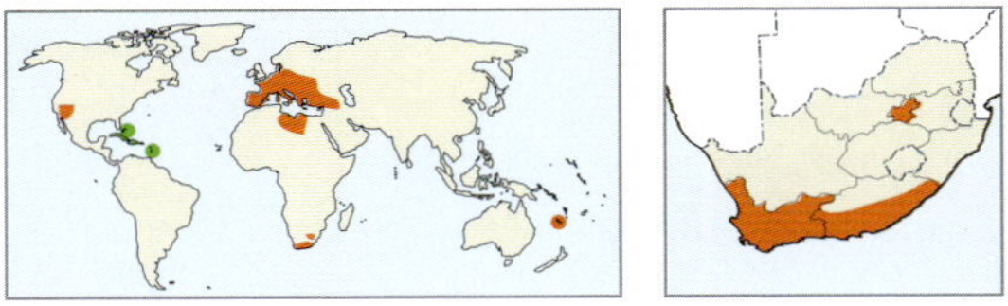

Habitat: Orchards and gardens. **Identification and biology:** Immatures are tiny (about 1 mm), oval and sedentary, occurring in clusters covered in white wax filaments, forming a characteristic woolly mass on the undersides of leaves. Adults are white, very small (about 1 mm) and resemble tiny moths. Often on citrus trees, especially lemon, but quite polyphagous and also recorded from crops like coffee and mango. **Introduction and spread:** Originally from the Caribbean, now widespread, occurring in numerous countries on several continents, but not yet in Australia. Detected in South Africa in 2007. **Impacts:** A pest of citrus, and also occurs on various other subtropical fruit crops. Heavy infestations cause leaf drop. Secretes honeydew, which coats leaves and provides a substrate for sooty mould growth. **Control:** Several insecticides are reportedly effective. Parasitic wasps belonging to the genus *Eretmocerus* are present in South Africa, and cause mortality.

Key reference: Giliomee, J.H. & Millar, I.M. 2009. The woolly whitefly, *Aleurothrixus floccosus* (Maskell) (Hemiptera: Aleyrodidae), a potentially serious citrus pest, recorded from South Africa. *African Entomology*, 17:232-233.

C. Griffiths

Greenhouse whitefly

Trialeurodes vaporariorum

Group: Sternorrhyncha: Aleyrodidae

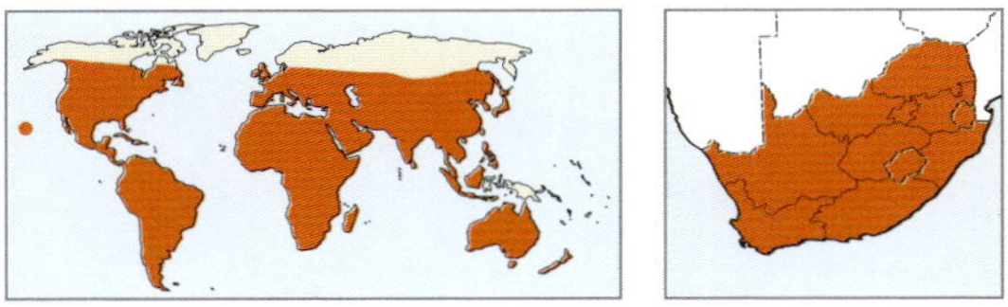

Habitat: Greenhouses, vegetable fields and gardens. **Identification and biology:** Immatures are tiny (up to about 1 mm), pale, yellowish, with translucent waxy filaments arising from the dorsum. Adults are yellow with white wings, very small (about 1 mm) and resemble tiny moths. At rest the folded wings hide the abdomen. Polyphagous, occurring particularly on herbaceous plants, where eggs are laid near the growing shoots. While early stages are active, later nymphs are sessile. **Introduction and spread:** Probably native to North America, now cosmopolitan. First record in Europe in 1856. Has occurred in South Africa since at least 1923. **Impacts:** A pest of vegetable crops such as tomatoes, and also some ornamentals. Can cause yellowing and wilting of leaves, although does not transmit viruses to any extent. **Control:** Certain parasitic wasp species belonging to the genus *Encarsia* kill this whitefly. Insecticides are available but chemical control is difficult.

Key reference: Visser, D. 2009. *A complete guide to vegetable pests in South Africa*. Pretoria: Agricultural Research Council, Roodeplaat Vegetable and Ornamental Plant Institute.

D. Visser

Pomegranate whitefly/ Ash whitefly

Siphoninus phillyreae

Group: Sternorrhyncha: Aleyrodidae

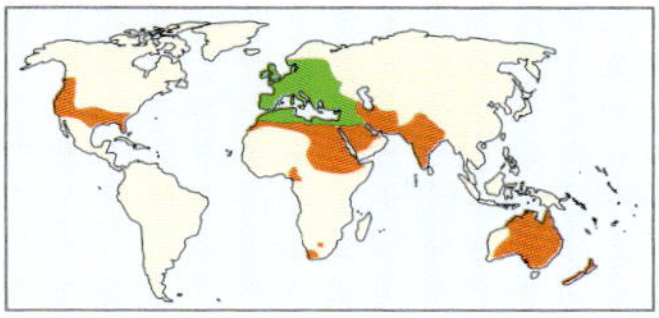

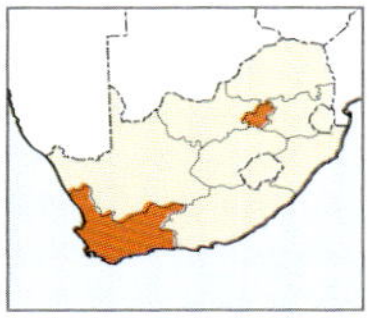

Habitat: Orchards and gardens.
Identification and biology: Immatures are tiny (length about 1 mm), oval with numerous short siphons or tubes on the back. They produce clumps of white wax along a wide strip of the central surface of the back. Nymphs sedentary, occurring in dense clusters on the underside of leaves, sometimes covering the entire leaf undersurface. Abundant on olive, pomegranate and ash trees and also found on rosaceous crops such as pear and apple**. Introduction and spread:** Native to the Mediterranean area; it has spread to parts of most continents, including West and East Africa. First found in South Africa in 2009. **Impacts:** A pest of crops such as pomegranate, ornamental olives and ash trees. Secretes honeydew, which provides a substrate for sooty mould growth. **Control:** Certain parasitic wasp species, and a predatory lady beetle (*Clitostethus arcuatus*), are important biocontrol agents. Parasitoids have been found in South Africa and their impact warrants investigation.

Key reference: Giliomee, J.H. & Millar, I.M. 2010. Pomegranate or ash whitefly, *Siphoninus phillyreae* (Haliday) (Hemiptera: Aleyrodidae), recorded from South Africa. *African Entomology*, 18:200-202.

C. Griffiths

Sugarcane planthopper

Perkinsiella saccharicida

Group: Auchenorrhyncha: Delphacidae.

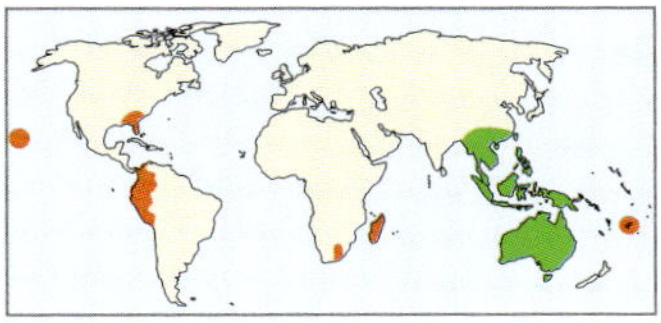

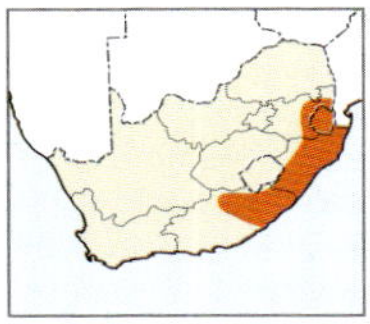

Habitat: Sugarcane fields. **Identification and biology:** Narrow, elongated, medium-sized (5–6 mm long), rounded head with ridges, pale marking on parts of thorax (pronotum and scutellum). Forewings marked with brown. Spur (calcar) on the apex of the hind tibia, and antenna with two enlarged basal segments. The complete life cycle takes *c.* 45 days. **Introduction and spread:** Origin probably Southeast Asia and Australia, now widespread in tropics and subtropics. Has caused damage to sugarcane in Pacific islands such as Hawaii. Other records include Australia, China, Fiji, Java, Malaysia, Mauritius, Reunion, Madagascar, Philippines, Africa and North America (detected in Florida since 1982). Confirmation of the identity of this species in South Africa is required, and it can easily be confused with *Perkinsiella manilae*.
Impacts: Potential transmitter of plant diseases. It is a known vector of Fiji disease of sugarcane in some countries, but this disease is not yet known in South Africa. **Control:** Host plant resistance or tolerance. A range of other insects such as mirid bugs (*Tytthus* species) assist in population regulation, largely by preying on eggs.

Key reference: Annecke, D.P. & Moran, V.C. 1982. *Insects and mites of cultivated plants in South Africa*. Durban: Butterworths.

C. Griffiths, M. Picker

Tamarix leafhopper

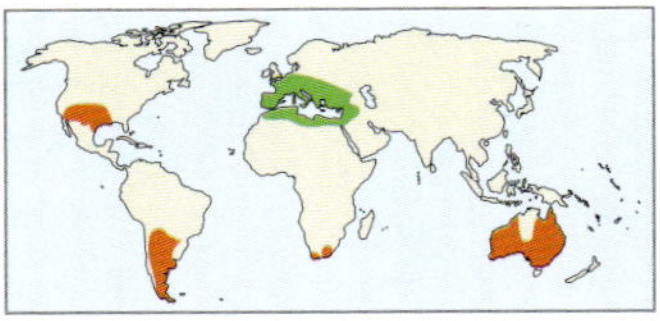

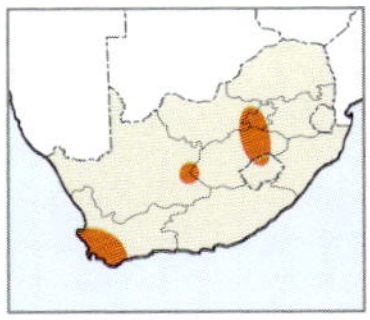

Opsius stactogalus

Group: Auchenorrhyncha: Cicadellidae

Habitat: Cosmopolitan on Saltcedar (*Tamarix*) or Sheoak (*Casuarina*) species, possibly as a result of the use of these plants for soil stabilisation. These plants are native to Eurasia and North Africa, and are generally found on sandy soils along riverbanks. **Identification and biology:** Medium-sized (4–5 mm long), adults green with head variegated, rounded, with yellow markings, and tip of wings brown. Likely to have other host plants such as grasses, as well as Wild tamarisk (*Tamarix usneoides*). **Introduction and spread:** European origin, now in North America, Australia (first record 1977) and Argentina.
Impacts: May shift hosts, and acts as a vector for plant diseases. Aggregated feeding known to reduce growth of host. **Control:** Biocontrol measures applied in some countries.

T. Bantock

Key references: Nickel, H. 2003. *The leafhoppers and planthoppers of Germany (Hemiptera, Auchenorrhyncha): Patterns and strategies in highly diverse group of phytophagous insects*. Sofia: Pensoft.

Virla, E.G., Logarzo, G.A. & Paradell, S.L. 2010. Occurrence of the tamarix leafhopper, *Opsius stactogalus* Fieber (Hemiptera: Cicadellidae), in Argentina. *Journal of Insects Science*, 10:23.

Beet leafhopper

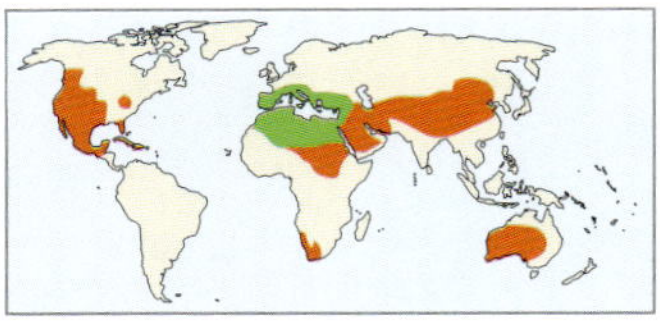

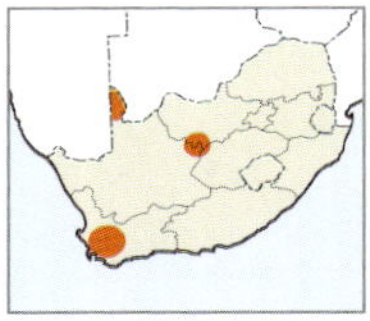

Circulifer tenellus

Group: Auchenorrhyncha: Cicadellidae

Habitat: Overwinters on weeds in disturbed areas and in spring and summer moves onto tobacco, tomato, chilli, pepper and eggplant. **Identification and biology:** Medium-sized (3–4 mm long), brown, sometimes with mottled pattern. Species determination is difficult as there are other species with similar appearance. Examination of internal male genitalia is required to confirm the species. **Introduction and spread:** Originally from North Africa and southern Europe, it now occurs in warm, dry parts of the world, e.g. western North America and Mexico, Mediterranean region and some neighbouring regions in Asia, Namibia and Australia. Known in South Africa but not common. Capable of long-distance migration. **Impacts:** In North America it is a vector of curly top virus and other viruses, causing leaf deformation. South African records suggest that this species has a wide host and habitat range. **Control:** Chemical; horticultural methods, such as companion planting with marigolds, which repel them. Can be drawn away from crops by planting pelargoniums and petunias nearby, the latter functioning as 'trap' plants. Removal of overwintering host weeds assists control.

M. Stiller

Key reference: Nault, L.R. & Rodriguez, J.G. 1985. *The leafhoppers and planthoppers*. New York: John Wiley and Sons.

Orosius leafhopper

Orosius albicinctus

Group: Auchenorrhyncha: Cicadellidae

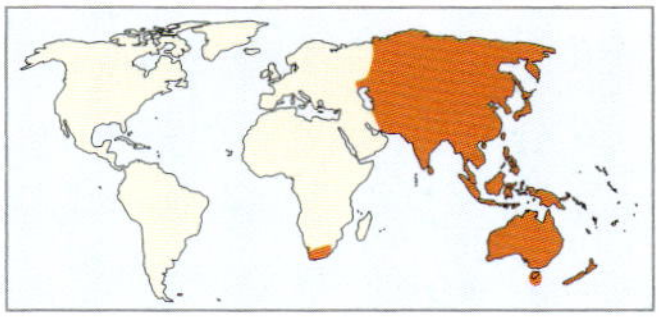

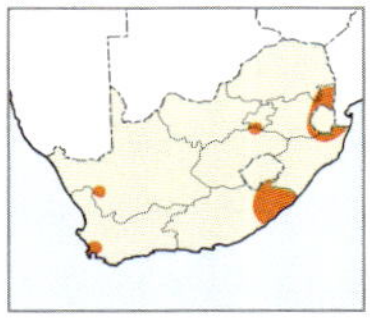

Habitat: Common on a wide variety of plants in many families, although it appears to be particularly associated with members of the Salt-bush family (Chenopodiaceae). **Identification and biology:** Medium-sized (3–5 mm long), triangular head, grey-brown-white colour pattern distinctly variegated. **Introduction and spread:** Origin difficult to determine, now widespread in Asia and Australasia. **Impacts:** This is one of the best-known vectors of plant phytoplasma diseases; affected leaf edges curl and turn brown. In South Africa it has wide host and habitat utilisation, but is never common. Virus transmission in South Africa is unknown to date. A serious pest of sesame in India, where it transmits *Sesamum* phyllody virus. Also damages chickpea crops. **Control:** Systemic insecticides and cultivation methods such crop rotation, intercropping, host resistance or tolerance.

Key reference: Ghauri, M.S.K. 1966. Revision of the genus *Orosius* Distant (Homoptera, Cicadelloidea). *Bulletin of the British Museum (Natural History)*, 18:231-252.

P. Weintraub

P. Weintraub

AUTHOR: M. STILLER

THRIPS Class Insecta, Order Thysanoptera

Very small insects (commonly 0.5–5 mm, some tropical species up to 12 mm); adults have two pairs of hair-fringed wings (folded at rest) and piercing mouthparts. They feed on a range of mostly plant-based material, including leaves, buds, blossoms and fruit. Since the mouthparts are very short, they damage the superficial layers of the plant. Apart from this direct destructive feeding, they are known vectors of plant viruses, bacteria and fungi. Populations can grow very rapidly, especially in monoculture crops. Thrips are regularly transported across political boundaries in fresh horticultural products, and the likelihood of alien thrips species occurring in South Africa was recognised as early as 1964. For quarantine authorities of importing countries, plants and plant products pose considerable problems as quarantine officers are expected to identify potential pest or invasive species. Exporting countries experience difficulties in certifying that exports are free of thrips pests, as required by trade agreements. This is complicated by the difficulty of correctly identifying various thrips species.

Greenhouse thrips

Heliothrips haemorrhoidalis

Group: Thripidae

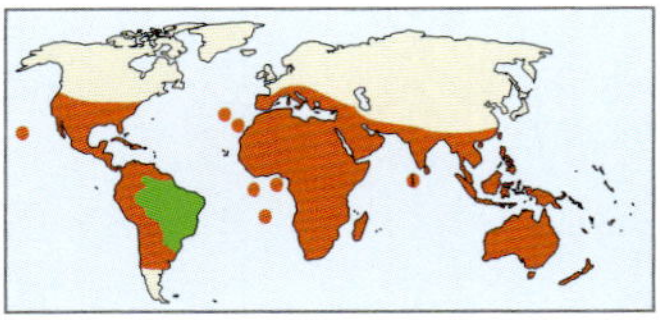

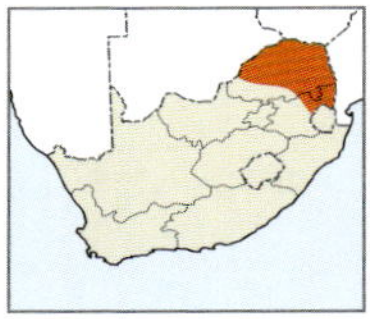

Habitat: Cultivated garden and indoor plants. **Identification and biology:** A small species (1–1.5 mm long). Three species are assigned to this genus – one species in South Africa, one in South America, and the third occurring worldwide but originally from South America. Species recognition requires the preparation of a microscope slide of a 'cleared' specimen (one that has been immersed in a potassium hydroxide solution). Larvae curve the abdomen forward over the body when disturbed and produce a dark secretion. Reproduces parthenogenetically. Feeds on many plants, usually those with a hard cuticle, including pines (*Pinus*), conebushes (*Leucadendron*), kiwi fruit and tea. **Introduction and spread:** Native to South America. Records in the thrips database of the *South African National Collection of Insects* suggest a long presence in South Africa (first record in 1912, many between 1940 and 1960, and recent records from 2000 to the present). This species has been intercepted on imported plant material from countries such as Portugal and New Zealand. **Impacts:** Direct feeding on leaves. Problematic in greenhouses and glasshouses in more temperate regions. **Control:** Quarantine, chemical control and horticultural manipulation.

David Cappaert/www.insectimages.org

Key reference: Mound, L.A. & Kibby, G. 1998. *Thysanoptera. An identification guide*. Wallingford, UK: CAB International.

Onion thrips

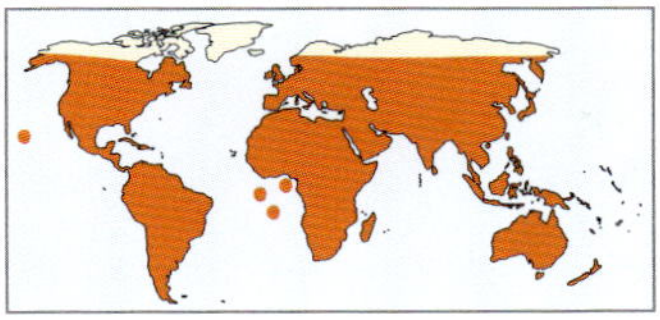

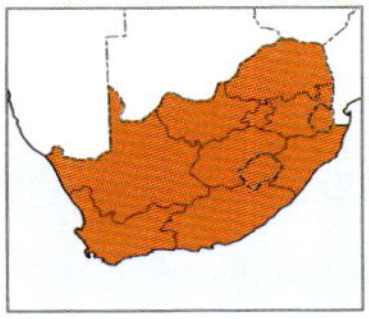

Thrips tabaci

Group: Thripidae.

Habitat: Agricultural fields and gardens. **Identification and biology:** Species recognition requires the preparation of a microscope slide of a 'cleared specimen'. Species identification has now been facilitated by the key to thrips species of the Afro-tropical region. This common pest species breeds in flowers and on leaves of a wide range of plants, but also feeds on the larvae of other thrips or mites. Adults (2 mm long) are yellow or brown, with colour determined by conditions during larval development. **Introduction and spread:** Cosmopolitan, origin uncertain. Readily exported and imported on a large range of plant parts. Some populations have the ability to transmit tospoviruses. **Impacts:** Potential as a virus vector, but also damages plant parts by direct feeding. Records in the thrips database of the *South African National Collection of Insects* reflect a long pest history in South Africa and a wide range of host plants. **Control:** Quarantine, chemical control and horticultural practices.

Alton N. Sparks, Jr., University of Georgia, Bugwood.org

Key references: Lewis, T. 1997. *Thrips as crop pests*. Wallingford, UK: CAB International.

Mound, L.A. 2010. Species of the genus *Thrips* (Thysanoptera, Thripidae) from the Afro-tropical region. *Zootaxa*, 2423:1-24.

Sugarcane thrips

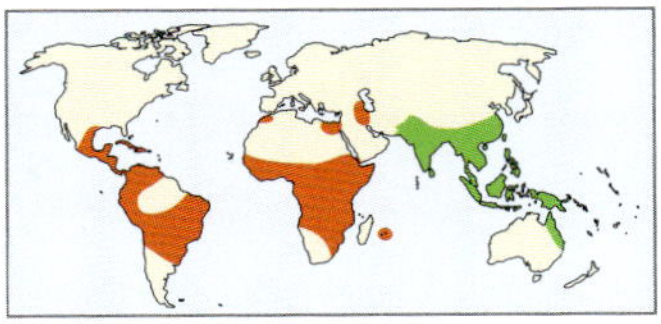

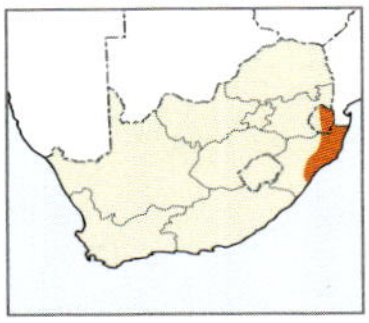

Fulmekiola serrata

Group: Thripidae

Habitat: Temperate and tropical sugarcane production regions.
Identification and biology: Small, 1.5 mm in length. Identification by morphological and molecular techniques. Morphological features include the position of the setae on the head and the serrated margin on the abdominal segments. Shares some similar features with *Stenchaetothrips* species. Damages young leaves causing stunting and knotting of the spindle (unrolled) leaves and silvering of older leaves. **Introduction and spread:** Native to Indian subcontinent and Indo-Pacific. Apparently spread from Mauritius to South Africa, where it has become a major sugarcane pest since 2004. **Impacts:** Caused large-scale damage to sugarcane in Taiwan during the 1950s. This grass-living member of the family Thripidae has spread through most sugarcane-growing areas of the world, and is a relatively recent introduction to South Africa. **Control:** Mainly varietal resistance or tolerance.

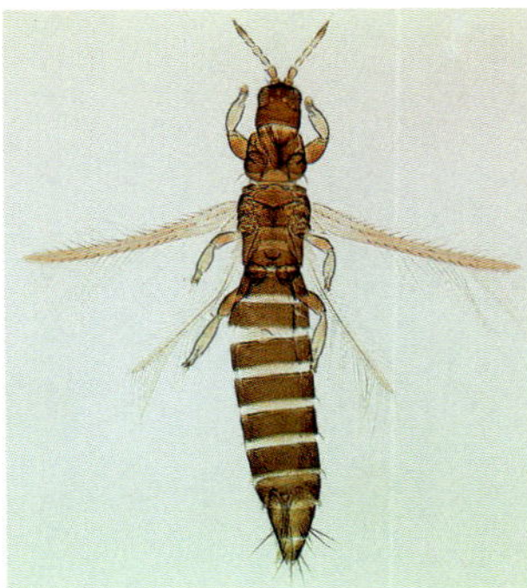

L. Mound, CSIRO Ecosystem Services, Canberra

Key references: Lewis, T. 1997. *Thrips as crop pests*. Wallingford, UK: CAB International.

Way, M.J., Stiller, M., Leslie, G.W., Conlong, D.E., Keeping, M.G. & Rutherford, R.S. 2006. *Fulmekiola serrata* (Kobus) (Thysanoptera: Thripidae), a new pest in southern African sugarcane. *African Entomology*, 14:401-403.

Melon thrips

Thrips palmi

Group: Thripidae

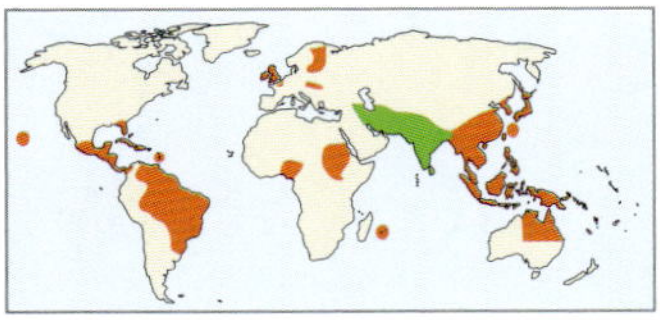

Habitat: Croplands, gardens and indoors on house plants. **Identification and biology:** Molecular markers (COI, ITS2) and morphological methods are required for a positive identification. This small (1 mm long) species is similar to a number of other species of *Thrips*, and therefore requires taxonomic expertise for accurate identification. Damages plants not only by direct feeding, but also by the transmission of a number of viruses. Feeds on a wide range of cultivated plants, but is most abundant on members of the cucumber and potato families. Eggs inserted into the leaf, flower and fruit tissue; larval instars 1 and 2 feed on the leaves, flowers and fruit, and the two pupal stages develop in the soil. **Introduction and spread:** Originated from southern Asia, now widespread in Asia, Central America and Caribbean. Introduced to, and occurs patchily in North and South America, Oceania and West Africa, but an economically important pest that may establish in South Africa. **Impacts:** Feeding and disease damage to crops, and potential effect on import and export commodities. **Control:** Quarantine protocols.

Key reference: Moritz, G., Mound, L.A., Morris, D.C. & Goldarazena, A. 2004. *Pest thrips of the world: an identification and information system using molecular and microscopical methods*. CD-ROM. Brisbane: University of Queensland, Centre for Biological Information Technology.

Hide beetles/ Larder beetles

Dermestes species

Group: Dermestidae

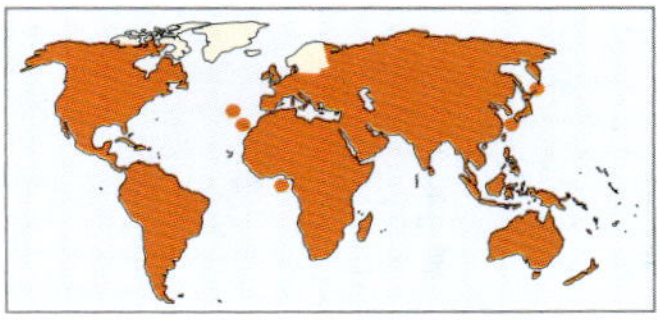

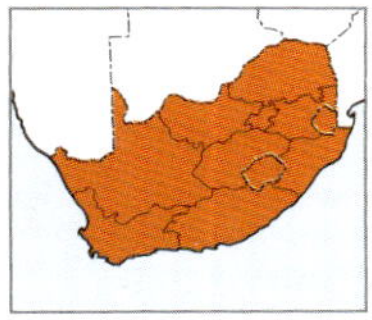

Habitat: Terrestrial. Usually associated with carcasses, on which it feeds, otherwise feeds on detritus mainly of animal origin. **Identification and biology:** Six alien species of *Dermestes* in South Africa, the Hide beetle (*D. maculatus,* shown) being the commonest. Sides of body nearly parallel, 5.5–10 mm long; brown or black, densely covered in fine hairs; light-coloured hairs form pale patches on body, patterning being species-specific. Distinct three-segmented antennal club. Larva elongate, tapering to rear; dark brown with longitudinal pale stripe on top; dense hair cover. **Introduction and spread:** Several *Dermestes* species are virtually cosmopolitan and cryptogenic; some may have African origins. Worldwide spread with human activity since early times, thus unrecorded. **Impacts:** Adults and larvae are pests of many dry substances of animal origin, including dried fish, biltong, skins, wool, leather, dog biscuits, Egyptian mummies; vegetable matter at times. Before pupation, larvae often bore into inedible substrates, e.g. woodwork, mortar, even lead. A recent pest in poultry houses. Commonly found on animal carcasses in the wild; ecologically beneficial. **Control:** Destruction of colonies; insecticides; hygiene.

C. Griffiths, M. Picker

Key reference: Mound, L. (ed.). 1989. *Common insect pests of stored food products*. British Museum (Natural History), Economic Series No. 15. 7th ed. London: British Museum.

Lesser mealworm beetle

Alphitobius diaperinus

Group: Tenebrionidae

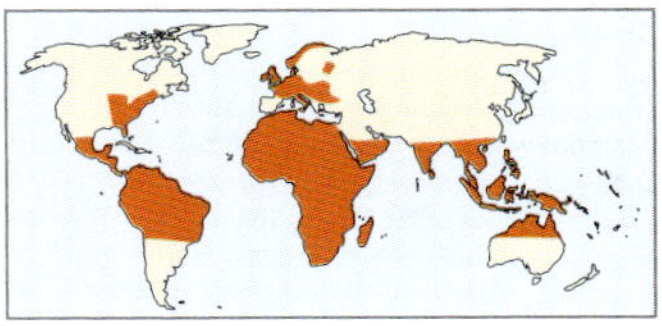

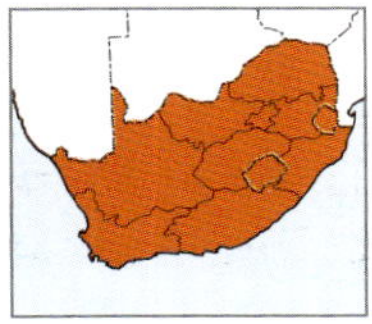

Habitat: Terrestrial. Omnivorous scavenger, particularly in mouldy detritus. **Identification and biology:** Shiny, dark brown to black, broadly oval, moderately convex darkling beetle (Tenebrionidae), 5.5–7 mm long; eye partly divided, elytra rounded. Larval abdomen ends in distinct single process. In tropics found in the wild in bird nests, mammal burrows, bat caves. The Black fungus beetle (*Alphitobius laevigatus*) is very similar, but less prevalent. **Introduction and spread:** Cosmopolitan pest; cryptogenic, but origin thought to be African. **Impacts:** Serious pest in deep-litter poultry farms, adults and larvae abundant in litter. Vector of several avian diseases and parasites, also of human pathogens; fowl infected by feeding on beetles and larvae. Known to attack and feed on live vertebrates in confined spaces. Burrowing larvae cause serious damage to structures, including insulation of poultry houses. Also a contaminant of stored damp and mouldy grain, cereals and animal feed. **Control:** Selection of appropriate poultry house construction materials; sanitation; appropriate cultural control and use of insecticides.

G. Goergen, IITA

Key reference: Barké, H.E. & Davis, R. 1969. Notes on the biology of the lesser mealworm, *Alphitobius diaperinus* (Coleoptera: Tenebrionidae). *Journal of the Georgia Entomological Society*, 4:46-50.

Spotted maize beetle

Astylus atromaculatus

Group: Melyridae

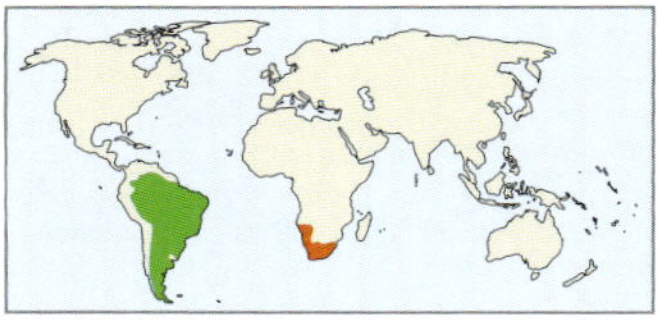

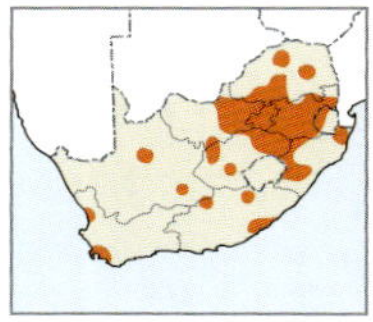

Habitat: Terrestrial. Various open natural habitats, croplands, gardens. **Identification and biology:** Adult medium-sized (8–11 mm long), soft-bodied, somewhat elongate. Head, antennae, pronotum and legs black, sometimes with greenish surface sheen. Head elongate, elytra yellow, with 10 (sometimes 12) irregular black markings. Back with sparse, stiff, erect black hairs. The number and appearance of larval instars varies; late instars with long brown to black hairs and two conspicuous abdominal hooks posteriorly; 15–20 mm long when fully grown. Pupation in soil, with one generation only per year. Adults most abundant January/February; occasionally extremely numerous, frequently huddle together, especially at night and on cool days. Adults primarily pollen feeders, but also take nectar, honeydew and fungi. Numerous host plants, but partial to amaranths and grasses/grains, frequently maize tassels. Eggs laid in clusters among plant debris near soil. Larvae very hardy; burrow in soil, feed opportunistically on vegetable and animal matter. **Introduction and spread:** Native to South America. Discovered in Pretoria in 1916; introduction mode unknown. Widespread and abundant in east-central interior and all maize and sorghum-producing areas. **Impacts:** Adults seldom truly harmful; can accidentally damage flowers, hampering pollination; may damage maize ears and sorghum heads by feeding on young kernels. Larvae more injurious, may destroy planted seed of maize, sorghum and other crops; also potatoes, asparagus, strawberries. Possibly also agriculturally beneficial, with adults as fungus feeders and pollinators, larvae as predators. Adults highly toxic to livestock; cattle deaths recorded after accidental ingestion of large beetle numbers. May be expanding range, agricultural importance may increase. True environmental impact unknown; research required. **Control:** Control through appropriate cultural control important. Adults rarely require control through contact insecticides. Guineafowl beneficial as they are unaffected by beetle poisons. Chemical treatment of seed before or during planting provides protection. No parasitoids ever found in South Africa; predators and fungal disease have little effect. Search for natural enemies in South America in 1950s was unsuccessful; renewed attempts justified.

C. Griffiths, M. Picker

Key reference: Bedford, E.C.G., de Wet, P. & Odendaal, J.S. 1974. Spotted maize beetle, *Astylus atromaculatus* Blanch., Dasytidae. In: Matthee, J.J. (ed.). Pests of graminaceous crops in South Africa. *RSA Entomology Memoir*, 40:7-11.

Harlequin lady beetle

Harmonia axyridis

Group: Coccinellidae

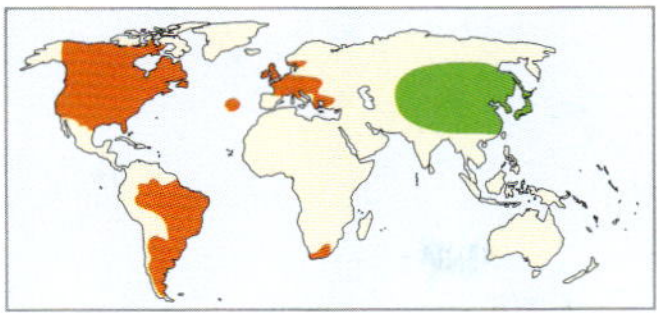

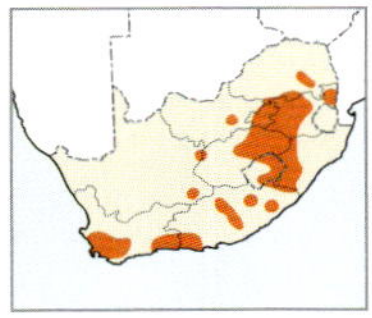

Habitat: Terrestrial. Various habitats, both natural and transformed.

Identification and biology: Medium-sized (5–8 mm, almost circular), 'typical' lady beetle, convex, shiny. Coloration extremely variable, elytra yellow, orange or red, with 0–19 variable black spots; immediately distinguishable from all other South African lady beetles by black '**M**' (or inverted '**W**') (see arrow) on white pronotum and white 'cheeks' (pronotal margins). Yellowish eggs laid in small clusters. Mature larva black, with pink markings and sharp projections on abdomen. Pupa exposed, on tree trunks, rocks, walls. Both adults and larvae effective predators of aphids and other smallish, soft-bodied arthropods, including many beneficial species and other lady beetle larvae; adult diet includes fruit. **Introduction and spread:** Origin central and eastern Asia, currently almost simultaneously invasive in Europe, the Americas and southern Africa. In Europe and Americas purposefully released as aphid biocontrol agent, invasive populations arising from 1988 to 2001. Invasion pathway to South Africa unknown, not deliberately released here. Earliest South African records 2001, in south-western Cape. By 2010, present almost throughout more moist, temperate southern, eastern and east-central South Africa; also Lesotho. Since 2009, additionally recorded from hotter and drier localities; northernmost record (2010) at Tropic of Capricorn. Further spread inevitable. **Impacts:** Emerged since 1990s as one of world's worst invasive, harmful animals. Impacts in both natural and transformed landscapes, including pristine grassland, various agricultural settings, plantation forestry, urban areas. Voracious generalist predator, feeding on a range of insects including other lady beetles and always outcompeting other species, potentially causing serious decline of native lady beetle species. Negative cascading effects throughout ecosystems anticipated. Economically disastrous when present among harvested grapes as may taint wine. Large overwintering aggregations frequently in buildings, significant domestic nuisance. Some people highly allergic to their secretions. **Control:** No effective means available. Various biocontrol options being researched in northern hemisphere, none yet applied; insecticides not feasible. Legislation currently before parliament will accord this beetle highest status of alien invasive species.

Key reference: Stals, R. & Prinsloo, G. 2007. Discovery of an alien invasive, predatory insect in South Africa: the multicoloured Asian ladybird beetle, *Harmonia axyridis* (Pallas) (Coleoptera: Coccinellidae). *South African Journal of Science*, 103:123-126.

C. Griffiths

Variegated lady beetle/ Spotted amber lady beetle

Hippodamia variegata

Group: Coccinellidae

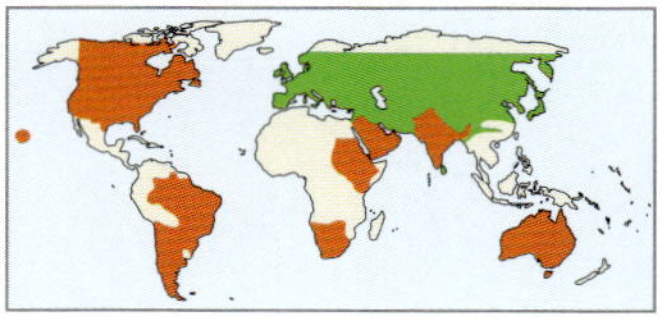

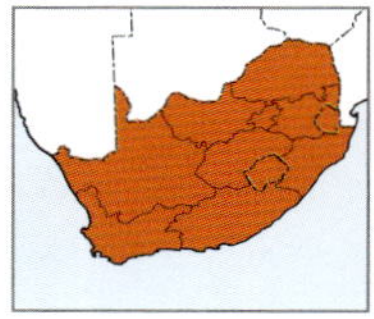

Habitat: Terrestrial. Feeds on various aphids in natural and agro-ecosystems. **Identification and biology:** Medium-sized lady beetle (length 4–6 mm, elongate-ovate), moderately convex; black pronotum with whitish margin; elytra yellowish to reddish, 9–13 variable black spots. Both larvae and adults are aphid predators. **Introduction and spread:** Palaearctic origin. Spread to numerous parts of the world, usually by unknown means; deliberate biocontrol introductions to a few countries. First known South African record Cape Town, April 1965; introduction pathway unknown. Widespread by 1990, but in 2000–2002 additionally introduced against Russian wheat aphid (p. 94), by Small Grain Institute of South Africa. Presently common and widespread in South Africa; familiar, rarely appreciated as an alien species. **Impacts:** Historical impact in South Africa unrecorded, present environmental impact not known. Beneficial aphid predator in agro-ecosystems, e.g. in wheat. Anecdotal evidence suggests that it caused decline of indigenous aphid-feeding lady beetle, *Lioadalia flavomaculata*. **Control:** Unwarranted; most likely impossible.

C. Griffiths

Key reference: Franzmann, B.A. 2002. *Hippodamia variegata* (Goeze) (Coleoptera: Coccinellidae), a predacious ladybird new in Australia. *Australian Journal of Entomology*, 41:375-377.

Vedalia beetle

Rodolia cardinalis

Group: Coccinellidae

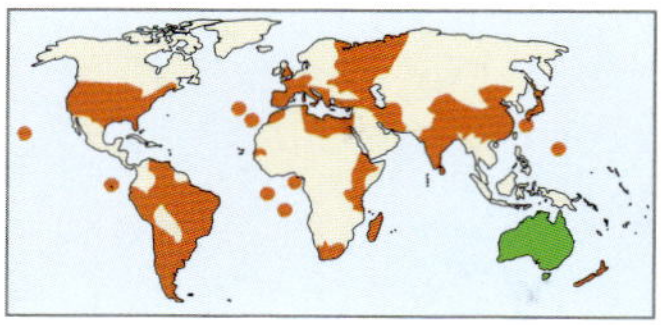

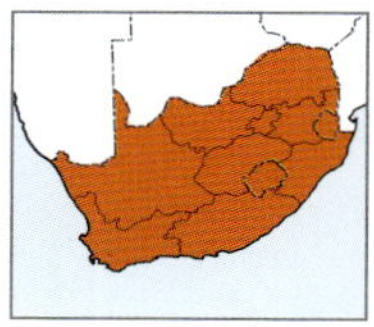

Habitat: Terrestrial. Spectacularly successful biocontrol agent of Australian bug (p. 98); usually found in agro-ecosystems, orchards and gardens. **Identification and biology:** Smallish lady beetle (width 2–3 mm, ovate), crimson with bold black markings, upperside densely hairy. Feeds primarily on *Icerya* species. Eggs laid on plants or onto egg sac of Australian bug; tiny pinkish larvae feed inside on bug's eggs, larger larvae on bug itself. **Introduction and spread:** An Australian native. First instance of biocontrol in South Africa, where deliberately released since 1892; now occurring widely. Introduced to over 70 other countries. **Impacts:** Renders complete biocontrol of Australian bug, with vast economic benefit. Highly susceptible to insect growth regulator hormone analogues when these are injudiciously used against Red scale (p. 96). Original success (California, late 1800s) initiated modern biocontrol of alien pests and invasive plants. **Control:** Unjustifiable. No non-target impacts known.

M. Hoddle

Key reference: Bedford, E.C.G, van den Berg, M.A. & de Villiers, E.A. 1998. *Citrus pests in the Republic of South Africa*. 2nd ed. Nelspruit: Institute for Tropical and Subtropical Crops.

Bamboo longhorn beetle/ Bamboo tiger longicorn

Chlorophorus annularis

Group: Cerambycidae

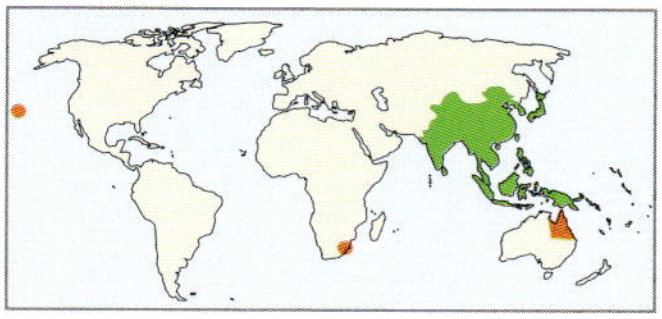

Habitat: Terrestrial. A destructive borer primarily infesting dry bamboo. **Identification and biology:** Unmistakeable beetle; upperside yellowish with characteristic dark markings, body slender (10–16 mm long), antennae very long. Larvae pale, 'round-headed' borers. Related to indigenous *Chlorophorus capensis*, which lacks any colour pattern and is a borer of indigenous trees. **Introduction and spread:** Native to Southeast Asia. Readily translocated inside ornamental bamboos and bamboo artefacts. Caused extensive damage to reed structures in Hluhluwe-iMfolozi Park (KwaZulu-Natal) in 2002. Present status or subsequent spread unknown; research urgently required. Intercepted in Durban (1953) and Pretoria (2004), but no evidence of establishment. **Impacts:** Has potential to develop new host association with indigenous reeds. Potentially a serious invader of natural reedbeds and serious pest of building material. An important species searched for worldwide in goods transported across political boundaries (phytosanitary regulation). **Control:** If a localised population is present at Hluhluwe, targeted eradication should be undertaken.

Key reference: Grobbelaar, E. 2003. Should we be concerned about the South-East Asian bamboo longhorn, *Chlorophorus annularis* (Coleoptera: Cerambycidae), an invader recently identified in South Africa? *Proceedings of the 14th Entomological Congress*. p. 32.

N. Stott

Eucalyptus longhorn borers

Phoracantha species

Group: Cerambycidae

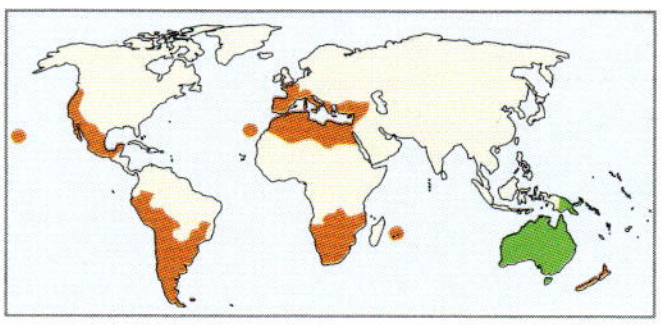

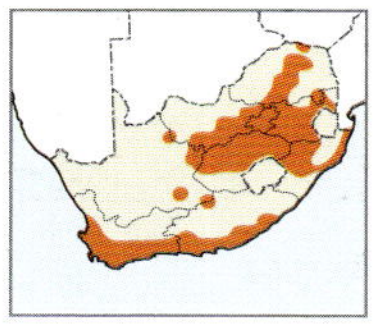

Habitat: Terrestrial. Urban, agricultural and other areas where eucalypts (gum trees) occur. **Identification and biology:** These two large black-brown and straw-coloured longhorn beetles have body lengths of 15–30 mm, and antennae even longer. *Phoracantha recurva* (shown) has elytra anteriorly yellowish with dark patches. The closely related *P. semipunctata* has a narrower yellowish band that includes a zigzag line in it. Egg clusters under loose bark. Larvae fleshy 'round-headed' borers, strong jaws. Pupal cell in heartwood. **Introduction and spread:** Australian species; widely spread globally and locally with infested wood. Entered South Africa around 1906, probably in freshly cut railway sleepers. Nocturnal adults moderately long-lived, capable of significant dispersal through strong flight. **Impacts:** Economically important pests of most cultivated *Eucalyptus* species. Mainly stressed trees attacked, but also freshly felled trees. Larvae tunnel deep into trunk, effectively ring-barking tree; trees killed rapidly, timber badly damaged. **Control:** Chemicals ineffective. Forestry cultivation measures, sanitation and tree selection may reduce infestations. Imported and indigenous natural enemies exert limited control.

Key reference: Drinkwater, T.W. 1975. The present pest status of eucalyptus borers *Phoracantha* spp. in South Africa. *Proceedings of the First Congress of the Entomological Society of Southern Africa*: pp. 119-129.

C. Griffiths, M. Picker

Prosopis seed beetles

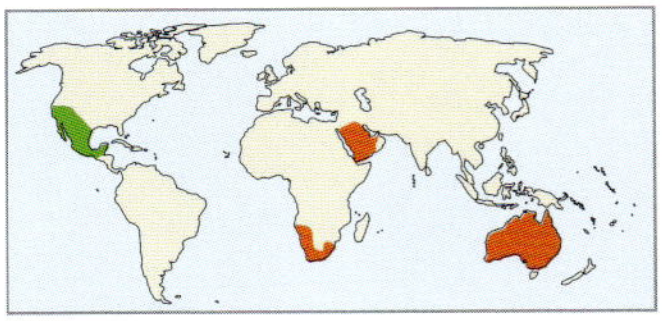

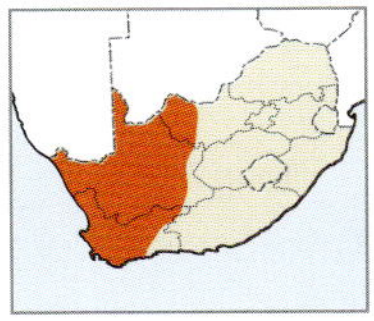

Algarobius prosopis and *Neltumius arizonensis*

Group: Chrysomelidae

Habitat: Terrestrial. Biocontrol agents of Mesquite trees (*Prosopis* species). **Identification and biology:** Smallish, hump-backed beetle with small head; elytra shortened to expose last abdominal segment (pygidium) on the back. *Neltumius arizonensis* (shown) is 2.5–4 mm long, elytra with distinctive black/brown-and-white pattern; *Algarobius prosopis* is 2–4 mm long, elytra brown/reddish-brown. Larvae develop in seeds of Mesquite trees, four species of which are serious declared plant invaders widespread in dry western regions of South Africa**. Introduction and spread:** Host trees and beetles all originated in south-western North America. Beetles purposely released in South Africa since 1987; locally established, widespread. (Maps apply to both species.) **Impacts:** Destroy *Prosopis* seeds, to an extent limiting spread of invaders. Use of seed-feeders for invasive tree biocontrol pioneered in South Africa. **Control:** Not relevant in the case of biocontrol agents.

C. Griffiths

Key reference: Impson, F.A.C., Moran, V.C. & Hoffmann, J.H. 1999. A review of the effectiveness of seed-feeding bruchid beetles in the biological control of mesquite, *Prosopis species* (Fabaceae), in South Africa. *African Entomology Memoir*, 1:81-88.

Satansbos leaf beetles

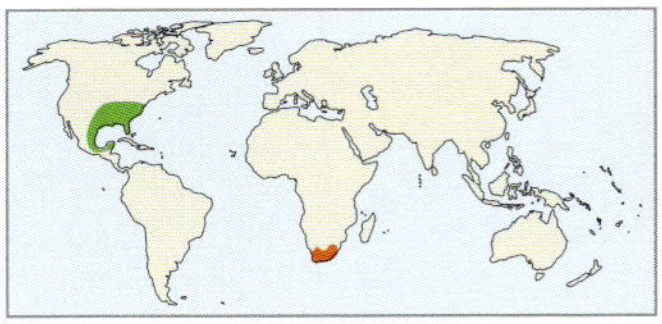

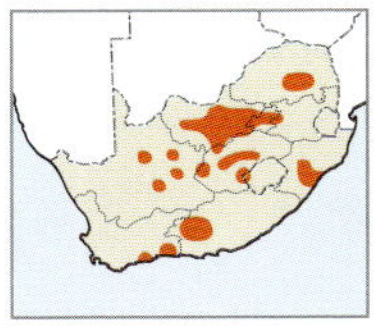

Leptinotarsa texana and *Leptinotarsa defecta*

Group: Chrysomelidae

Habitat: Terrestrial. Biocontrol agents of Silverleaf nightshade/Satansbos (*Solanum elaeagnifolium*), occurring in disturbed and agricultural settings. **Identification and biology:** Striking yellow-and-black, shiny, convex beetles, 10 mm long; *Leptinotarsa texana* (shown) with four lines per wing cover, *L. defecta* with only two lines. Adults and larvae feed on leaves, flowers, bark; skeletonise host plants. Closely related to Colorado potato beetle (*L. decemlineata*), one of the world's most notorious insect pests (not known from Africa). **Introduction and spread:** Origin North America; since 1992 purposely released in South Africa, now established widely. World's first implementation of biocontrol against weeds in the potato family (Solanaceae). (Maps apply to both species.) **Impacts:** Substantial control of *S. elaeagnifolium*, a major agricultural and environmental weed from north-eastern Mexico and south-western USA. Damage to plants mainly by *L. texana*; causes stunted growth and reduced fruiting, allowing natural vegetation to regain dominance. **Control:** Not relevant in the case of biocontrol agents.

E. Holm

Key reference: Olckers, T., Hoffmann, J.H., Moran, V.C., Impson, F.A.C. & Hill, M.P. 1999. The initiation of biological control programmes against *Solanum elaeagnifolium* Cavanilles and *S. sisymbriifolium* Lamarck (Solanaceae) in South Africa. *African Entomology Memoir*, 1:55-63.

Northern pine weevil/ Deodar weevil

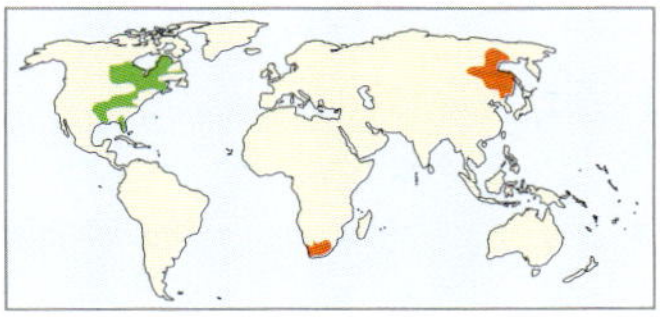

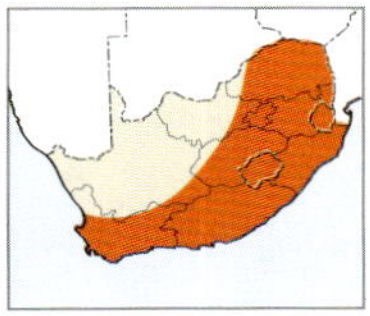

Pissodes nemorensis

Group: Curculionidae

Habitat: Terrestrial. Pine trees, largely in plantations. **Identification and biology:** Typical weevil appearence, with conspicuous snout; cylindrical body reddish to grey-brown, 5–6 mm long, with pale patches. Feeds on and breeds in cambium/phloem of all pine species grown in South Africa. Larvae yellowish-white, legless grubs. Pupation (autumn) underneath bark, in oval 'chip cocoons' constructed from wood fragments. Single generation per year. **Introduction and spread:** Origin eastern North America. First noticed in South Africa in 1942; introduced via infested timber, now in all pine-growing areas. **Impacts:** Principally a secondary pest associated with stressed trees; in some areas damage is primary. Ring-barks growth shoots, causing dieback, retarded growth, poor tree form and decreased wood yield. Nursery trees killed by heavy larval infestations. Potential vector of pitch canker, a serious fungal disease. Of phytosanitary concern. **Control:** Cultural management in plantations, sanitation most important; no insecticidal control locally; host resistance and biocontrol are future possibilities.

C. Ernst

Key reference: Gebeyehu, S. & Wingfield, M.J. 2003. Pine weevil *Pissodes nemorensis*: threat to South African pine plantations and options for control. *South African Journal of Science*, 99:531-536.

Azolla weevil

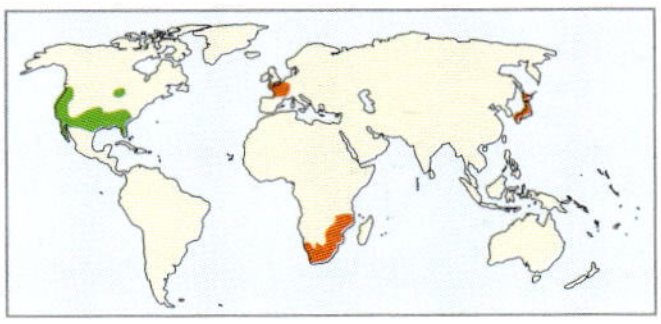

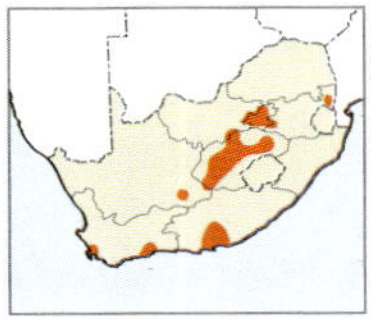

Stenopelmus rufinasus

Group: Curculionidae

Habitat: Semi-aquatic (freshwater). Biocontrol agent of Red water fern (*Azolla filiculoides*). **Identification and biology:** Small weevil (1.7 mm long), grey-black, with variably scattered red, black and white scales; snout shorter than head, its tip reddish. A water beetle, always in association with host plant (Red water fern), a serious floating weed of still and slow-moving waters in both natural and transformed habitats. Older larvae can consume several plants per day. **Introduction and spread:** Both weed and weevil originated from the Americas. Since 1997, Azolla weevil purposely released in South Africa by Plant Protection Research Institute; it established rapidly and widely. **Impacts:** One of the most successful weed biocontrol agents ever employed. Most weed infestations completely controlled in a matter of months, locally even causing 100% clearance. **Control:** Not relevant in the case of biocontrol agents.

M. Hill

Key reference: Hill, M.P. 1999. Biological control of red water fern, *Azolla filiculoides* Lamarck (Pteridophyta: Azollaceae), in South Africa. *African Entomology Memoir*, 1:119-124.

AUTHOR: M. PICKER

BUTTERFLIES, MOTHS Class Insecta, Order Lepidoptera

Lepidoptera are among the largest insect groups and easily recognised by their coiled, tubular mouthparts; minute flattened, overlapping scales covering the body; and two pairs of wings. The eggs hatch into elongated larvae (caterpillars) that have short abdominal 'prolegs' in addition to the three pairs of thoracic legs. Adults usually feed on nectar, but caterpillars are typically voracious herbivores, many being important crop pests that have been spread with their food-plants across the planet.

Large cabbage white butterfly

Pieris brassicae

Group: Pieridae

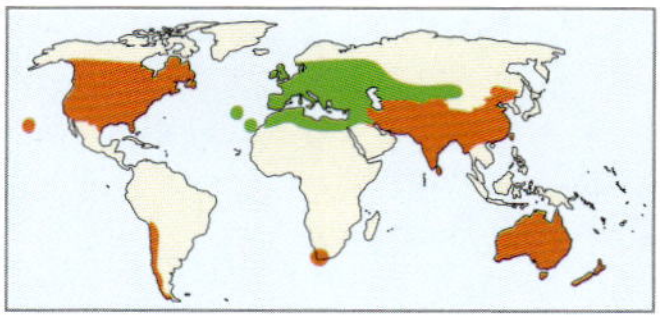

Habitat: Gardens, vacant land, agricultural fields. **Identification and biology:** Large white butterfly (wingspan 55–60 mm), forewings with black marginal band; females have two black spots on upperside, males have only underside spots. Between 150 and 300 yellow eggs laid under host plant leaves, hatching into yellowish-black striped caterpillars. Mature larvae skeletonise leaves, and after approximately one month form a greenish-yellow pupa, often suspended against walls, trunks and branches by silken girdle. **Introduction and spread:** Introduced from Europe to Canada in 1860, reaching California by 1883 and Hawaii by 1893. In 1930 reached New Zealand, and by 1943 occurred across Australia, where early attempts at biological control using the wasp *Pteromalus* and a virus, failed. In the 1970s invaded Chile. First recorded in Cape Peninsula in 1994. **Impacts:** Caterpillars are pests on cabbage and other cultivated Brassiceae (mustard family), also on wild Brassiceae. May feed on onions and nasturtiums (*Tropaeolum majus*), sustaining populations in the Western Cape, where nasturtiums are a weed. All known food-plants have a common defensive compound, a mustard glycoside, which is stored as a mustard oil by the Large cabbage white butterfly. Population numbers fluctuate annually, probably from varying levels of parasitism by *Apanteles* wasps (which kill up to 80% of caterpillars) and native *Pteromalus puparum* (which attack pupae). Not currently a serious pest of brassicaceous crops in the region. **Control:** Regular crop rotation, deep winter ploughing with removal of all plant debris, and spraying with insecticides, plus removal of nasturtiums in vacant land, all reduce populations.

C. Griffiths

C. Griffiths, M. Picker

Key reference: Visser, D. 2009. *A complete guide to vegetable pests in South Africa*. Pretoria: Agricultural Research Council.

Prickly pear moth/ Cactus moth

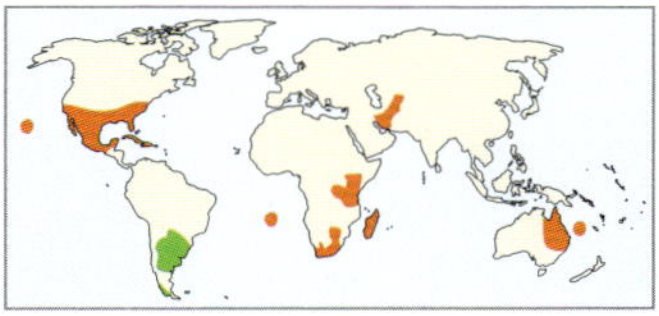

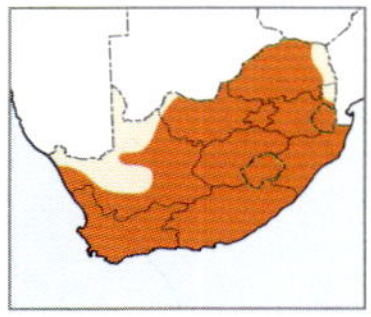

Cactoblastis cactorum

Group: Pyralidae

Habitat: Arid parts infested with prickly pear cactus. **Identification and biology:** Small, short-lived moths (wingspan 23–40 mm) which wrap their pale brown forewings and whitish hindwings around the body when at rest. Egg sticks resembling cactus spines laid on leaf pads and hatch in 18–28 days. Orange, black-banded larvae feed internally in leaf pads, with yellowing pads oozing sap and caterpillar droppings. After hollowing out and destroying leaf pad, mature larvae pupate at base of the cactus for 21–28 days, before emerging as adults. **Introduction and spread:** First introduced from Argentina to Australia in 1926 as a biocontrol agent, where it successfully controlled several weedy species of prickly pear cactus (*Opuntia*). From Australia introduced to South Africa (1933), and from there to Mauritius and Hawaii (1950), the Caribbean (1957) and various other countries. **Impacts:** After successfully controlling weedy *Opuntia* species in the Caribbean, has spread to Florida, where it began attacking native cacti. Now poses a serious threat to native North American and Mexican *Opuntia* species, as well as a diverse community of *Opuntia*-dependent vertebrates and invertebrates. Also threatens the economy and livelihood of owners of Mexican *Opuntia* plantations (from which fruit, pads and the Cochineal bug (*Dactylopius coccus*) are harvested). Erect prickly pear (*O. stricta*) invaded the Kruger National Park in the 1960s, and now covers 35 000 ha, but is not effectively controlled there by the Cactus moth. In other parts of South Africa, the woody stems of the tree-like Barbary fig (*O. ficus-indica*) are never killed by the Cactus moth, which only destroys terminal branches and young cladodes. **Control:** In South Africa the Cactus moth is mostly a desirable biocontrol agent, although less effective than elsewhere, possibly since its numbers may be controlled by native braconid wasps, ants that feed on the egg sticks, and baboons and Vervet monkeys that open the cladodes to eat larvae. Together with the Cochineal bug, it is a pest in South African plantations of Barbary fig and Wheel cactus (*O. robusta*) grown for fruit and livestock fodder. Here integrated chemical and mechanical methods are used to regulate its numbers, and include hand-picking of egg sticks. Use of sterile males may hold promise for future regulation.

H. Robertson

C. Griffiths, M. Picker

Key reference: Zimmermann, H., Bloem, S. & Klein, H. 2004. *Biology, history, threat, surveillance and control of the Cactus moth,* Cactoblastis cactorum. Vienna: International Atomic Energy Agency.

Sorghum stem borer/ Spotted stem borer

Chilo partellus

Group: Pyralidae

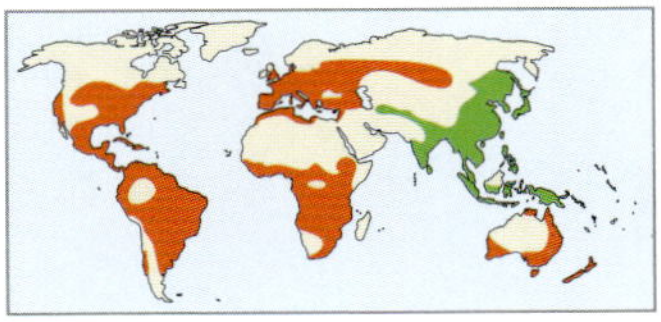

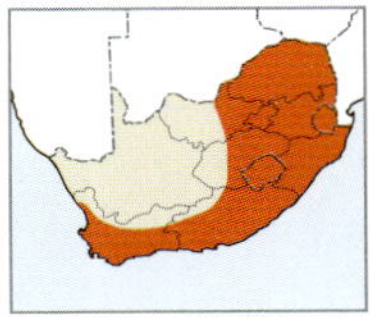

Habitat: Agricultural and disturbed land. **Identification and biology:** Larvae of these pale-brown moths (wingspan 20–30 mm) feed on various grasses, including agricultural crops such as sorghum, maize, millet, rice, sugarcane and Job's tears. **Introduction and spread:** Origin likely Asia, but now widespread in Africa, including South Africa (after likely introduction to the Springbok Flats in the 1950s). There are two generations in South Africa each summer. **Impacts:** In South Africa, primary hosts are sorghum and sugarcane. Young larvae damage leaf whorls, eventually burrowing into and hollowing grass culms. May also attack maize cobs. **Control:** Insecticides generally ineffective in mid-summer (as larvae have already bored into grass stems), and should be employed in mid-October, after the eggs have been laid. Similarly, planting after mid-October bypasses the first cohort of moths. In late summer the wasp *Dentichasmias busseolae* is an effective population regulator.

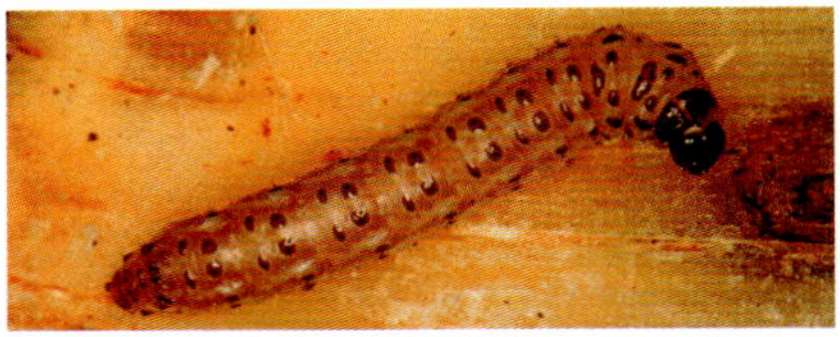

D. Visser

D. Visser

Key reference: Annecke, D.P. & Moran, V.C. 1982. *Insects and mites of cultivated plants in South Africa.* Durban: Butterworths.

Codling moth/Apple codling moth

Cydia pomonella

Group: Tortricidae

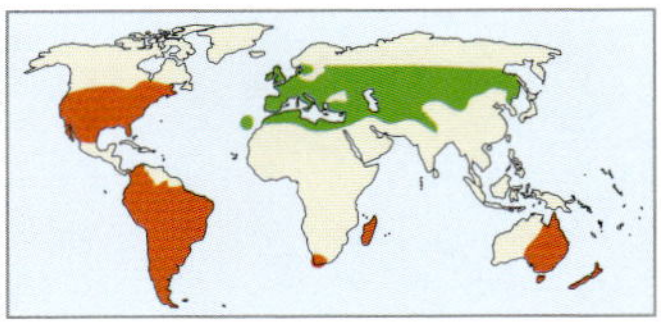

Habitat: Orchards, gardens.
Identification and biology: Small squat moths (wingspan 16 mm) with coppery tip to grey wings. Eggs laid on young fruit or leaves; caterpillars feed on various pome (apple, pear, quince) and *Prunus* fruits (peach, apricot, almond, plum), also walnuts, pupating after 3–6 weeks. **Introduction and spread:** Originated in temperate Asia, but widely invasive. First confirmed report from South Africa in Graaff-Reinet in 1892. In 1898 it had reached the Western Cape, possibly via an unrelated apple import. **Impacts:** Although not impacting peach crops as severely as apple and pear, the former act as a reservoir for later infestations. There are 1–3 generations in the Western Cape on apple crops. **Control:** Biocontrol by a native egg parasite (the wasp *Trichogrammatoidea lutea*) needs to be supplemented with the application of insecticides. Monitoring infestation levels with pheromone-baited traps informs frequency of insecticide application.

C. Griffiths, M. Picker

Key reference: Annecke, D.P. & Moran, V.C. 1982. *Insects and mites of cultivated plants in South Africa.* Durban: Butterworths.

Diamondback moth/ Lesser cabbage moth

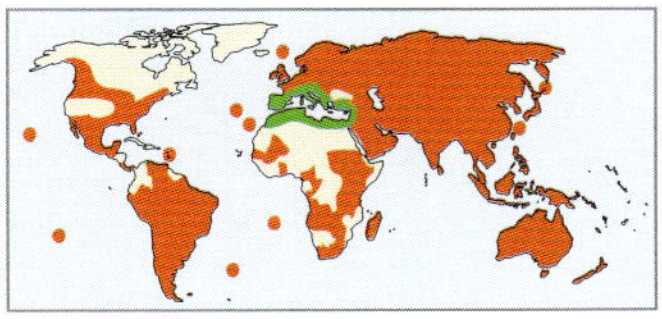

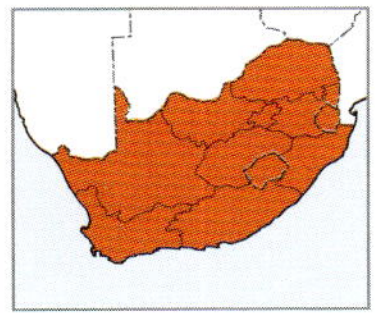

Plutella xylostella

Group: Plutellidae

Habitat: Agricultural and disturbed land. **Identification and biology:** Adults small (wingspan 12 mm), with brown and buff wings that form diamond-shaped patterns above when at rest. Bright green larvae restricted to cultivated and wild species and strains of Cruciferae (mustard family), especially *Brassica* (which includes many weeds). Larval duration 9–32 days and many generations produced each year. **Introduction and spread:** Origins probably Mediterranean, now widespread on all continents. Adults capable of large-scale migration, but eliminated by very cold winters. **Impacts:** Larvae strip lower leaf epidermis, producing clear patches, that later develop into holes. Seedlings may be killed, or head formation in cabbage or Brussels sprouts disrupted. **Control:** Natural mortality is high, but insecticidal application is required for effective control, especially at the seedling stage. The use of insecticidal 'cocktails' retards the development of insecticidal resistance. However, this compromises the regulatory role of natural parasites.

C. Griffiths, M. Picker

Key reference: Visser, D. 2009. *A complete guide to vegetable pests in South Africa*. Pretoria: Agricultural Research Council.

Potato tuber moth

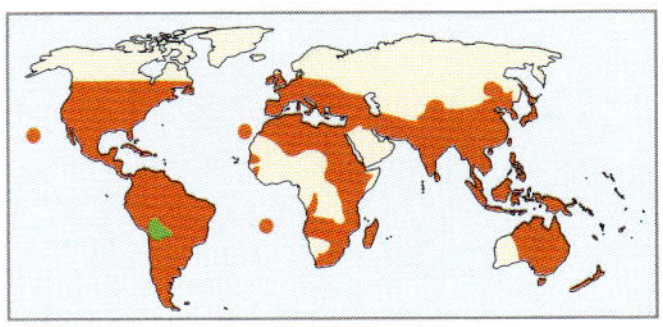

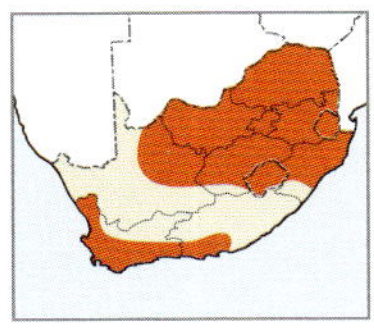

Phthorimaea operculella

Group: Gelechiidae

Habitat: Agricultural lands, storage sheds. **Identification and biology:** Small (wingspan 9 mm) greyish-brown moth with six overlapping generations in South Africa, the life cycle taking 6–8 weeks. Caterpillar mines leaves, eventually attacking potato tubers via cracks in the soil, forming black tunnels in the potato filled with faecal pellets. Caterpillar occasionally feeds on introduced Thorn apples (*Datura* species) and other solanaceous plants (tomato, Cape gooseberry, tobacco and eggplant). **Introduction and spread:** Native to Bolivia, present in South Africa for more than 100 years. **Impacts:** Late summer crop most susceptible. Has the potential to transfer to native Solanaceae. **Control:** Integrated management (including insecticides) requires pest-free seed potatoes, removal of old tubers from fallow fields before planting, crop rotation and deep planting. Soil should be mounded up around growing plants. Control also required during storage. Native parasites and virus, aided by imported parasitic wasps (*Copidosoma koehleri*, *Apanteles subandinus*) may reduce infestation by 98%.

D. Visser

Key reference: Visser, D. 2009. *A complete guide to vegetable pests in South Africa*. Pretoria: Agricultural Research Council.

French grain moth

Sitotroga cerealella

Group: Gelechiidae

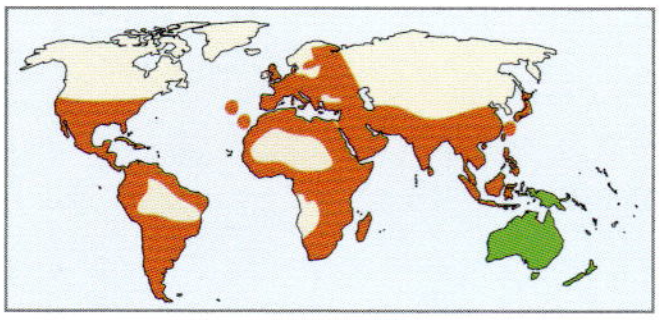

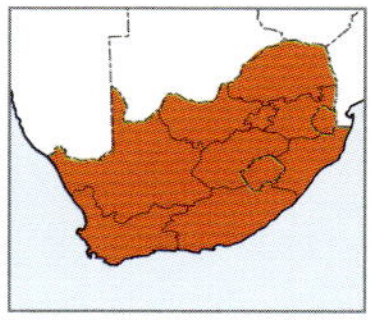

Habitat: Agricultural and disturbed fields. **Identification and biology:** Small moth (wingspan 13–20 mm) with dull yellow forewings and grey hindwings. Caterpillars develop inside seed of Asian rice, pearl millet, sorghum, wheat and maize, rarely on dry plant matter. In warmer countries, more than one generation is produced annually. Caterpillars enter soft maize kernels in the field, then continue to develop on dried kernels. **Introduction and spread:** Native to Australia. Virtually global pest in temperate and tropical countries, translocated via grain shipments. **Impacts:** Grain usually infected in storage, where up to 50% of wheat grains may be infested. Infested grain has a bad odour and taste and is unpalatable. **Control:** Heating to 60°C kills eggs and larvae, but germination capacity is reduced and the grain cannot be used for flour. Fumigation of silos and mills with various insecticides eliminates infestations.

Key reference: Harney, M. 1993. *A guide to the insects of stored grain in South Africa*. Pretoria: Plant Protection Research Institute.

Clemson University, USDA Cooperative Extension Slide Series, Bugwood.org, Wikimedia Commons

Indian meal moth/ Flour moth

Plodia interpunctella

Group: Pyralidae

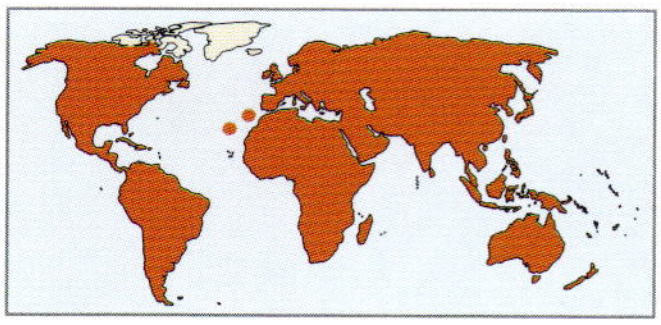

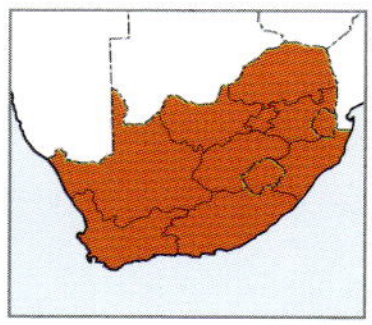

Habitat: Households, warehouses, mills, food-processing plants. **Identification and biology:** Small moths (wingspan 16–20 mm) with slow fluttering flight, last half of forewings bronze, grey or copper-coloured, separated by dark band from the yellow remainder. The 12 mm long, cream-white, dark-headed larvae ('waxworms') infest cereal and related products and take 2–41 weeks to complete development, depending on temperature. **Introduction and spread:** Cosmopolitan, origin not known. **Impacts:** Major pest of stored food products, including flours, pasta, rice, dried fruit, nuts and chocolate; binds the food together with loose webbing. **Control:** Store products in airtight containers, and discard contaminated food. Pheromone-laced sticky traps remove males, and conventional sticky traps ('fly tape') reduce population. Parasitic wasps *Bracon hebetor* and *Trichogramma* species regulate numbers. Exposure to low (-18°C) and high temperatures (54°C) for a few days kills both larvae and eggs.

Key reference: Harney, M. 1993. *A guide to the insects of stored grain in South Africa*. Pretoria: Plant Protection Research Institute.

C. Griffiths, M. Picker

Mediterranean flour moth

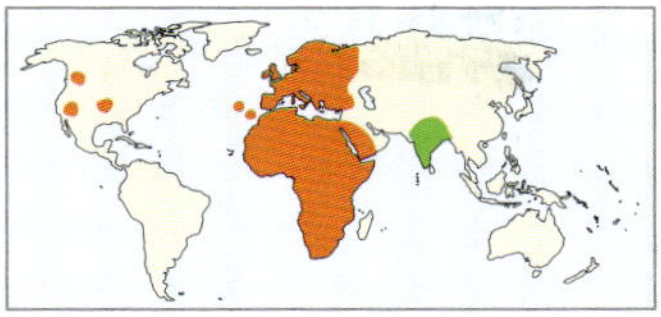

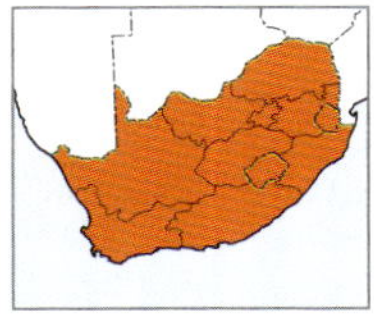

Ephestia kuehniella

Group: Pyralidae

Habitat: Households, factories, flourmills, shops. **Identification and biology:** Small, pale grey moth (wingspan 16–20 mm) with faint dark markings on forewing. Adults live around 14 days, active at dusk and dawn. Cream-coloured larvae with dark heads infest cereals, dried fruit, nuts, cocoa and baked goods. May be confused with similar Almond moth (*Cadra cautella*) and Warehouse moth (*Ephestia elutella*). **Introduction and spread:** Native to India, reached Europe by 1877 in imported American wheat, now cosmopolitan in temperate and subtropical regions. **Impacts:** Silk webbing produced by larvae contains faeces and moulted skins, making infected foodstuff inedible and blocking sieves in flourmills. Pupation occurs within food product. **Control:** Good sanitary measures (tightly sealed containers, eliminating spilt produce) block infestations. In severe infestations, fumigation with pesticides controls larvae and eggs residing in crevices. Biological control, using mites and fungi, currently being explored.

C. Griffiths, M. Picker

Key reference: Harney, M. 1993. *A guide to the insects of stored grain in South Africa*. Pretoria: Plant Protection Research Institute.

Case-making clothes moth

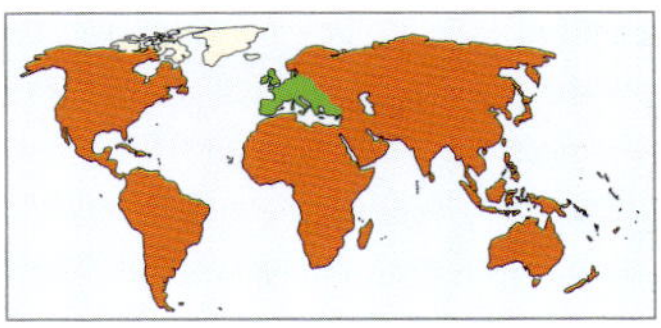

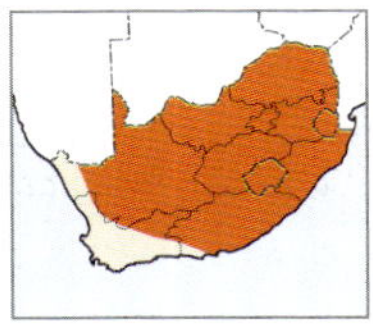

Tinea pellionella

Group: Tineidae

Habitat: Houses, warehouses. **Identification and biology:** Small (wingspan 10 mm), inconspicuous buff-coloured moth, three dark spots on forewings. Larvae feed from within silken case covered in faecal pellets and discarded fibres; only head and legs visible. Caterpillar moults 5–45 times, life cycle lasting two months to several years. Caterpillars typically feed on materials containing animal protein, including wool, fur, leather, hair, feathers and fishmeal; also tobacco. Larvae produce characteristic furrows on surface of woollen garments. Another similar and very common introduced clothes moth is the straw-coloured Common clothes moth (*Tineola bisselliella*). **Introduction and spread:** Common across its native Europe, but introduced to many other parts including Australia, North America and Africa. **Impacts:** Damages stored clothing, carpets, felt and furs. **Control:** Careful vacuum cleaning of floors, crevices and cupboard floors reduces populations. Residual approved insecticides can be applied to drawers and cupboards, and affected clothing steam-cleaned.

C. Griffiths, M. Picker

Key reference: Cheema, P.S. 1956. Studies on the bionomics of the case-bearing clothes moth, *Tinea pellionella* (L.). *Bulletin of Entomological Research*, 47:167-182.

Greater wax moth

Galleria mellonella

Group: Pyralidae

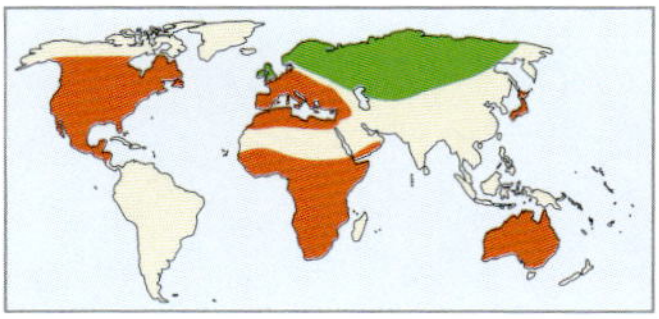

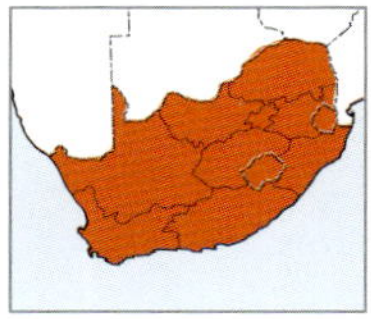

Habitat: Beehives, honeycomb storage rooms. **Identification and biology:** Medium-sized moth (wingspan 30–45 mm), wings greasy grey with dull purple markings. Females enter hives of honeybees (*Apis* species) at night to lay eggs. Greyish-white caterpillars feed on honeycomb, although also infest bumblebee and wasp nests, and dried figs. In South Africa there are three generations per year, the late summer brood having the greatest impact. **Introduction and spread:** Widely distributed from its presumptive native range, Eurasia. **Impacts:** Most important wax moth in South Africa. Heavy losses occur in sub-Saharan Africa, with stored and deserted combs (especially those still containing pollen) reduced to a mass of webbing and caterpillar droppings. Larvae bore depressions/holes in woodwork of hive to form pupae. **Control:** Strong bee colonies able to depress infection levels – providing colonies with too much comb can also result in uncontrolled infestations.

dhobern, Wikimedia Commons

Key reference: Buys, B.A. 1982. A survey on honeybee pests in South Africa. *South African Bee Journal*, 54:86-90.

Lesser wax moth

Achroia grisella

Group: Pyralidae

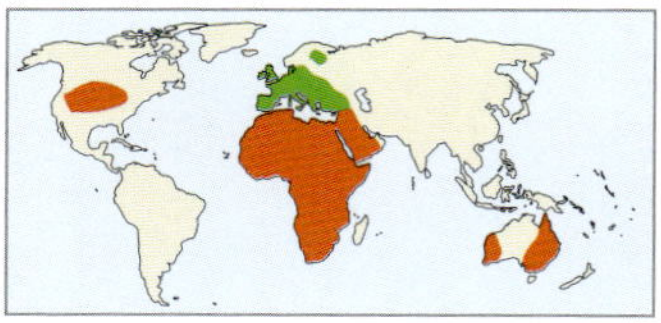

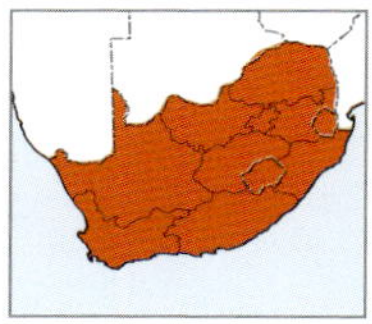

Habitat: Beehives, honeycomb storage rooms. **Identification and biology:** Larvae of this plain grey moth (wingspan 20 mm) bore through honeycomb using silk-lined tunnels, feeding on beeswax, pollen, honey and occasionally bee brood. Occasionally feed on vegetable matter and keratin. Very large numbers can build up in weakened colonies. Males attract females using ultrasound, mating inside or near beehives. **Introduction and spread:** Spread from its native Europe to most parts of the world, except the coldest parts. **Impacts:** Very destructive to honeycomb, with considerable economic losses to beekeepers. **Control:** Since levels of infestation reflect colony weakness, cleanliness of hives essential. Bees in healthy colonies regulate numbers of larvae in the hive. Storage of honeycomb at temperatures below freezing, or fumigation of empty combs with para-dichlorobenzene kills larvae (combs filled with honey can be fumigated with carbon dioxide, frozen or heated).

R. Underhill, Peace Bee Farm, Arkansas, USA

Key reference: Buys, B.A. 1982. A survey on honeybee pests in South Africa. *South African Bee Journal*, 54:86-90.

AUTHOR: A. KIRK-SPRIGGS

FLIES

Class Insecta, Order Diptera

'True flies' (Diptera) are one of the largest orders of insects, with 154 853 species (20 308 in the Afrotropical region). Most have a single pair of wings, the hind pair being reduced to tiny balancing organs, the 'halteres'. Some highly adapted ectoparasitic/island groups have the wings reduced or absent. The applied significance of Diptera cannot be overestimated. On the negative side, flies are especially significant in Africa, as vectors of deadly insect-borne diseases and other pathogens to humans and their livestock, including malaria, trypanosomiasis (and the equivalent animal disease, nagana), leishmaniasis and African horse sickness. Others are serious agricultural pests. On the positive side, the role of Diptera in pollination has received increasing attention in recent years, with studies indicating that Diptera may be far more important in pollination biology than previously thought. Many parasitoid species, especially the Tachinidae (tachinids), are potential agents in biocontrol, while others are now used routinely in forensic investigations. The Diptera are also probably the most significant group in terms of the degeneration and decomposition of animal and other organic matter. The majority of alien invasive species among the Diptera appear to have resulted directly from importations with ship cargo or other imported foodstuffs.

Window-waisted soldier fly

Hermetia illucens

Group: Stratiomyidae

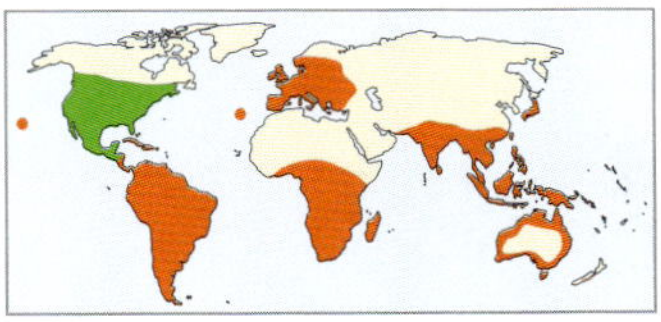

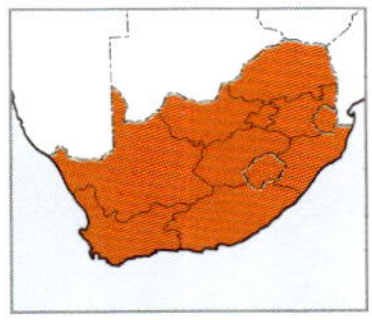

Habitat: Terrestrial, in dead organic matter and faeces. **Identification and biology:** Medium to large fly (wingspan 16 mm), sleek, robust, elongate wasp-mimic, with smokey wings. Body predominantly black, with two pale translucent markings at base of abdomen (dorsally and ventrally). Legs black with white bands. Adults frequently enter houses. Larvae large, with sclerotised body segments, covered in short, sparse spinules. Develop in decaying organic matter in compost heaps and livestock waste. **Introduction and spread:** Originally New World, now cosmopolitan. Purportedly introduced into KwaZulu-Natal from the New World, now virtually ubiquitous in South Africa. **Impacts:** Reduces animal manure in commercial swine and poultry facilities. Adult flies are a potential mechanical vector of various pathogens. Known to develop in intestines of humans and other animals (myiasis) following accidental ingestion of larvae. **Control:** Well established and beneficial. Does not appear to compete with any indigenous species, therefore no control measures required.

S. Marshall

Key reference: James, M.T. 1947. The flies that cause myiasis in man. *USDA Miscellaneous Publications*, No. 631.

Common coffin fly

Megaselia scalaris

Group: Phoridae (scuttle flies)

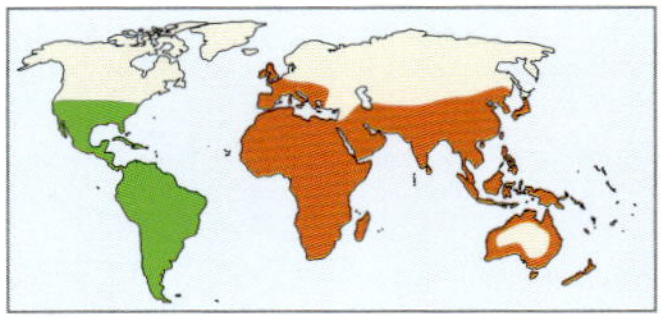

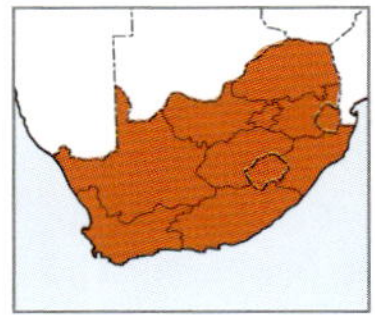

Habitat: Terrestrial, on decaying matter, corpses and faeces.
Identification and biology: Tiny fly (wingspan 4 mm), predominantly pale yellow with darker bands on abdominal segments. Head with a series of strong backwardly directed spines. Wings have spines along leading edge and greatly reduced venation. Identification of *Megaselia* species based largely on microscopic characteristics of the male terminalia. Scuttle flies are easily recognised by their erratic scuttling movements. Larvae breed and complete their development in a range of media. Many are specialised predators or parasitoids. The normal larval food of the Common coffin fly is decaying plants and fungi, dead arthropods and molluscs. Maggots have, however, been reared from the pools of oil that develop below sugarcane-crushing machines in Africa, where they feed on other entrapped insects, and from shoe polish and emulsion paint. **Introduction and spread:** A well-known 'tramp species', probably originally native to North and South America. Has been distributed worldwide by its association with humans, mainly in ship cargo. Also known to have been transported 800 km by aeroplane, and adult flies have been transported from North Africa to Spain in the plumage of migrating birds. Recently recorded from Antarctica breeding in a consignment of damaged chicken eggs. **Impacts:** Maggots reported to contaminate foodstuffs such as pasta, dates, soya flour, cheese, dried fish and rotting potatoes; regularly infests milk bottles. Infested food may be ingested by humans, leading to development of the larvae in the intestine (myiasis). Maggots may infest open cuts and wounds, the urogenital tract and nasal sinuses. Frequently infests foodstuff in homes and common around urinals. Larvae also feed on human corpses and therefore feature in forensic investigations. **Control:** Good sanitation and freezing of infested items for several weeks at -4°C often proves effective.

C. Griffiths, M. Picker

M. Mostovski

Key reference: Disney, R.H.L. 2008. Natural history of the scuttle fly, *Megaselia scalaris*. *Annual Review of Entomology*, 53:39-60.

Common kelp fly

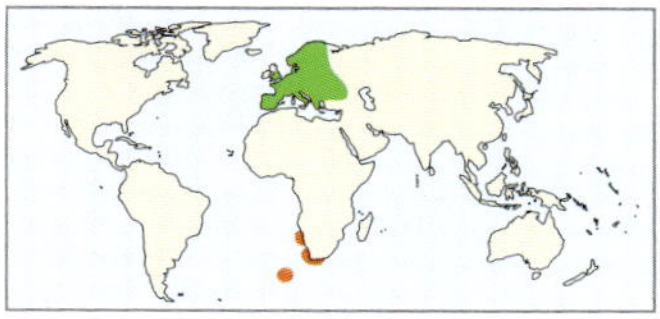

Fucellia tergina

Group: Anthomyiidae

Habitat: Coastal, on decaying seaweeds and faeces. **Identification and biology:** Medium-sized (wingspan 12 mm), uniformly grey-black, with characteristic head shape and spinose tubercle on hindleg. Three *Fucellia* species occur in southern Africa; only one (*F. capensis*) is indigenous and occurs widely along the south-western African coast, associated with kelp beds. Specific identification based largely on microscopic characteristics of legs and male terminalia. Occupy the intertidal zone, principally associated with marine algae (wrack), often swarming in enormous numbers on tide-washed kelp. Frequently blown far inland, but other northern hemisphere species truly associated with the margins of saline lakes. Larvae cylindrical, grey-white maggots that normally develop between blades of partially buried kelp washed up during high tides, forming pupae in the sand beneath. Larvae may also develop in other plant and animal material, including marine molluscs thrown up by the tide. Other rearing records (for the Common kelp fly) include peas, living snails and the spawn of the Caplin fish (*Mallotus villosus*). **Introduction and spread:** Native to Europe. Earliest known record is from Muizenberg, Cape Town (1949). Subsequently from Walvis Bay (1974) and Swakopmund (1976). A second species of the genus, *F. maritima*, first recorded from West Coast National Park in 2001, probably via the guano trade from native Europe in the 1840s. **Impacts:** Introduction of these two species unlikely to affect the indigenous species *F. capensis* negatively, as the latter is better adapted to local conditions; however, some competition may take place. **Control:** No control measures recommended as these insects are largely beneficial in the degeneration of shoreline kelp.

Key references: Kirk-Spriggs, A.H. 2003. The immature stages of *Fucellia capensis* (Schiner) (Diptera: Anthomyiidae) from the southwestern seaboard of Africa, with a review of the known immature stages and ecology of the genus. *Cimbabasia*, 19:95-109.

Kirk-Spriggs, A.H., Ismay, J.W., Ackland, M., Roháček, J., Mathis, W.N., Foster, G.A., Pape, T., Cranston, P.S. & Meier, R. 2001. Inter-tidal Diptera of southwestern Africa (Chironomidae, Canacidae, Chloropidae, Milichiidae, Tethinidae, Ephydridae, Sphaeroceridae, Coelopidae, Sarcophagidae and Anthomyiidae). *Cimbebasia*, 17:85-135.

S. Marshall

Common greenbottle

Lucilia sericata

Group: Calliphoridae (blow-flies)

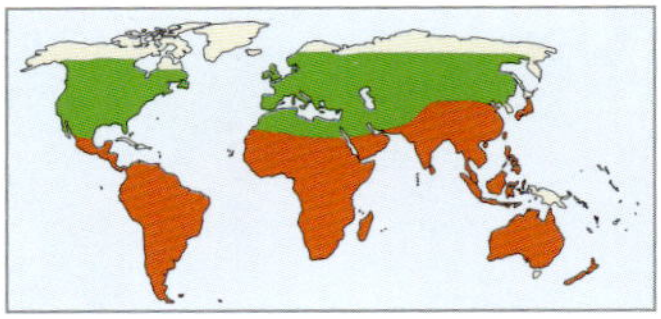

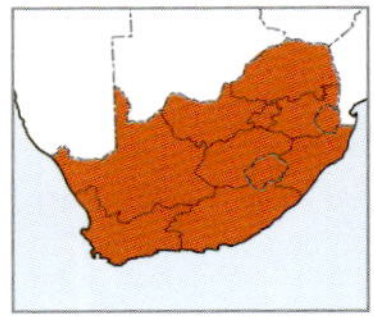

Habitat: Terrestrial, breeding in decaying animal and plant matter and faeces. **Identification and biology:** Medium-sized blow-fly (wingspan 10–14 mm) with a brilliant, metallic, blue-green coloration. Frequently confused with the Australian sheep blow-fly (*Lucilia cuprina*), which has similar habits. Male terminalia and relative positions of three setae behind the eyes are the features most reliably used to distinguish the species. Prefers temperate habitats. Adults attracted to carrion, open wounds, soiled or wet fleece of sheep and, to a lesser extent, faeces, in all of which larvae are able to complete their development. Has been recorded as breeding in kelp in South Africa, and in mink dung in the USA. Adults attracted to fermenting liquids and frequently encountered on flowering plants. **Introduction and spread:** Indigenous to the Holarctic region, now widespread in the Palaearctic and Oriental regions, Australia, New Zealand, the Nearctic and Neotropical regions. Transported by humans (probably recently) to almost all parts of the world. Probably introduced into South Africa over a century ago in ship cargo or with transported sheep or other livestock. **Impacts:** In Great Britain and the Netherlands a primary causal agent of myiasis in sheep ('sheep strike'), either alone, or in combination with other species, but in South Africa the related Australian sheep blow-fly is the main culprit. Also a wound parasite of cattle, horses and humans, although may be beneficial and stimulate the healing process. As a result, the larvae are the main maggots used in human surgery (maggot therapy) to clean infected wounds. The forensic significance of the species is uncertain, mainly due to confusion with Australian sheep blow-fly at crime scenes, but the role of the two species is similar. **Control:** The best method of controlling blow-flies is via proper sanitation, but various commercially produced Pyrethrin-based insecticides are available to deal with extreme cases.

Key references: Kurahashi, H. & Kirk-Spriggs, A.H. 2006. The Calliphoridae of Namibia (Diptera: Oestroidea). *Zootaxa*, 1322:1-131.

Tourle, R.A., Downie, D.A. & Villet, M.H. 2009. Flies in the ointment: a morphological and molecular comparison of *Lucilia cuprina* and *L. sericata* (Diptera: Calliphoridae) in South Africa. *Medical and Veterinary Entomology*, 23:6-14.

C. Griffiths, M. Picker

European bluebottle

Calliphora vicina

Group: Calliphoridae

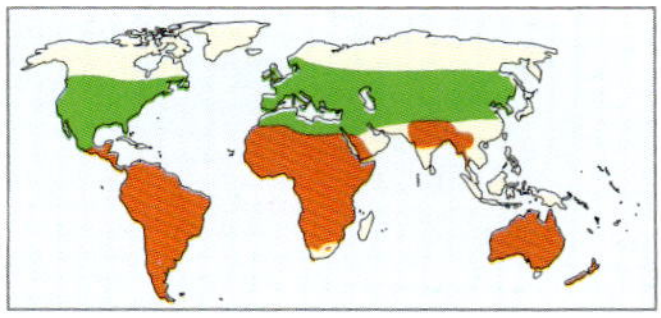

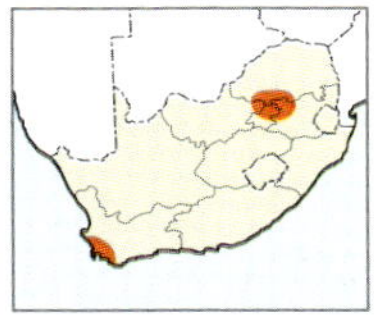

Habitat: Terrestrial, breeding in decaying animal matter. **Identification and biology:** Large (wingspan 30 mm), conspicuous blue-black blow-fly, frequently attracted to carrion. In South Africa, frequently misidentified as the indigenous and widespread Highveld blow-fly (*Calliphora croceipalpis*) but differs in the coloration of the second wing vein and cheek (genae) of the head. Larvae develop exclusively in carrion, on which the eggs are laid. **Introduction and spread:** Originating from, and widespread in the Holarctic region, and has followed humans as an invasive species to northern India, Africa, South America, Australia and New Zealand. First recorded from South Africa (Johannesburg) in 1965, and later in 1967 and 1969 in Benoni, near O.R. Tambo International Airport. Later collected in Cape Town in 1976 and Witbank (about 100 km east of O.R. Tambo) in 2001. Other species of blow-fly have been found in the luggage holds of aircraft and larvae have been found in dried meat and fish imported by air into Texas (USA). This invasive blow-fly species has, therefore, been introduced via South Africa's airports, probably as a direct result of the increased trade between South Africa and other countries in recent decades. **Impacts:** In Europe, very common in urban areas, and one of the most significant species in forensic investigations due to its consistent time of arrival and colonisation of the body following death. It therefore has potential to become a forensically useful fly in South Africa if it becomes fully established. Whether its establishment would have a detrimental effect on the indigenous Highveld blow-fly remains uncertain, but as the main factor regulating the number of generations per year is temperature, it is likely that conditions in South Africa would be highly conducive to its establishment. The two species could therefore potentially compete for resources. **Control:** The best method of controlling blow-flies is via proper sanitation, but various commercially produced Pyrethrin-based insecticides are available to deal with extreme cases.

Key reference: Williams, K.A. & Villet, M.H. 2006. A new and earlier record of *Chrysomya megacephala* in South Africa, with notes on another exotic species, *Calliphora vicina* (Diptera: Calliphoridae). *African Invertebrates*, 47:347-350.

S. Marshall

Oriental latrine fly

Chrysomya megacephala

Group: Calliphoridae

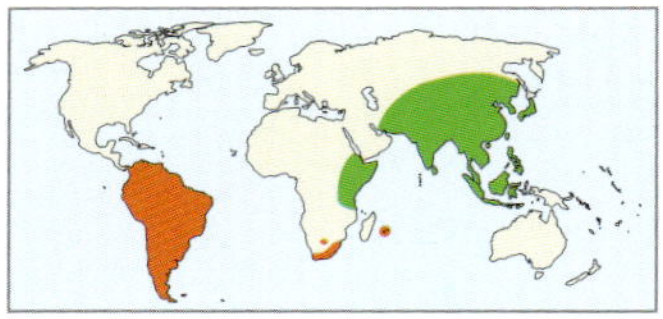

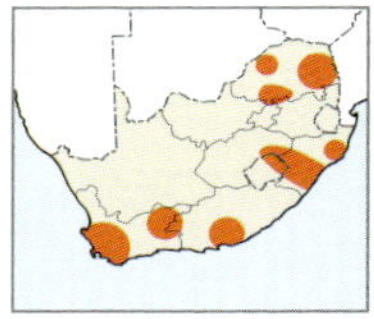

Habitat: Terrestrial, breeding in decaying animal and plant matter and faeces. **Identification and biology:** Medium-sized to large blow-fly (wingspan 24 mm), with a greenish-blue metallic coloration, darker bands on the abdomen and very large, bright red eyes. Adults may be confused with other similar species of *Chrysomya* blow-flies, but the males have eyes that are distinctively divided into small facets ventrally and large facets dorsally, with a sharp transition between the two zones. **Introduction and spread:** Originating in Southeast Asia and East Africa, it was introduced to South America (via Brazil) in the 1970s. The earliest recorded specimen from South Africa was collected in Amanzimtoti on the south coast of KwaZulu-Natal in 1971. The most likely means of introduction into South Africa is transportation in ship cargos from Mauritius (where the species was introduced in 1962) or from East Africa via several South African ports. In the past 35 years since the first South African record, the species has extended its range into the Western Cape, KwaZulu-Natal, the Eastern Cape and Gauteng provinces. **Impacts:** A forensically and medically important species recorded as breeding in faeces, meat and fish. It is therefore an important vector of a range of human pathogens, such as *Escherichia coli*. If larvae of other species are present in the food resource, competition takes place, which reduces the amount of time spent feeding, leading to the early formation of pupae and therefore smaller adults. Adult flies actively seek out flowers for nectar and are therefore instrumental in pollination. In Taiwan, flies are actively encouraged as the main pollinators of commercially grown mangoes. **Control:** The best method of controlling blow-flies is via proper sanitation, but various commercially produced Pyrethrin-based insecticides are available to deal with extreme cases, including specific treatments for 'sheep strike'.

Key reference: Williams, K.A. & Villet, M.H. 2006. A new and earlier record of *Chrysomya megacephala* in South Africa, with notes on another exotic species, *Calliphora vicina* (Diptera: Calliphoridae). *African Invertebrates*, 47:347-350.

S. Marshall

European cheese fly ('Cheese skipper')

Piophila casei

Group: Piophilidae

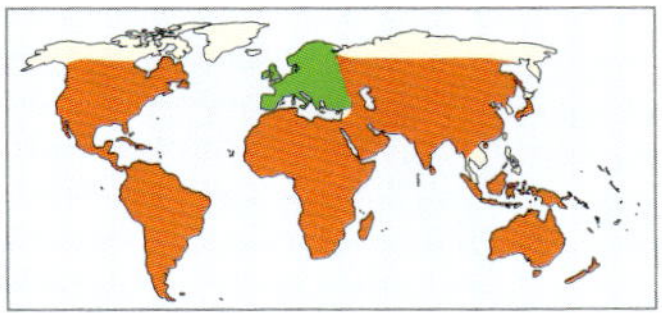

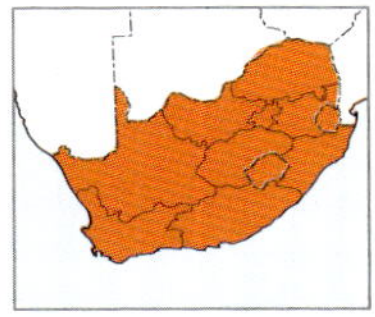

Habitat: Terrestrial, breeding in decaying organic material and corpses. **Identification and biology:** Small (wingspan 4 mm) metallic-coloured, usually black/bluish-black with bronze-coloured tints on head, thorax and abdomen; reddish-brown eyes and iridescent wings. Larvae often referred to as 'Cheese (or ham) skippers', due to their habit of 'jumping', achieved by bending the body double, grasping the anal body compartment with the mouthhooks, tensing the muscles, and then releasing the mouthhooks, thus flicking the larva into the air. **Introduction and spread:** Probably native to Europe and transported around the world in ship cargos, now virtually cosmopolitan. **Impacts:** Larvae infest a range of high-protein foodstuffs, including cured meats, smoked fish, cheeses and decaying animal produce. Signs of European cheese fly infestation include small whitish-coloured eggshells and grooves or creases on the surface of cheeses, made by first-instar larvae. Infested cheeses usually have soft or sunken areas and meats may have a shiny grease-like liquid dripping from infested areas. Larvae may be inadvertently ingested with contaminated foodstuffs and begin developing in the intestine. The European cheese fly is the leading cause of intestinal myiasis in humans, with larvae causing serious intestinal lesions as they attempt to bore through the intestinal walls. Living or dead larvae may pass in the stool. Larvae may also infest the chest and nasal passages of human patients. Due to its habit of infesting mammalian carrion, the species is also significant in forensic investigations. Larvae normally infest corpses during the later stages of decomposition when the body is desiccated, 3–6 months after death. Larvae then form pupae in soil beneath the body or among the folds of dried skin, bones, etc. They have even been found on the exhumed remains of Egyptian mummies. **Control:** Sanitation is considered the most important aspect in management of Cheese skippers. Exclusion of European cheese fly from all levels of food production (processing, curing and storage) ranges from proper sanitation techniques to proper wrapping/protection of food products. Light larval infestations can be removed individually, but the use of fumigation techniques is often required for more severe infestations.

Key reference: Martín-Vega, D., Gómez-Gómez, A., Baz, A. & Díaz-Aranda, L.M. 2010. New piophilid in town: the first Palaearctic record of *Piophila megastigmata* and its coexistence with *Piophila casei* in central Spain. *Medical and Veterinary Entomology*, 25:64-69.

S. Marshall

Moth fly/Drain fly

Clogmia albipunctata

Group: Psychodidae

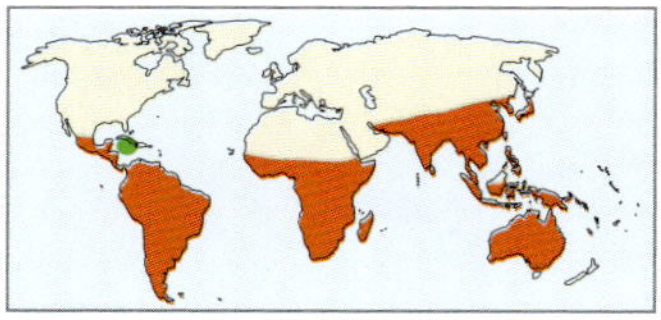

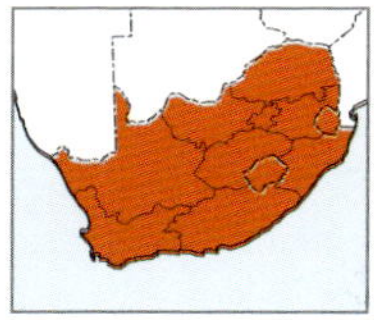

Habitat: Terrestrial, feeding on decaying organic matter. **Identification and biology:** Small fly (wingspan 3–4 mm), easily identified by the distinctive colour patterning of the wings, with white spots at the tips of the wing veins. Body and wings usually black, but may be brown. If crushed, leaves a powdery smudge. Larvae probably breed naturally in rot holes in trees and in stagnant water collected in natural vegetation (phytothelmata), but adults are now most commonly found in association with humans. **Introduction and spread:** Originally described from Cuba, now distributed throughout the tropical and subtropical regions of the world and one of the most common synanthropic species in the family Psychodidae. **Impacts:** In human habitations, found mostly in bathrooms and kitchens, where they breed in drains, sewers and septic tanks. They also occur in soil that has been contaminated with sewage, urinals, water treatment plants and any other areas where waste water containing organic debris collects. The larvae and pupae occur in the thin gelatinous film found in drains, septic tank field lines or filter stones, etc., where they feed on sediment, fungi, decaying vegetation and microscopic plants and animals. **Control:** Adult flies are virtually harmless in terms of disease transmission, as they generally do not stray far from the larval breeding source and are not attracted to settle on foodstuffs. In some cases, however, the number of emergent flies may become troublesome, in which case the source of the infestation must be located. The presence of numerous adult flies inside and around a drain is a good sign that the drain is a breeding site. Once the breeding site has been located, various drain cleaners and gels are commercially available for treatment.

Key reference: Hribar, L.J., Vlach, J.J., DeMay, D.J., James, S.S., Fahey, J.S. & Fussell, E.M. 2004. Mosquito larvae (Culicidae) and other Diptera associated with containers, storm drains, and sewage treatment plants in the Florida Keys, Monroe County, Florida. *Florida Entomologist*, 87:199-203.

S. Marshall

Asian tiger mosquito

Aedes albopictus

Group: Culicidae

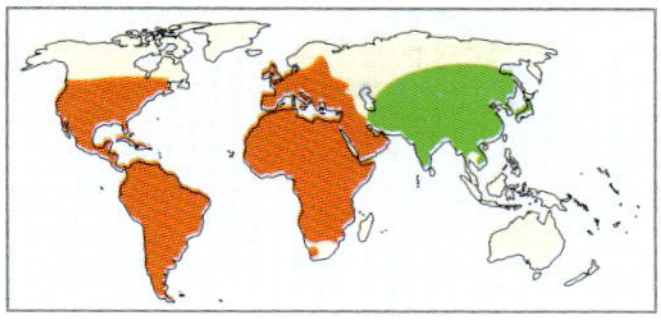

Habitat: Terrestrial, blood and nectar-feeding. **Identification and biology:** Small to medium-sized (wingspan 5.5 mm) mosquito with striking white and black patterning. In Africa, the species may be confused with the Yellow fever mosquito (*Aedes aegypti*), which belongs to the same subgenus, and has similar body coloration and markings. As with other mosquitoes, only the adult female requires a blood meal for maturation of the developing eggs, normally feeding on nectar and other sweet plant juices, as do the males. The larvae are aquatic and breed in stagnant pools. Any open container containing water will suffice for larvae development, even if containing less than 30 ml of water. It can also breed in running water, so breeding sites are not confined to stagnant pools. **Introduction and spread:** Originally from Southeast Asia, from where it has spread to Europe, the Americas, the Caribbean, Africa and the Middle East, via the international tyre trade (due to rainwater retained in tyres when stored outdoors). Listed as one of the world's 100 Worst Invasive Species by the IUCN (International Union for Conservation of Nature) Global Invasive Species Programme database (www.issg.org/database). First detected in Cape Town in 1990; found entering South Africa in used tyre casings from Japan (on three separate occasions). At that time it was suggested that the species may already be established in the country. **Impacts:** Known to transmit pathogens and viruses, including West Nile virus, Yellow fever virus, as well as those causing St. Louis encephalitis, dengue fever and Chikungunya fever. Also relevant in veterinary medicine as a transmitter of Heartworm (*Dirofilaria immitis*), a parasitic roundworm that causes heartworm disease in dogs and cats. Also significant, as it can out-compete, and even eradicate, other species with similar breeding habitats from the very onset of its establishment in other regions and biotopes. **Control:** The Asian tiger mosquito has proven to be very difficult to suppress or control due to its remarkable ability to adapt to various environments, its close association with humans, and its reproductive biology. Monitoring or surveillance of imported plants and stockpiles of tyres at ports, airports and warehouses is an essential intervention in preventing its spread and establishment.

Key reference: Cornel, A.J. & Hunt, R.H. 1991. *Aedes albopictus* in Africa? First records of live specimens in imported tires in Cape Town. *Journal of the American Mosquito Control Association*, 7:107-108.

S. Marshall

European drone fly

Eristalis tenax

Group: Syrphidae

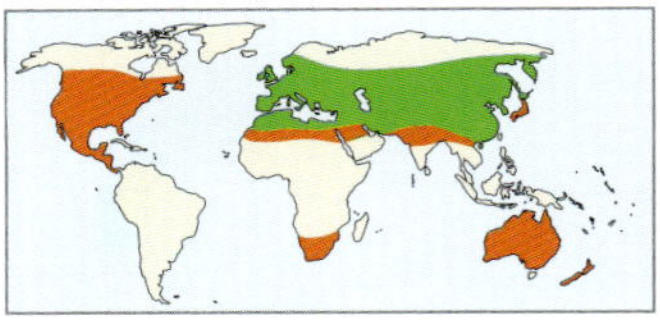

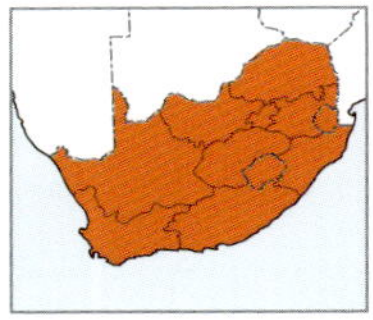

Habitat: Terrestrial, feeds on nectar and decaying matter. **Identification and biology:** Large (wingspan 15 mm), stocky, bee-mimicking in appearance and behaviour. The only species of the genus recorded from South Africa, superficially resembling the Band-eyed drone fly (*Eristalinus taeniops*). The eyes of the European drone fly are hairy, the hairs forming two vertical parallel bands; one brown, the other brown-black. Often observed hovering. The large, grey larvae are the well-known 'rat-tailed maggots', which have a long siphon that can be extended to the water surface for respiration. They inhabit drainage ditches, pools around manure piles, sewage, and similar places containing water badly polluted with organic matter. In southern Africa they are, for example, commonly encountered in pools around permanent springs that have been soiled with baboon dung. **Introduction and spread:** A widespread Palaearctic species, now virtually cosmopolitan in temperate and subtropical areas. Does not survive well in more tropical countries. Arrived in North America in the 1870s and in New Zealand over the same period. It was not known from South Africa in 1860, so was probably introduced subsequently. The means of introduction are uncertain, but may have been via ship cargos or from other outside sources of polluted and infested water. **Impacts:** Adult flies are beneficial, regular flower visitors and are highly important pollinators, especially of carrot and fennel crops. There have occasionally been documented cases of human intestinal myiasis by larvae, which result from humans drinking polluted water or ingesting contaminated salad. Larvae frequently appear in cisterns, baths and basins in South Africa, causing major public consternation. **Control:** Unlikely to be in direct competition with any indigenous species and control measures are therefore not required.

C. Griffiths

S. Marshall

Key reference: Bartsch, H. 2009. Tvåninger: Blomflugor. Diptera: Syrphidae: Eristalinae & Microdontinae. *National Encyclopedia of the Swedish Flora and Fauna*. DH 53b.

Invasive fruit fly

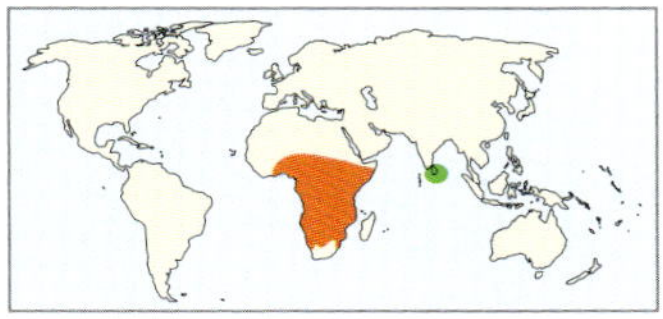

Bactrocera invadens

Group: Tephritidae

Habitat: Terrestrial, plant-feeding. **Identification and biology:** Medium-sized (wingspan 6 mm), strikingly marked fly, with stripes and lines on the abdomen, yellow patches on the thorax and two dark stripes on the wing. Highly variable and may be easily confused with other species of the genus. One of four *Bactrocera* species of Asian origin now established as invasive alien species on the African continent. The three other species, Melon fly (*B. cucurbitae*), Malaysian fruit fly (*B. latifrons*) and Peach fruit fly (*B. zonata*), are not yet recorded from South Africa, but remain serious phytosanitary risks. **Introduction and spread:** Originating in Sri Lanka, it was first recorded in Kenya in 2003 and later from Tanzania in 2004. Although initially thought to be a variant form of the Oriental fruit fly (*B. dorsalis*), it was described as a distinct species in 2005. Since its first discovery in Africa, it has spread rapidly and now occurs in several countries in East, central and West Africa. It was reported from Namibia in 2008, from the Chobe area of Botswana in 2010, from the Beitbridge border post on the Zimbabwean/South African border and Groblersbrug border post in 2010. At this stage a quarantine and eradication programme was immediately implemented by the South African government in a successful attempt to prevent its spread and establishment. The larvae develop in fruit and are major pests of cultivated crops. **Impacts:** Recorded attacking plants in several unrelated families, but particularly injurious to mango and guava. Several studies on the host plant range of the Invasive fruit fly are ongoing and it is to be expected that the host list will increase in future. **Control:** High-risk entry points into South Africa include Johannesburg, Cape Town and Durban (all of which have international airports), and the Beitbridge and Lebombo border posts (which receive the most vehicular traffic from other African countries). If established, *Bactrocera* fruit flies can be successfully managed on a local scale by bagging fruit, field sanitation, the use of protein baits, male annihilation trapping (MAT), cultivation of fruit fly-resistant genotypes, augmentation of biocontrol agents and soft insecticides.

S. Marshall

Key references: De Meyer, M., Mohamed, S. & White, I.M. 2010. Invasive fruit fly pests in Africa. Available at www.africamuseum.be/fruitfly/AfroAsia.htm (last accessed 4 November 2010).

Drew, R.A.I., Tsuruta, K. & White, I.M. 2005. A new species of pest fruit fly (Diptera: Tephritidae: Dacinae) from Sri Lanka and Africa. *African Entomology*, 13:149-154.

White-footed lesser house fly

Fannia albitarsis

Group: Fanniidae

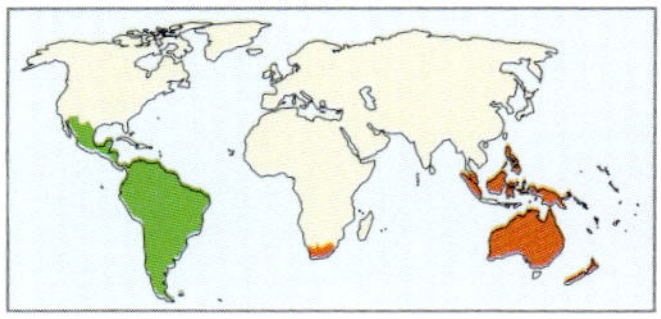

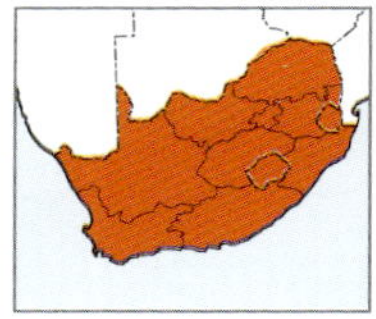

Habitat: Terrestrial, feeding on decomposing organic matter. **Identification and biology:** Medium-sized (wingspan 10.5 mm), uniformly shiny, black fly, superficially resembling small House fly (*Musca domestica*). Particularly characteristic are the ornamentations of the fore- and hindlegs of males. One of four species, the others being the Lesser house fly (*Fannia canicularis*), Chicken dung fly (*F. pusio*) and Latrine fly (*F. scalaris*), all of which are cosmopolitan and may be invasive in Africa, having originated in the northern hemisphere. The larvae can be found in a range of decaying organic matter and are easily recognised by the many well-developed feathery lateral and dorsal processes on each body segment. Larvae have been reared from bird nests, poultry and sheep dung, and the carcasses of a stingray, fish and bird (genus shown). **Introduction and spread:** Origin Neotropical. A 'tramp species' widespread in South America (Peru, Bolivia, southern Brazil, Chile, Argentina, Falkland Islands) and introduced to Amsterdam Island, Australia, Fiji, New Zealand and South Africa, probably through commerce. First recorded from South Africa in 1953, but already well established by then. **Impacts:** An important pest in poultry houses in Argentina, New Zealand and South Africa. **Control:** In poultry houses, the use of mists, fogs or baits may be necessary for fly control. Insecticides should not be applied to the manure for maggot control. Manure should be kept dry and removed only during winter.

Key references: Hulley, P.E. 1983. A survey of the flies breeding in poultry manure, and their potential natural enemies. *Journal of the Entomological Society of Southern Africa*, 46:37-47.

Paterson, H. 1953. *Fannia albitarsis* Stein – a species new to the Ethiopian Region. *Journal of the Entomological Society of Southern Africa*, 16:79.

Sanja565658, Wikimedia Commons

AUTHOR: C. GRIFFITHS

FLEAS

Class Insecta, Order Siphonaptera

Fleas are a distinctive group of small, completely wingless and laterally flattened insects with piercing and sucking mouthparts. They feed on the blood of birds and mammals, some being restricted to a single host, but most infesting several hosts. The hindlegs are greatly enlarged and capable of spectacular leaps, assisted by a sudden release of energy stored in elastic proteins in the cuticle. Females require a blood meal before laying their large eggs, which hatch into maggot-like larvae that undergo three moults before pupating. Pupae can remain dormant for months, hatching in response to vibrations caused by the presence of suitable hosts. Flea bites cause irritation and significant blood loss in some hosts. Fleas are also important vectors of diseases, including plague, and act as intermediate hosts for parasitic tapeworms. They are therefore of considerable medical and veterinary importance.

Jigger flea

Tunga penetrans

Group: Tungidae

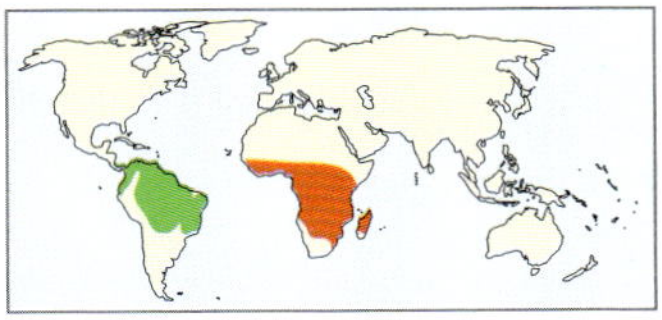

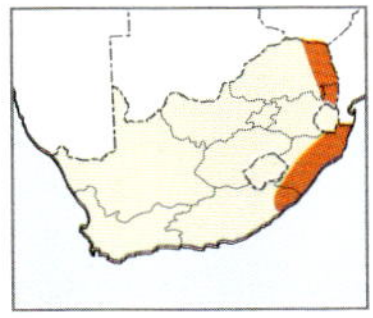

Habitat: Feed on blood of a variety of hosts in dry, sandy tropical areas, but fertilised females burrow into the soft skin of hosts, especially between the toes and under toenails. **Identification and biology:** One of the smallest fleas (1 mm long), easily recognised by the unusual burrowing habits of fertilised females, which maintain a small opening in the skin, through which they obtain air and void faeces and eggs. As the eggs develop the female's abdomen gradually swells until she reaches the size of a small pea. When mature, some 100 eggs are released through the hole in the host's skin and fall to the ground, where they hatch into maggot-like larvae. The flea then dies and is sloughed off with the growing skin. **Introduction and spread:** Introduced to Africa from South America in the 17th century and now widespread in more tropical regions. **Impacts:** Causes intense itching, inflammation and infection, sometimes leading to painful sepsis and ulceration, and in the absence of treatment, even loss of the infected toe. **Control:** Wearing of shoes greatly reduces chances of infection. Once present, female fleas should be removed intact by carefully enlarging the hole with a sterile needle and extracting the entire flea, then treating the wound with antiseptics. Soaking infected areas daily in hydrogen peroxide or commercial disinfectants is also effective.

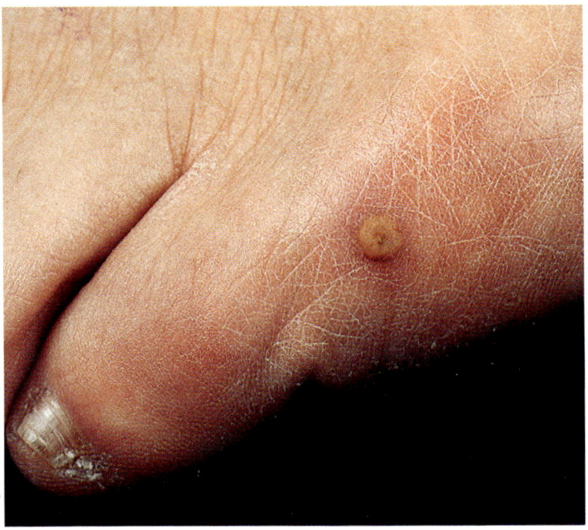

J. Segerman

Key references: Jeffreys, M.D.W. 1952. *Pulex penetrans.* The jigger's arrival and spread in Africa. *South African Journal of Science*, 48:249-255.

Skaife, S.H. 1979. *African insect life.* Cape Town: Struik Publishers.

Human flea

Pulex irritans

Group: Pucilidae

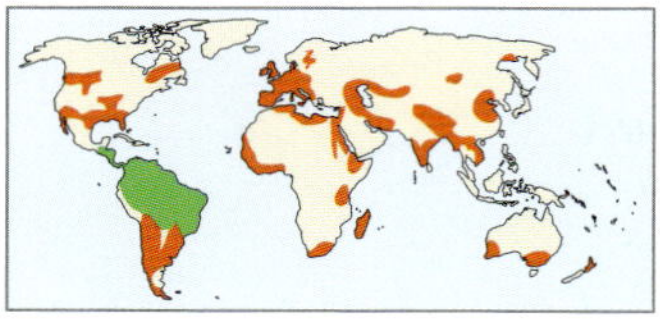

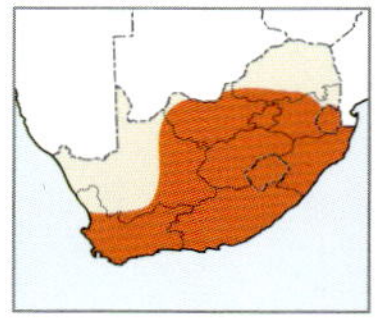

Habitat: Usually associated with human dwellings, especially in temperate regions, but less common in South Africa than the Cat flea (below). **Identification and biology:** Identified by smoothly rounded head and large eyes, as well as its association with humans and domestic animals. Males (2 mm long) smaller than females (4 mm long). **Introduction and spread:** Thought to have originated in South America, but probably reached South Africa via Europe, to which it was introduced centuries ago. Now present in most urban areas in the region. **Impacts:** Apart from the irritation caused by their bites, can act as intermediate hosts for dog tapeworm. **Control:** Various commercial products available for the control of fleas in the home.

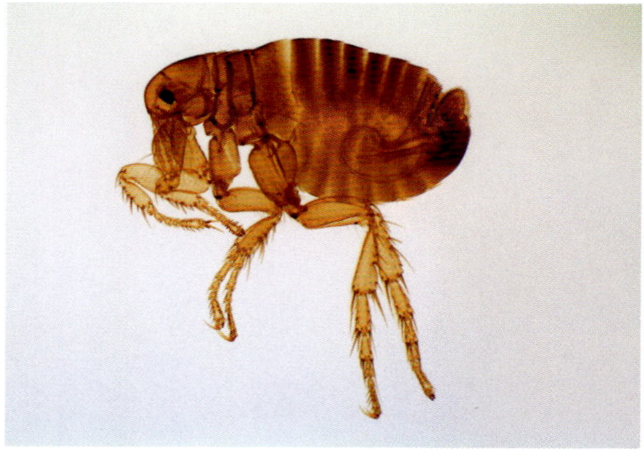

C. Griffiths

Key reference: Segerman, J. 1995. *Siphonaptera of South Africa: Handbook for the identification of fleas*. Johannesburg: South African Institute for Medical Research.

•Cat flea

Ctenocephalides felis

Group: Pucilidae

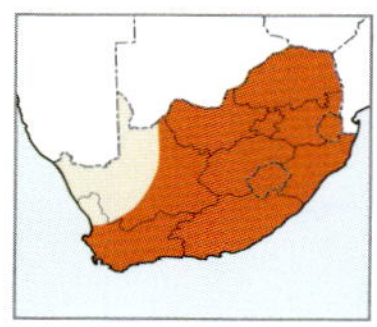

Habitat: Most common domestic flea in the region; despite its name, occurs equally on dogs and cats, also humans. Found almost exclusively on domestic hosts and only exceptionally in the wild. **Identification and biology:** A small flea (2–3 mm long), best recognised by the non-specialist on the basis of its host associations. Voracious blood feeder, injecting anti-coagulant saliva into bites made by the serrated and cutting mouthparts. **Introduction and spread:** First specimen was described from Egypt, now cosmopolitan and widely dispersed via domestic animals and found wherever there are dogs and cats. **Impacts:** One of the most disliked of all insects. Bites can cause intense irritation and allergic reactions and bloody faecal deposits can soil linen. Also intermediate host for cat and dog tapeworm. **Control:** Regular cleaning of carpets and upholstery reduces abundance of eggs and larvae in the home. A wide variety of commercial flea powders, collars and topical applications also available to control fleas on pets.

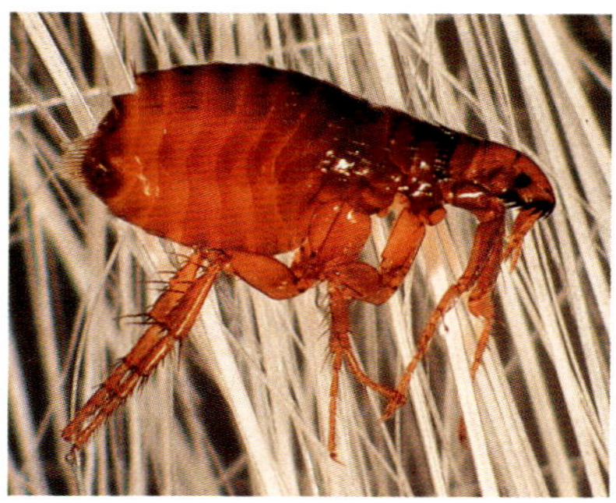

C. Griffiths

Key references: Segerman, J. 1995. *Siphonaptera of South Africa: Handbook for the identification of fleas*. Johannesburg: South African Institute for Medical Research.

Skaife, S.H. 1979. *African insect life*. Cape Town: Struik Publishers.

Stick-tight flea

Echidnophaga gallinacea

Group: Pucilidae

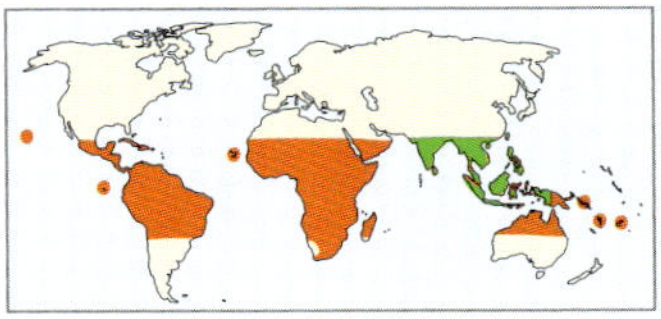

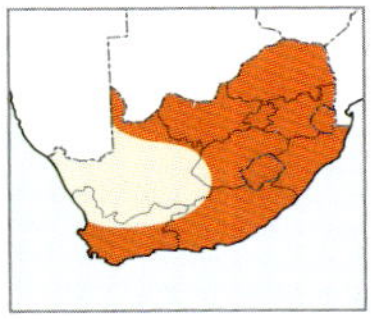

Habitat: Common parasite of domestic fowls; also infects other birds and mammals. **Identification and biology:** Tiny flea (2 mm long), best recognised by its unusual habits. Female attaches tightly to skin around the face, comb and wattles of chickens, where it is fertilised by free-living males. Eggs simply drop to the ground. **Introduction and spread**: Introduced from its native Asian range to tropics and subtropics around the world via domestic chickens. **Impacts**: Cause inflammation and discomfort and sometimes occur in such densities that they significantly reduce growth rate and egg production of hosts. **Control**: Best controlled by frequent cleaning of henhouses to remove eggs and larvae. To kill adult fleas, dust birds with insect powder obtained from vet or pet shop.

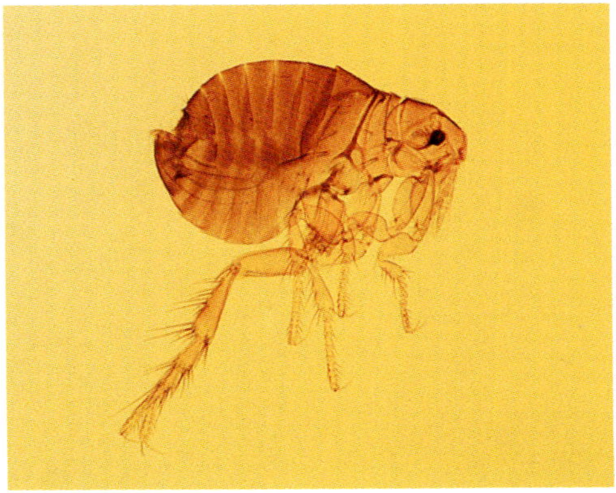

C. Griffiths

Key references: Segerman, J. 1995. *Siphonaptera of South Africa: Handbook for the identification of fleas*. Johannesburg: South African Institute for Medical Research.

Skaife, S.H. 1979. *African insect life*. Cape Town: Struik Publishers.

Oriental rat flea

Xenopsylla cheopis

Group: Pucilidae

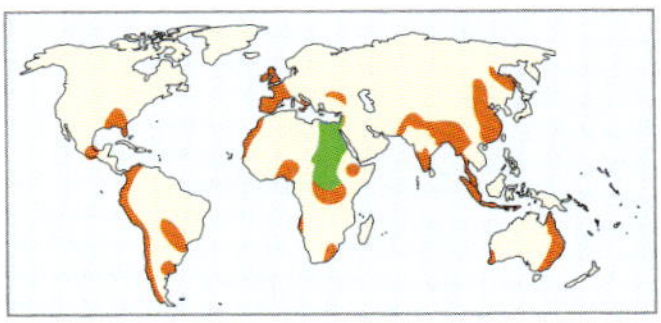

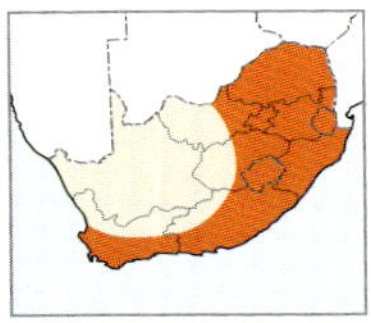

Habitat: Associated with domestic rats and mice and found mostly in urban areas. **Identification and biology:** One of at least four species of flea found associated with rats in South Africa; males (1.5 mm long) smaller than females (4 mm long). **Introduction and spread:** Introduced to South Africa either directly from its native North Africa, or secondarily via Europe by sea, in association with House rats (p. 26). Now the commonest rat flea in ports and coastal towns. Now cosmopolitan, especially in the tropics and subtropics. **Impacts:** One of the primary vector species responsible for spread of bubonic plague, which has killed millions in epidemics such as the 'Black Death' and which is caused by the bacterium *Yersinia pestis*, itself introduced to South Africa in 1900. **Control**: Best eliminated by control of host rodents, but can also be killed by fumigation or application of various commercial insecticides.

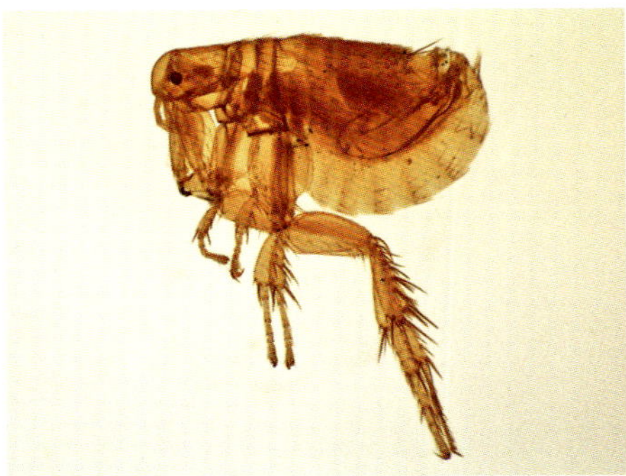

C. Griffiths

Key reference: Segerman, J. 1995. *Siphonaptera of South Africa: Handbook for the identification of fleas*. Johannesburg: South African Institute for Medical Research.

WASPS, BEES, ANTS Class Insecta, Order Hymenoptera

This is a large, common and widespread insect order, including wasps, bees and ants. They are best characterised by having fore- and hindwings coupled in flight by tiny hooks along the front edge of the hindwings. Nearly all have a 'wasp waist' formed by constricted first and second abdominal segments. Hymenoptera are unique among insects in terms of their diverse modifications of the egg-laying apparatus (ovipositor) and occurrence of haploid males (which have a single copy of each chromosome). Primitive members (sawflies) have caterpillar-like larvae and are plant feeding, while most of the others are parasitic, predaceous, or nectar and pollen-feeding species with maggot-like larvae. Many are important pollinators.

• Argentine ant

Linepithema humile

Group: Formicidae

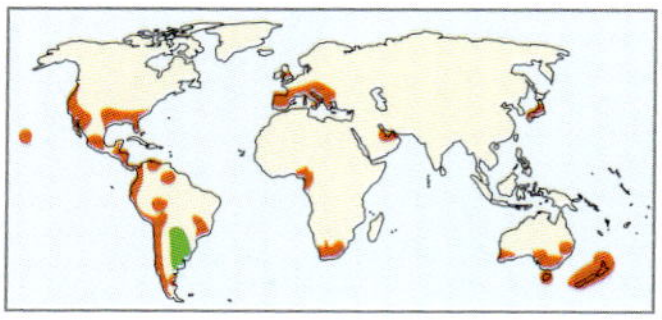

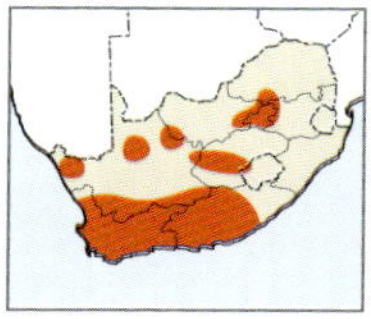

Habitat: Urban and agricultural land, plantations, forests. When in pristine vegetation, then usually associated with urban edge, habitat disturbance or watercourses. Also in undisturbed fynbos and southern Afro-temperate forest in Western Cape province. **Identification and biology:** Small (2–3 mm long), brown ant. Under microscope, mandibles show alternating small denticles and large teeth, and clypeus (upper lip) mostly concave towards middle of head. Typically observed in trails from relocating colonies. Less tolerant of high temperatures and dry conditions than native ants occupying semi-arid biomes. Feeds on honeydew (from aphids and mealybugs) and protein (dead insects, carrion). **Introduction and spread:** First recorded in 1908 from Western Cape, having been introduced with horse fodder imported from Argentina for British troops in the Anglo-Boer War. Introduced ('uniclonal') populations have lower genetic diversity than native populations, resulting in reduced territoriality between colonies, allowing fusion into huge, very dense super-colonies. Colonies have many queens and spread by budding. Spread globally by trade and industry. **Impacts:** Ranked among world's 100 Worst Invasive Species by the IUCN (International Union for Conservation of Nature) Global Invasive Species Programme database (www.issg.org/database); known to reduce native ant diversity causing collapse of various mutualisms between native ants and plants. In fynbos, interfere with burial of large proteaceous seeds by native ants, resulting in changes in plant community composition. Also reduce the number of seeds that are formed in proteas by dominating inflorescences and deterring pollinators. Biotic resistance offered by some native ant species, such as the Small black sugar ant (*Lepisiota capensis*). **Control:** Not easily achieved. Recommended preventive measures include diversifying agricultural activities to avoid monocultures, and limiting habitat disturbance and urban runoff within native vegetation.

H. Robertson

Key reference: Lach, L., Picker, M.D., Allsopp, M., Colville, J. & Griffiths, C.L. 2002. Alien invertebrate animals in South Africa. In: Pimental, D. (ed.). *Biological invasions. Economic and environmental costs of alien plant, animal and microbe species*. Florida: CRC Press.

Sirex woodwasp

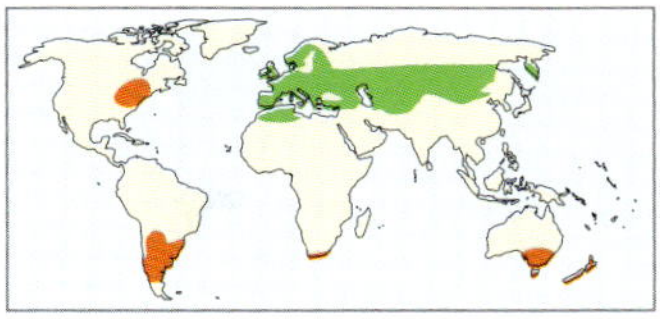

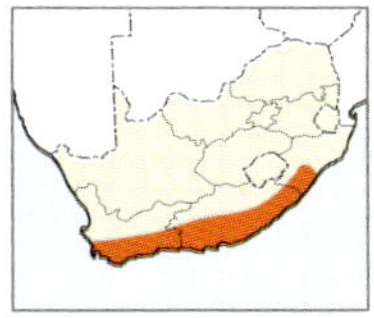

Sirex noctilio

Group: Siricidae

Habitat: Terrestrial, phytophagous. **Identification and biology:** Large, body length 10–40 mm. Female metallic bluish-black with orange legs, male with striking orange abdomen. Females lay eggs and symbiotic fungal spores into dead or dying pine trees. Larvae are wood-borers. **Introduction and spread:** Native to temperate Eurasia. First recorded in imported timber in Port Elizabeth in 1962, but did not become established. In 1994 found infesting pine plantations in Cape Town. Rapidly spread through Western Cape. Now also present in Eastern Cape and KwaZulu-Natal. **Impacts:** Pest in Monterey pine plantations (*Pinus radiata*), reducing production and leading to extensive loss of revenue (up to R300 million per annum). **Control:** The parasitoid Sirex ibaliid wasp (below) introduced in 2000 and has become established. Another parasitoid wasp, the Giant ichneumon (*Megarhyssa nortoni)* was released without success in 1998. Woodwasp nematode (p. 173) released in 1995, 1996 and 2004/5, with limited impact.

C. Griffiths

Key reference: Tribe, G.D. & Cillié, J.J. 2004. The spread of *Sirex noctilio* Fabricius (Hymenoptera: Siricidae) in South African pine plantations and the introduction and establishment of its biological control agents. *African Entomology*, 12:9-17.

Sirex ibaliid wasp

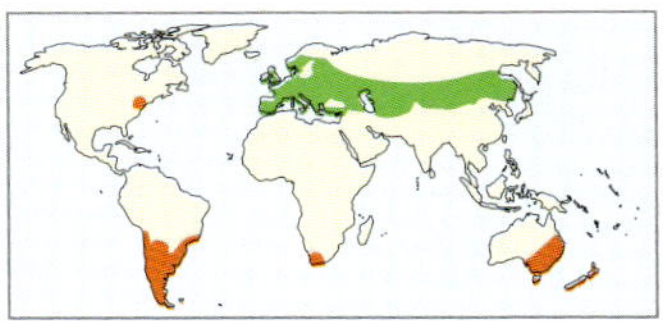

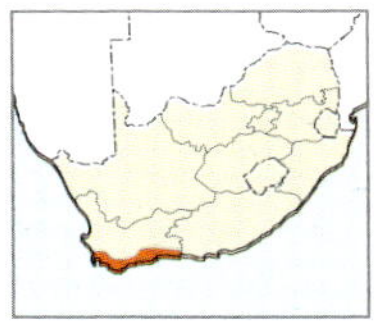

Ibalia leucospoides

Group: Ibaliidae

Habitat: Terrestrial, parasitoid. **Identification and biology:** Medium-sized wasp (body length 10–15 mm). Black, with tooth on front top end of thorax. Parasitoid of woodwasps. Males mate with females while they are laying eggs. Female inserts ovipositor down tunnel bored in pine trees by host woodwasp larva, to lay an egg either into egg of host or into young host larva. On hatching, Ibaliid wasp larva emerges from body of host and feeds externally. **Introduction and spread:** Native to the Holarctic region; introduced to Australia, New Zealand and South Africa (1995–2007) to control the Sirex woodwasp (above), a pest in Monterey pine plantations (*Pinus radiata*). **Impacts:** Positive, contributes towards control of pest Sirex woodwasp. **Control:** Not appropriate for biocontrol agent.

S. van Noort

Key reference: Hurley, B.P., Slippers, B. & Wingfield, M.J. 2007. A comparison of control results for the alien invasive woodwasp, *Sirex noctilio*, in the southern hemisphere. *Agricultural and Forest Entomology*, 9:159-171.

German wasp/ European wasp

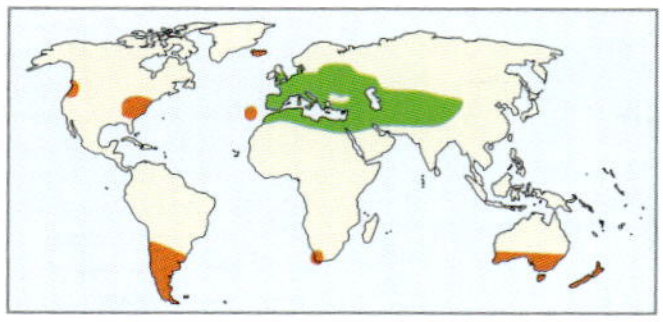

Vespula germanica

Group: Vespidae

Habitat: Terrestrial, predatory.
Identification and biology: Medium-sized wasp (body length 12–17 mm). Head and thorax mostly black with some yellow, legs yellow, abdomen yellow with black markings. Social, constructing concealed communal paper nests, often underground. Larvae fed on chewed-up, soft-bodied insects such as caterpillars. **Introduction and spread:** Introduced to Cape Peninsula in 1975 from Europe and although it has the potential to spread through much of the winter-rainfall areas of South Africa, has not dispersed yet beyond Stellenbosch and the Hottentots Holland mountain range. **Impacts:** Can be a pest and incidental threat to picnickers as they are protein scavengers, attracted to sugary substances, and sting readily if disturbed. Has potential impact on local ecosystems through prey competition: the honey industry as they hunt bees; and on wine and fruit industry through fruit-feeding damage and subsequent secondary fungal infection. **Control:** Destruction of nests.

C. Griffiths

Key reference: Tribe, G.D. & Richardson, D.M. 1994. The European wasp, *Vespula germanica* (Fabricius) (Hymenoptera: Vespidae) in southern Africa and its potential distribution as predicted by ecoclimatic matching. *African Entomology*, 2:1-6.

European paper wasp

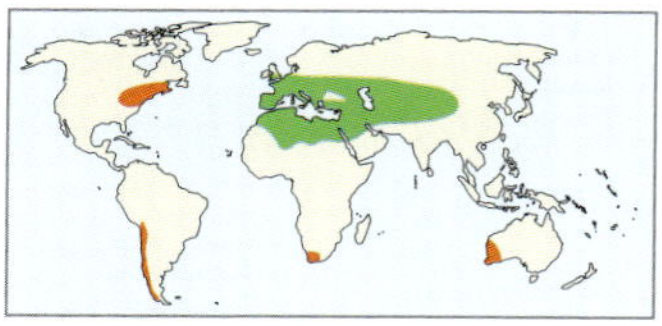

Polistes dominulus

Group: Vespidae

Habitat: Terrestrial, predatory.
Identification and biology: Medium-sized wasp (body length 12–16 mm). Yellow, with black barring on abdomen. Social, constructing exposed communal paper nests, which can reach a much larger size than indigenous paper wasp nests. Larvae fed on chewed-up, soft-bodied insects, such as caterpillars. Adults feed on nectar from flowers. **Introduction and spread:** Accidentally introduced to South Africa in 2008 from either Europe or America, where it was introduced in 1981. First observed in Cape Town in 2008, where it appears to be well established in the Kuils River area. Likely to spread to other areas of South Africa. **Impacts:** None yet recorded, but has potential to become an invasive species. **Control:** Destruction of nests.

Alvesgaspar, Wikimedia Commons

Key reference: Eardley, C., Koch, F. & Wood, A.R. 2009. *Polistes dominulus* (Christ, 1791) (Hymenoptera: Polistinae: Vespidae) newly recorded from South Africa: short communication. *African Entomology*, 17:226-227.

Megalyrid wasp

Megalyra fasciipennis

Group: Megalyridae

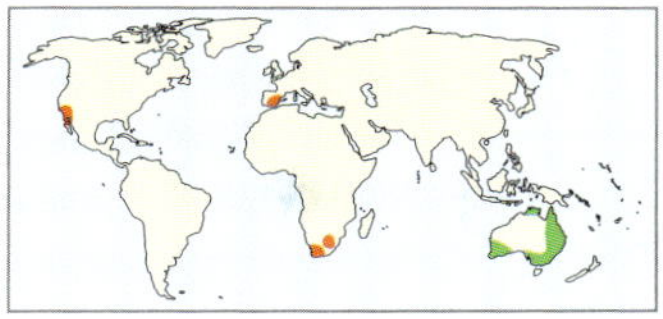

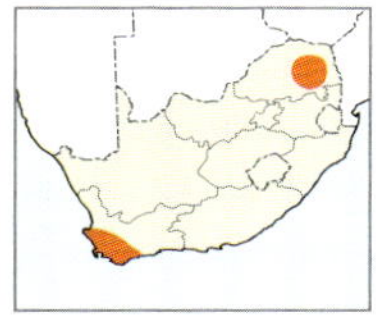

Habitat: Terrestrial, parasitoid. **Identification and biology:** Medium-sized wasp (body length 10–15 mm). Black and red with very long ovipositor (2–4 times body length). Pupal parasitoid of beetle larvae in genus *Phoracantha* (p. 115), which are wood-borers of *Eucalyptus* species. Female inserts long ovipositor through plug produced by the boring beetle larvae to block its feeding tunnel, and then lays an egg on pupa. **Introduction and spread:** Deliberately introduced to South Africa (Cape Town) in 1910 from Australia. Although rearing of wasps for release failed, presumably established itself as it was recovered in 1962 from infested logs. Appears not to have spread beyond the Western Cape, but has been introduced into the Tzaneen area. **Impacts:** Positive, controlling introduced invasive pest beetle species. **Control:** Not necessary, beneficial introduction.

S. van Noort

Key references: Gess, F.W. 1964. The discovery of a parasite of the *Phoracantha* beetle (Coleoptera: Cerambycidae) in the Western Cape. *Journal of the Entomological Society of Southern Africa*, 27:152.

Neuenschwander, P., Borgemeister, C. & Langewald, J. (eds). 2003. *Biological control in IPM systems in Africa*. Wallingford: CABI Publishing.

Acacia gall wasp

Trichilogaster acaciaelongifoliae

Group: Pteromalidae (Chalcidoidea)

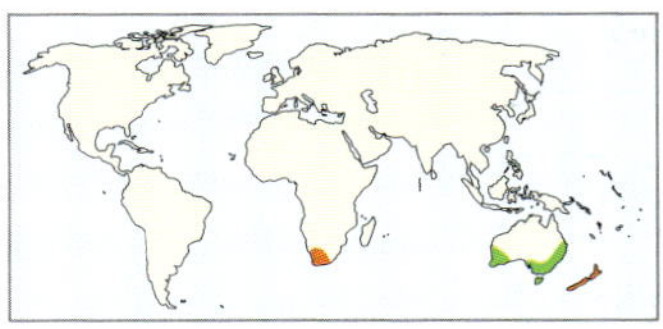

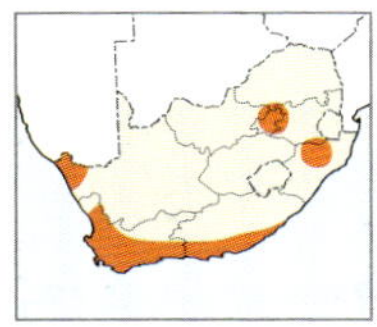

Habitat: Terrestrial, phytophagous. **Identification and biology:** Small wasp (body length 2–2.5 mm). Orange and black with reduced wing venation. Gall formers on the Long-leaved wattle (*Acacia longifolia*). Adults live only a few days. Eggs laid into flower buds. On hatching, larvae instigate a galling response in the plant and live and feed inside the galls. Adults chew their way out of gall before mating and searching for young flower buds. A related species, *Trichilogaster signiventris*, has been introduced to control the alien Golden wattle (*Acacia pycnantha*). **Introduction and spread:** Deliberately introduced to South Africa in the 1990s from Australia; successfully established. **Impacts:** Effective biocontrol agent of introduced invasive Australian Long-leaved wattle. Galling of flower buds reduces seed set and galls impact on and stunt plant growth. **Control:** Not necessary, beneficial introduction.

C. Griffiths

Key references: Dennill, G.B., Donnelly, D., Steward, K. & Impson, F.A.C. 1999. Insect agents used for the biological control of Australian *Acacia* species and *Paraserianthes lophantha* (Willd.) Nielsen (Fabaceae) in South Africa. *African Entomology Memoir*, 1:45-54.

Prinsloo, G.L. & Neser, O.C. 2007. Revision of the pteromalid wasp genus *Trichilogaster* Mayr (Hymenoptera: Chalcidoidea): gall-inducers on Australian acacias. *African Entomology*, 15:161-184.

Eucalyptus gall wasp (Blue gum chalcid)

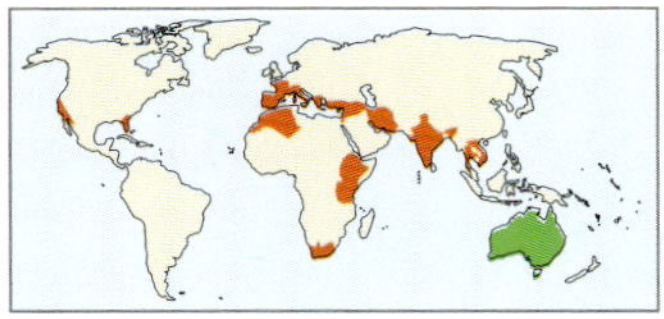

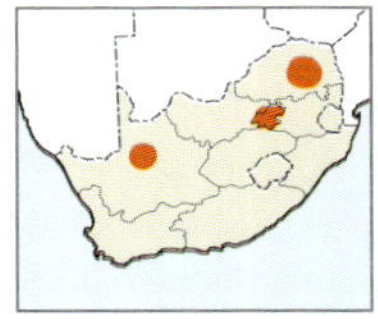

Leptocybe invasa

Group: Eulophidae (Chalcidoidea)

Habitat: Terrestrial, phytophagous.
Identification and biology: Tiny wasp (body length 1 mm). Blackish-brown, with metallic refringence, pale legs and antennae. **Introduction and spread:** An Australian species first recorded in the Middle East and Mediterranean region in 2000; has since spread down through Africa, reaching South Africa in 2007. **Impacts:** Devastates eucalyptus plantations by forming galls on leaves, deforming foliage and stunting growth. **Control:** No effective measures established. Insecticides have some effect in preventing infestations on nursery saplings. There is potential for implementation of biocontrol through introduction of suitable parasitoid agents, and/or selection of resistant tree genotypes.

Z. Mendel

Key reference: Mendel, Z., Protasov, A., Fisher, N. & La Salle, J. 2004. The taxonomy and natural history of *Leptocybe invasa* (Hymenoptera: Eulophidae) gen & sp. nov., an invasive gall inducer on Eucalyptus. *Australian Journal of Entomology*, 43:101-113.

Erythrina gall wasp

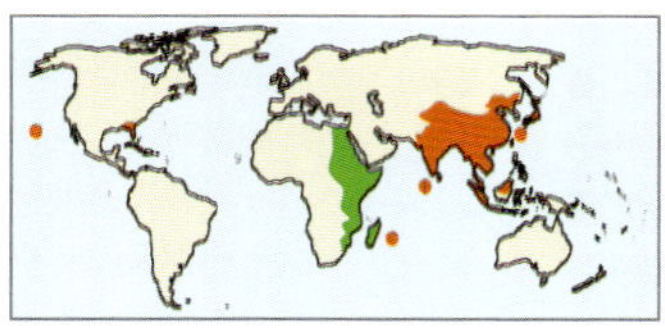

Quadrastichus erythrinae

Group: Eulophidae (Chalcidoidea)

Habitat: Terrestrial, phytophagous.
Identification and biology: Tiny wasp (body length 1.4–1.6 mm). Dark brown with yellow markings. Larvae hatching from eggs laid in young leaf tissue induce formation of leaf and stem galls on *Erythrina* species (Coral trees). Leaves curl and deform and petioles become swollen (shown). Wasp larvae feed on plant tissue; pupate inside gall and on emerging chew an exit hole to exterior. **Introduction and spread:** Probably native to East Africa. Not yet recorded from South Africa but has been identified as a potential threat to Coral trees here. Introduced to various parts of the world. **Impacts:** Gall infestation results in reduction in growth and may cause complete defoliation and death of tree. **Control:** No effective measures established. Insecticides of limited use. Biocontrol potentially feasible using parasitoid wasps of the genus *Eurytoma* (Eurytomidae).

C. Griffiths

Key references: Prinsloo, G.L. & Kelly, J. 2009. The tetrastichinae wasps (Hymenoptera: Chalcidoidea: Eulophidae) associated with galls on *Erythrina* species (Fabaceae) in South Africa, with the description of five new species. *Zootaxa*, 2083:27-45.

Kim, I., Delvare, G. & La Salle, J. 2004. A new species of *Quadrastichus* (Hymenoptera: Euphidae): A gall-Inducing pest on *Erythrina* (Fabaceae). *Journal of Hymenoptera Research*, 13:243-249.

Cockroach egg parasitoid (Ensign wasp)

Evania appendigaster

Group: Evaniidae

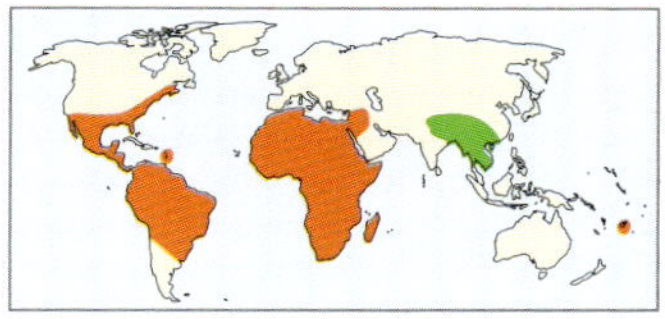

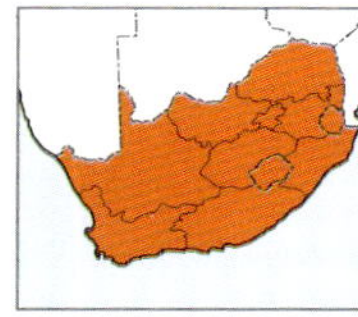

Habitat: Terrestrial, parasitoid.
Identification and biology: Small wasp (body length 5–7 mm). Black, clear wings, with characteristic flag-like abdomen. Parasitoid of cockroach eggs, particularly those of the American cockroach (p. 82) and the Oriental cockroach (*Blattella orientalis*). Female penetrates tough cockroach egg case with her ovipositor and lays a single egg inside. Larva eats cockroach eggs and pupates inside the case. On emerging from the pupa, adult wasp chews an exit hole in the side of the case. Adults occasionally found inside homes and buildings. **Introduction and spread:** Probably originally an Oriental species, but now widespread throughout tropical and temperate areas. **Impacts:** Positive, controls pest cockroach species, but impact on indigenous cockroach species not established. **Control:** Not necessary, beneficial introduction.

Key reference: Cameron, E. 1957. On the parasites and predators of the cockroach. II. *Evania appendigaster* (L.). *Bulletin of Entomological Research*, 48:199-209.

C. Griffiths

Pear slug

Caliroa cerasi

Group: Tenthridinidae

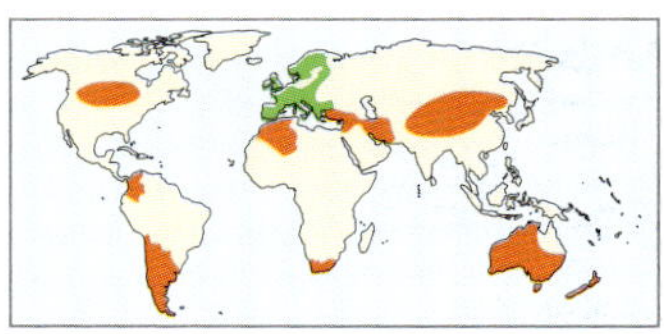

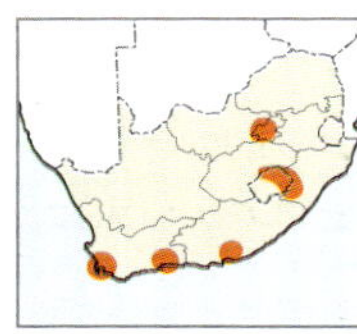

Habitat: Terrestrial, phytophagous.
Identification and biology: Small wasp (body length 4–6 mm). Black, with transparent wings; sluggish fliers. Larvae (shown) resemble small slugs (greenish-brown with a large head-end tapering to the tail), cover themselves with slime as a protection against predation, and feed on epidermis of leaves, in the process skeletonising them, and pupate in the soil beneath host plant. Adult wasps emerge, mate, and female uses her saw-like ovipositor to lay eggs in leaves. A related species, the European Alder leafminer (*Fenusa dohrnii*), is a leaf miner of Alder trees (*Alnus* species) and was accidentally introduced to South Africa from Eurasia and North America. **Introduction and spread:** Originated in Eurasia, but accidentally introduced to North and South America, South Africa, Australia and New Zealand. **Impacts:** Pest of fruit industry, damaging leaves of cherry, plum, pear and apple trees. Heavy infestations reduce fruit yield. **Control:** Manual tillage of soil beneath fruit trees to kill cocoons. Insecticide application.

Key reference: Kaiser, C. & Sheard, A.G. 2001. *Integrated pest and disease management (IPDM) of apples and pears in KwaZulu-Natal.* KZN Agri Report No. NA-2001-06.

C. Griffiths

Diamondback moth parasitoid wasp

Cotesia plutellae

Group: Braconidae

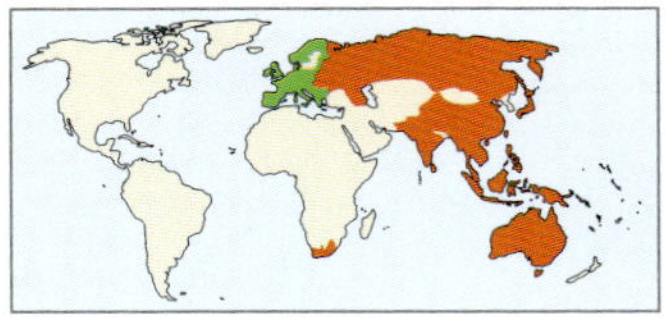

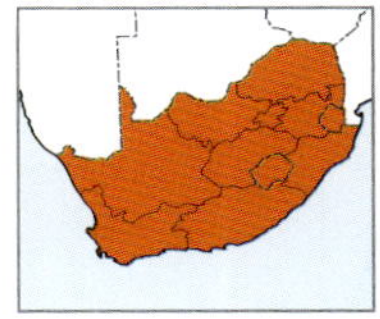

Habitat: Terrestrial, parasitoid.
Identification and biology: Small wasp (body length 3–4 mm). Black, with yellow markings on abdomen, clear wings. Eggs laid in early instar moth caterpillars; wasp larvae feed inside living caterpillars and exit from dying, mature caterpillars to spin a small, white, silken cocoon, from which adult wasps emerge. **Introduction and spread:** Deliberately introduced from Europe as a biocontrol agent for the Diamondback moth (p. 121). **Impacts:** Extremely important biocontrol agent of the Diamondback moth, the most significant pest on cabbages and other Cruciferaceae. The moth is resistant to insecticides. **Control:** Not necessary, this parasitoid wasp is a beneficial introduction.

B. Rowell

Key references: Nofemela, R.S. & Kfir, R. 2005. The role of parasitoids in suppressing Diamond Back moth, *Plutella xylostella* (L.) (Lepidoptera: Plutellidae), populations on unsprayed cabbage in the North West Province of South Africa. *African Entomology*, 13:71-83.

Sarfraz, M., Keddie, A.B. & Dosdal, L.M. 2005. Biological control of the diamondback moth, *Plutella xylostella*: A review. *Biocontrol Science and Technology*, 15:763-789.

Potato tuber moth parasitoid wasp

Apanteles subandinus

Group: Braconidae

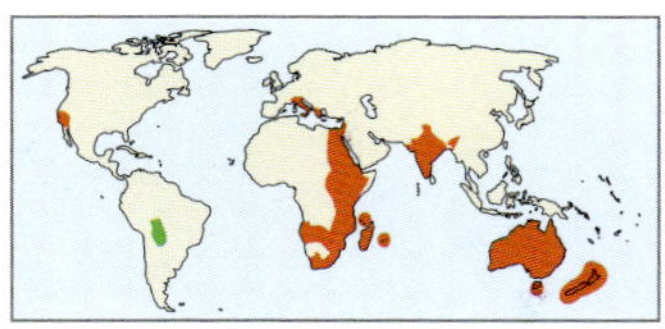

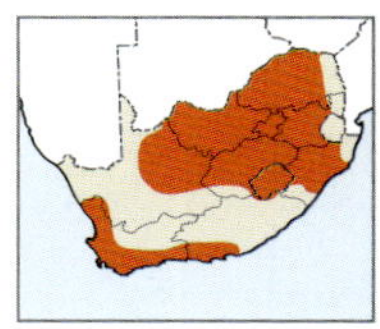

Habitat: Terrestrial, parasitoid.
Identification and biology: Small wasp (body length 3–4 mm). Black, with pale areas on abdomen, clear wings. Eggs laid into early instar caterpillars. On hatching, wasp larvae feed inside the living caterpillars, exiting from dying, mature caterpillars to spin a small, silken cocoon, from which adult wasps emerge. Overwinter within their host inside un-harvested potatoes. **Introduction and spread:** South American species deliberately introduced to South Africa in 1960s to control the Potato tuber moth (p. 121). Subsequently introduced to other potato-growing areas of Africa, Madagascar and surrounding islands. **Impacts:** One of a cohort of effective biocontrol agents of the Potato tuber moth, a pest originating from South America. **Control:** Not necessary, this parasitoid wasp is a beneficial introduction.

G. Goergen, IITA

Key references: Neuenschwander, P., Borgemeister, C. & Langewald, J. (eds). 2003. *Biological control in IPM systems in Africa*. Wallingford: CABI Publishing.

Watmough, R.H., Broodryk, S.W. & Annecke, D.P. 1973. The establishment of two imported parasitoids of potato tuber moth (*Phthorimaea operculella*) in South Africa. *Biocontrol*, 18:237-249.

AUTHOR: C. GRIFFITHS

CENTIPEDES, MILLIPEDES Subphylum Myriapoda

Centipedes and millipedes are easily recognised by their elongated, segmented bodies and large number of legs. Centipedes are active, often fast-moving carnivores equipped with fangs and poison glands and have one pair of legs on each body segment. Millipedes are slow-moving herbivores that feed mostly on decaying plant material. Their body segments are fused into pairs, so each appears to have two short pairs of legs. Both groups are easily transported with trade goods as a result of their tendency to shelter in crevices.

Portuguese millipede

Ommattoiulus moreleti

Group: Diplopoda: Julidae

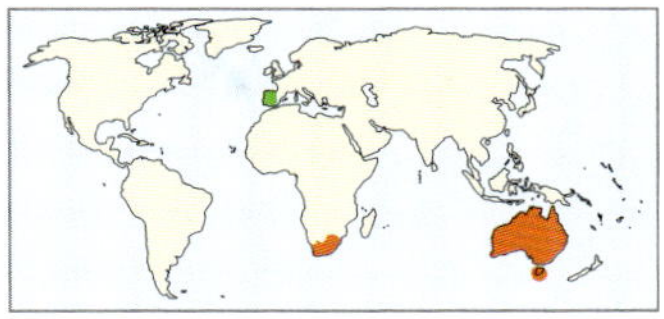

Habitat: Terrestrial, found under rocks, bark, logs, compost heaps and in houses. **Identification and biology:** Relatively small, stout, shiny black millipede, reaching 40 mm in length. Adults have 40–50 body segments, each with two pairs of legs. Dorsally the posterior margin of each body segment has a crimped pattern. Curl into a tight, flat spiral when disturbed. Seldom found far from human dwellings. Eggs are laid in the soil and newly hatched juveniles have fewer body segments, moulting many times before reaching maturity after about two years. **Introduction and spread:** Probably introduced from Portugal with timber or other commercial produce. Common in Cape Town and in the south-western Cape, especially in gardens and homes. **Impacts:** Can damage fruit and vegetables, especially those that lie on the ground, such as strawberries, squashes and melons. Attracted to light and can be a nuisance in homes when abundant. Release a pungent, mildly toxic yellow secretion when handled, which stains clothing and skin. Reach formidable numbers in Australia, where they have been reported to stop trains by swarming over the lines! **Control:** Chemical control is possible, but can also be attracted to traps and disposed of. Keep compost heaps away from the house and avoid lighting the walls to limit numbers in the home.

Key references: Hamer, M.L. 1998. Checklist of southern African millipedes (Myriapoda: Diplopoda). *Annals of the Natal Museum*, 39:11-82.

Lawrence, R.F. 1984. *The centipedes and millipedes of Southern Africa: a guide*. Cape Town: A.A. Balkema.

C. Griffiths

Hothouse millipede

Orthomorpha gracilis

Group: Diplopoda: Paradoxosomatidae

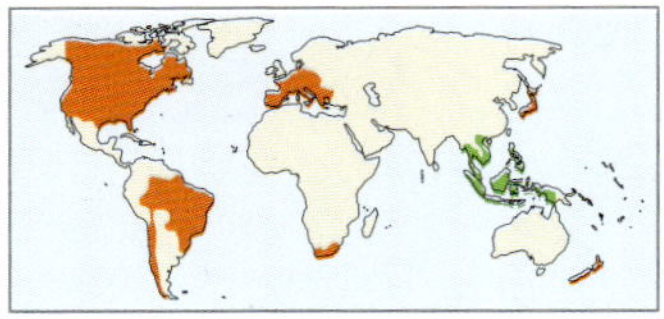

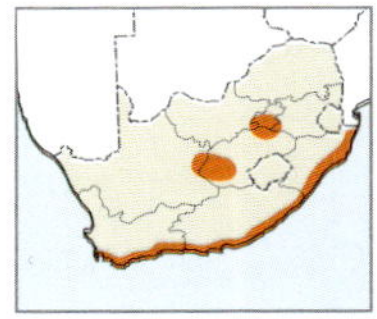

Habitat: Under logs, bark or stones; emerging at night to feed. Also underground in caves and mines. **Identification and biology:** Represents a primitive group of millipedes with only 19 or 20 rounded segments resembling beads. Prominent flange along the side partially conceals legs, giving body a square cross-section. Length 20–30 mm, head small, eyes absent. Brown, segments bordered with yellow. Pale after moulting, becoming darker as it approaches next moult. Feeds on decaying wood and leaves. **Introduction and spread:** Originated from the East Indies, but now widely distributed. Originally described from a hothouse in Austria, hence the common name. **Impacts:** Firmly established as a minor garden pest in Cape Town and other coastal towns and cities. Unusual in that it also occurs hundreds of metres underground, feeding on timber props of mines in Kimberley and Johannesburg. **Control:** Keep compost heaps away from the home, seal cracks and reduce humidity. Millipedes also dislike wood ash and diatomaceous earth.

Key reference: Lawrence, R.F. 1984. *The centipedes and millipedes of Southern Africa: a guide*. Cape Town: A.A. Balkema.

C. Griffiths

•House centipede

Scutigera coleoptrata

Group: Myriapoda: Scutigeridae

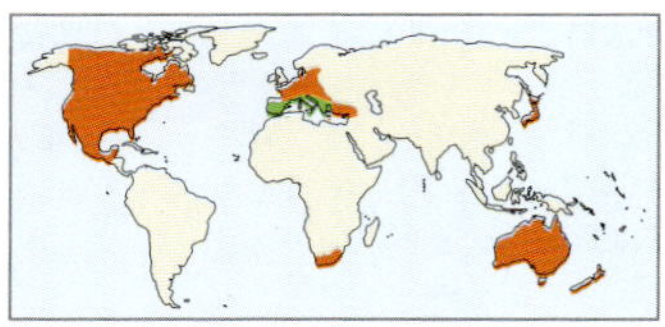

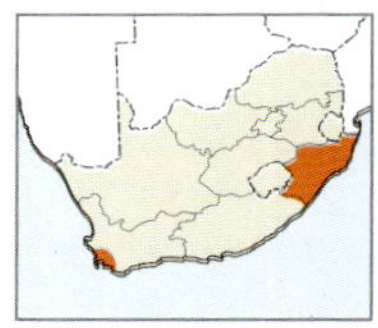

Habitat: Under rocks and logs, compost heaps and other humid locations, also houses, especially bathrooms. **Identification and biology:** Striking, easily identified yellowish centipede up to 50 mm long and with 15 pairs of remarkably long, delicate legs and three dark stripes along body, legs striped. Hindlegs resemble antennae, and when still, can be difficult to identify front end. Very fast runner, scurrying up walls and along ceilings. Feeds on insects and spiders, which are injected with venom from modified, pincer-like front legs. Eyes unusually well developed. Eggs laid in the ground and guarded by female. Newly hatched juveniles have only four pairs of legs and gain 1–2 pairs with each moult. Live 3–7 years. **Introduction and spread:** Native to southern Europe, but widely spread through association with wood and other traded produce. Common in Cape Town and KwaZulu-Natal. **Impacts:** Generally harmless to humans and can control insects in the home, although an intimidating houseguest! **Control:** Measures include sealing cracks in walls, minimising insect populations that act as food, and reducing damp.

Key reference: Lawrence, R.F. 1984. *The centipedes and millipedes of Southern Africa: a guide*. Cape Town: A.A. Balkema.

C. Griffiths

Stone centipedes

Lithobius obscurus and *L. peregrinus*

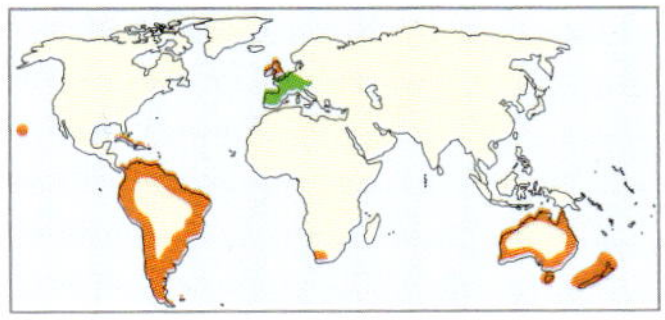

Group: Myriapoda: Lithobiomorpha

Habitat: Under stones and logs, or in leaf litter or compost heaps.
Identification and biology: Shiny brown to mauve centipedes typically 2.5–3 cm long and with 15 body segments; longer body segments characteristically alternate with much shorter ones. Fifteen pairs of legs, each longer than the width of the body and becoming progressively longer posteriorly, but all still much shorter than those of the House centipede (p. 148). Pores on the undersides of the last four pairs of legs secrete a sticky liquid used in defence. Last pair of legs not used when running, but held lifted in the air (shown). Feed on small flies and other soft prey, which are pierced with the large fangs and paralysed with poison where necessary. Two closely related species have been introduced to the region – the Purple stone centipede (*Lithobius obscurus*, shown) and Peregrine's stone centipede (*L. peregrinus*). Since these have similar distributions, and are difficult to identify to species in the field, they are combined into a single entry here. The species are best separated by the number of segments in the antenna, which varies from 26 to 28 joints in the Purple stone centipede, but reaches 40 or more in Peregrine's stone centipede. The eyes of the Purple stone centipede are also comprised of 10 ocelli, whereas those of Peregrine's stone centipede are made up of 15–18 ocelli. **Introduction and spread:** Purple stone centipede native to western Mediterranean region (distribution shown), and Peregrine's stone centipede to central Europe. To date, introduced populations of both species have only been reported from Cape Town, but their actual distribution is probably more widespread, especially in other urban areas. **Impacts:** Predatory on small insects, but unlikely to have major ecological impacts. Harmless to humans, as their fangs are too small and weak to pierce human skin. **Control:** None required.

Key references: Attems, C. 1928. The Myriapoda of South Africa. *Annals of the South African Museum*, 26:1-413.

Lawrence, R.F. 1984. *The centipedes and millipedes of Southern Africa: a guide*. Cape Town: A.A. Balkema.

C. Griffiths

AUTHORS: A. DIPPENAAR, E. UECKERMANN & M. PICKER

SPIDERS, MITES, TICKS Class Arachnida

Arachnids are a diverse group of eight-legged animals consisting of 12 orders, of which nine occur in South Africa. Of these, the mites and ticks (Acari) and the spiders (Araneae) most commonly move around with humans, either on plants, domestic animals or packing material. **Spiders** (Araneae) vary considerably in shape, size (0.8–120 mm) and colour. Their body is divided into two sections, a cephalothorax (fused head and thorax), bearing the mouthparts and legs, and an abdomen. The hard, top part of the cephalothorax is the carapace. Most spiders have eight eyes, but the number varies from zero to eight. Their piercing mouthparts (chelicerae) are strong and form fangs with a venom gland at the tip. They produce silk from well-developed spinnerets on the tip of the abdomen. **Mites** (Acari) are mostly less than 1 mm long, but include many economically important pests. Their body segments are fused into one unit that can be oval, sac- or worm-like in appearance. There are usually eight legs, but the number may vary from two to eight. **Ticks** (Ixodidae) have a similar body form to mites, but are larger. They are bloodsucking parasites of vertebrates, and are important transmitters of diseases.

Brown button spider

Latrodectus geometricus

Group: Araneae: Theridiidae

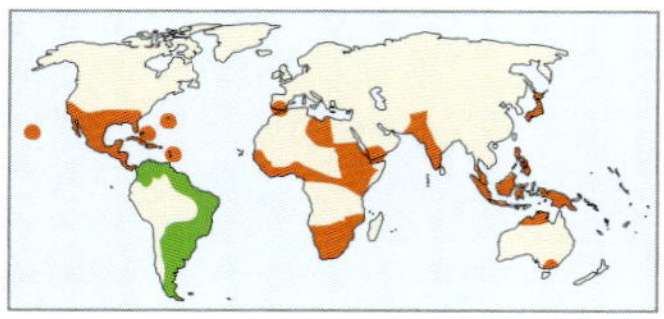

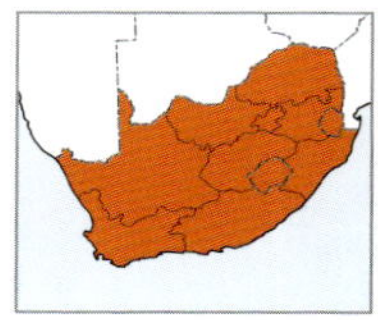

Habitat: Terrestrial, in cracks and under eaves. **Identification and biology:** Body length of female 9–10 mm; male much smaller. Abdomen globular; usually brown, but colour varies from cream to pitch black, with a geometric pattern visible dorsally in paler specimens; distinct orange-red hourglass pattern ventrally; legs moderately long and slender, third pair shortest. Construct three-dimensional webs in dark corners. The round egg sacs are decorated with silk tufts and attached to side of web. **Introduction and spread:** Native to South America, now cosmopolitan, first introduction unknown, widespread throughout South Africa. **Impacts:** One of the most common spiders in built-up areas. Of medical importance, producing a neurotoxic venom that affects the nervous system in humans. **Control:** A useful biocontrol agent, as it feeds on a wide range of pest species; however, removal from inside buildings is recommended.

W. Schmidt

Key references: Lotz, L.N. 1994. Revision of the genus *Latrodectus* (Araneae: Theridiidae) in Africa. *Navorsinge van die Nasionale Museum Bloemfontein*, 10:1-60.

Müller, G.J. 1993. Black and brown widow spider bites in South Africa. A series of 45 cases. *South African Medical Journal*, 83:399-405.

Long-fanged six-eyed spider

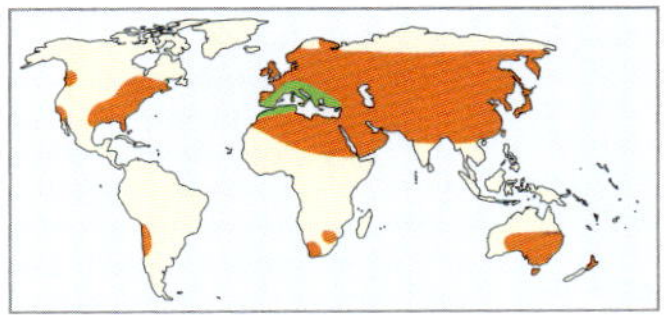

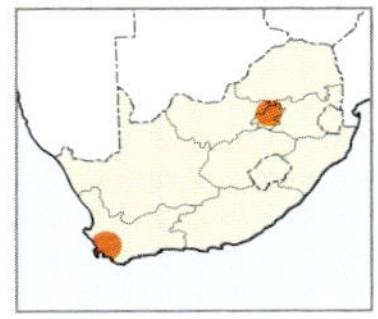

Dysdera crocata

Group: Araneae: Dysderidae

Habitat: Terrestrial. **Identification and biology:** Body length in both sexes 9–15 mm. Carapace and legs deep red to orange, abdomen pinkish-grey; only six eyes in a compact central group; chelicerae and fangs long and well-developed; abdomen elongated-oval, with light cover of short setae (short, stiff hairs); legs of moderate length, without strong setae. Free-living, nocturnal ground-dwellers; use their long fangs to catch prey. Spend the day in a silken retreat, which is also used for moulting, laying eggs, and raising young for up to a month. Known to prey on woodlice, but recent studies indicate that they also prey on insects. **Introduction and spread:** Native to North Africa and southern Europe, now cosmopolitan, first introduction unknown; not widespread in South Africa; known only from areas around Cape Town and the Witwatersrand. **Impacts:** Nasty bite, resulting in painful aching limbs and a necrotic wound. **Control:** Nothing registered to control spiders; removal from inside buildings recommended.

N. Larsen

Key reference: Jackson, R.R. & Pollard, S.D. 1982. The biology of *Dysdera crocata* (Araneae, Dysderidae): interspecific interactions. *Journal of Zoology*, 198:197-214.

Humped back daddy-long-legs spider

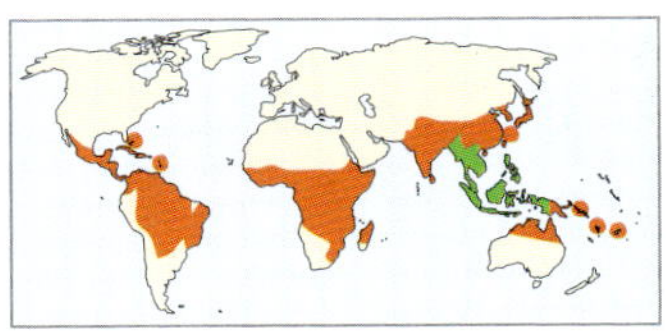

Crossopriza lyoni

Group: Araneae: Pholcidae

Habitat: Terrestrial. **Identification and biology:** Body length in both sexes 5–7 mm. Body cream with brown median markings; carapace short, broad and almost circular; eight eyes; abdomen flattened, triangular in lateral view, with back end tapering to a point; stridulation organs are present, which produce sound by scraping the carapace against the abdomen; legs extremely long with flexible end segments. Live in space-webs high up against walls near ceiling. Female carries eggs in her chelicerae, unprotected as they lack an enveloping sac. When young hatch, female carries them around for a short period. **Introduction and spread:** Native to Southeast Asia, now cosmopolitan, first recorded from South Africa in 2007, known so far only from buildings in Pretoria. **Impacts:** First record from Tshwane Fresh Produce Market, where their extensive webbing causes problems. **Control:** Nothing registered to control spiders; removal of webbing recommended.

R. Atkinson

Key reference: Huber, B.A., Deeleman-Reinhold C.L. & Pérez Gonzáles, A. 1999. The spider genus *Crossopriza* (Araneae, Pholcidae) in the New World. *American Museum Novitates*, 3262:1-10.

Spitting spider

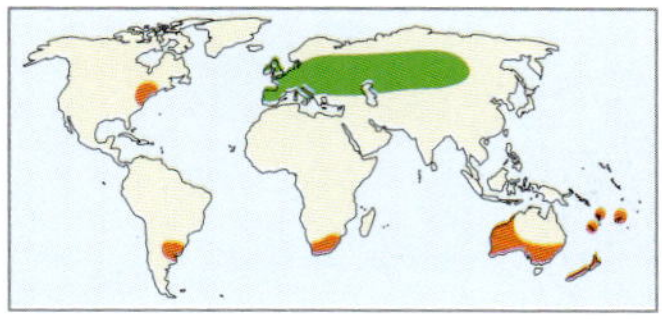

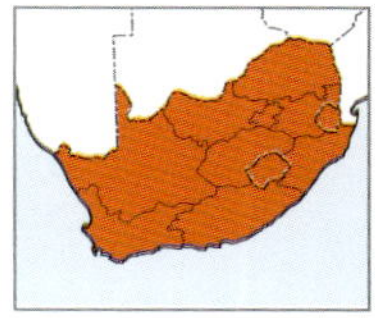

Scytodes thoracica

Group: Araneae: Scytodidae

Habitat: Terrestrial. **Identification and biology:** Body in both sexes 5–10 mm long, yellow to orange with dark stripes and spots joined to form symmetrical patterns; carapace domed to accommodate glue-producing glands; six small eyes, arranged in three well-separated groups; legs long and slender. Nocturnal wanderers. When prey is encountered, rapid contraction of carapace muscles squirts a mixture of venom and gluey silk from the chelicerae to stick prey to the substratum. Eggs held together by a few silk threads and carried in chelicerae. Common in houses, where they roam around in cupboards, stalking prey such as Grey silverfish (p. 81) and other small invertebrates. Frequently confused with Violin spiders (*Loxosceles* species). **Introduction and spread:** Introduced from Europe, date of first introduction unknown; widespread throughout South Africa. **Impacts:** Common in houses where it feeds on small invertebrates; not of medical importance. **Control:** Unnecessary.

Fritz Geller-Grimm, Wikimedia Commons

Key reference: Dippenaar-Schoeman, A.S. & Jocquè, R. 1997. African spiders, an identification manual. *Plant Protection Research Institute Handbook* 9. Pretoria: Agricultural Research Council.

Brown clover mite

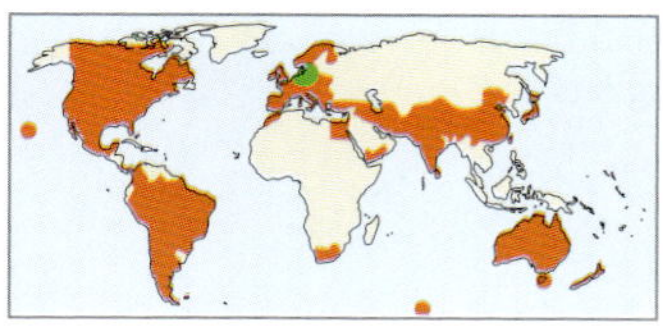

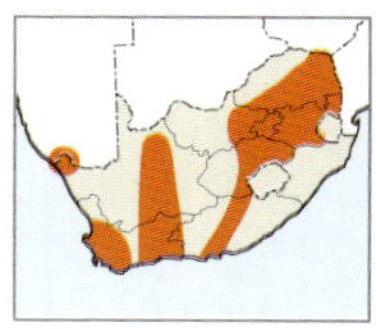

Bryobia praetiosa

Group: Acari: Tetranychidae

Habitat: On leaves of host plants. **Identification and biology:** Females broadly oval, flattened, 0.9 mm long, brownish-green, with pale red legs, first pair longer than body. Males smaller than females, tapered posteriorly. With exception of larvae, all stages overwinter. Overwintering eggs hatch in spring. Optimal egg hatching and ovipositing occur when relative humidity is high and at temperatures between 8°C and 18°C. Occasionally invade buildings, where can become a nuisance, just before mites are about to hibernate, or during spring, when overwintering eggs hatch. Eggs hatch into six-legged larvae, followed by eight-legged nymphal stages and adults. **Introduction and spread:** A European species, date of introduction into South Africa unknown; distributed worldwide. **Impacts:** Pest of clover, lucerne, cereals and occasionally vegetables. Initial symptoms lighter stipples and fine, meandering streaks on leaves; leaves subsequently become yellow or brown and wilt. **Control:** Controlled with miticides.

T. Gotoh

Key reference: Smith Meyer, M.K.P. 1996. Mite pests and their predators on cultivated plants in Southern Africa. Vegetables and berries. *Plant Protection Research Institute Handbook* 6. Pretoria: Agricultural Research Council.

Two-spotted mite and Tobacco spider mite

Tetranychus urticae and *T. evansi*

Group: Acari: Tetranychidae

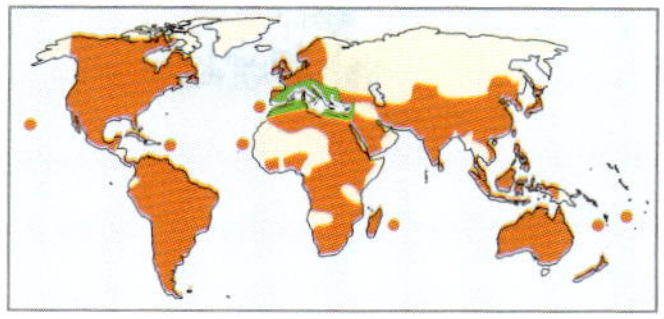

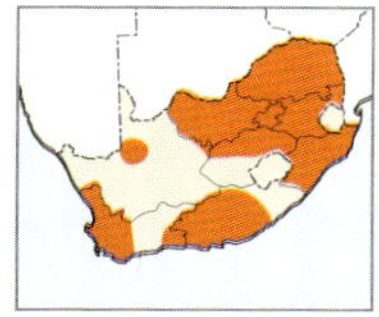

Habitat: On leaves of host plants.
Identification and biology: Two-spotted mite (shown and in maps) female brownish-red, straw-coloured or green, with dark blotch on each side; legs yellowish. Tobacco spider mite orange-red with an indistinct dark blotch on each side and with reddish legs. Male smaller than female, orange-coloured and tapered posteriorly. Both about 0.5 mm long. Two-spotted mite a common pest on most crops, and has been recorded on more than 200 host plants. Tobacco spider mite attacks mostly solanaceous (tomato and pepper family) plants. Both adults produce silk webbing that can serve as protection against enemies, pesticides and high temperatures, as well as for distribution and in mating rituals. Eggs pearly pink, light red or ivory-white; spherical, and hatch into six-legged larvae, followed by eight-legged nymphs and adults. Unfertilised eggs produce only males. Two-spotted mites feed and breed throughout the year, except in areas of extreme cold, where they remain quiescent in winter and hibernate as females on ground under leaves, in cracks, crevices and other sheltered places. Most numerous in hot, dry weather, with populations usually declining after rain. Oviposition of Tobacco spider mite increases under high humidities. **Introduction and spread:** Two-spotted mite introduced from its native range in North Africa and Europe to South Africa in 1970. Tobacco spider mite introduced to South Africa in 1980s, native range being South America; both have worldwide distributions, the latter mainly where tomatoes are grown. **Impacts:** First symptoms of injury are pale stipples on leaves. Large areas subsequently turn yellow, become bronzed, dry out and drop off. **Control:** Both species mostly chemically controlled, but have developed resistance to pesticides in parts of South Africa. In apple orchards in Western Cape, biocontrol has been implemented with great success by introducing the mite predator, *Neoseiulus californicus. Phytoseiulus persimilis*, a Mediterranean species, was also introduced into the Western Cape in 1977 on imported plant material, but has not yet proved itself an effective predator of the Two-spotted mite locally. Local mite species, namely *Euseius addoensis* and *Phytoseiulus longipes*, have also been shown empirically to be fairly good predators of the Two-spotted mite.

Key reference: Smith Meyer, M.K.P. 1996. Mite pests and their predators on cultivated plants in Southern Africa. Vegetables and berries. *Plant Protection Research Institute Handbook* 6. Pretoria: Agricultural Research Council.

C Griffiths

Citrus mite and European red mite

Panonychus citri and *P. ulmi*

Group: Acari: Tetranychidae

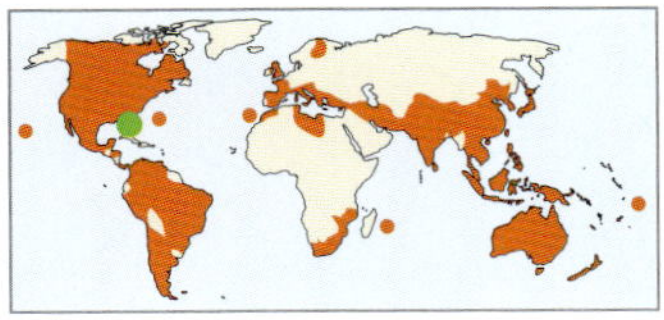

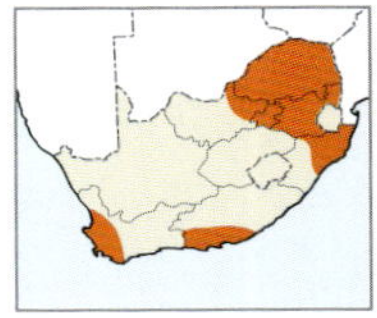

Habitat: On host plants in citrus orchards. **Identification and biology:** Citrus mite female oval, dark red with long dorsal hairs (shown). Eggs pale to dark red, somewhat flattened and with stalk. European red mite female spherical to oval, reddish-brown to blackish, dorsal hairs also long, eggs similar to those of the Citrus mite. Males smaller than females and tapered posteriorly. Body length similar in both species, about 0.3–0.5 mm. The Citrus mite is a pest of citrus fruit and prefers moderate climatic conditions; temperatures of 35–40°C and relative humidity below 25% are unfavourable for development and can cause death. The European red mite is a pest of apples and overwinters in the egg stage. Hatching of eggs only begins when temperatures rise above 7°C. High humidities (80–100%) during egg hatching reduce hatching percentage. First eggs usually hatch when first apple blossoms become pink. Both species spin rather thick threads, also known as 'guy ropes', with which they anchor their eggs, and disperse by 'ballooning', with the long threads pulling them into the air. These threads are spun when walking. Eggs hatch into six-legged larvae followed by eight-legged nymphs and adults. **Introduction and spread:** The Citrus mite (shown in maps) described from Florida, although its native range is not known. It has been a pest of citrus fruit since 1950 in South Africa, but also occurs in most citrus-growing areas worldwide. The European red mite was first recorded in South Africa in 1974 and spread to the deciduous fruit-producing areas of different countries by transport of eggs on plant material. It was first described from European material. **Impacts:** Citrus mite feeds on upper surface of leaves and bark of green branches of citrus trees. Leaves initially take on stippled appearance, developing in pale patches; as the population increases leaves become grey, silver or yellow. Feeding by European red mite causes leaves to become pale yellow and later bronze due to their withdrawing chlorophyll from the leaves. Continuous feeding results in leaves becoming completely brown and falling. Fruit remain small. **Control:** Both species mostly chemically controlled. Locally no effective biocontrol agent is known.

Nigel Cattlin/Visuals Unlimited

Key reference: Smith Meyer, M.K.P. 1981. Mite pests of crops in Southern Africa. *Science Bulletin, Department of Agriculture and Fisheries Republic of South Africa* No. 397.

Black sand mite

Halotydeus destructor

Group: Acari: Penthaleidae

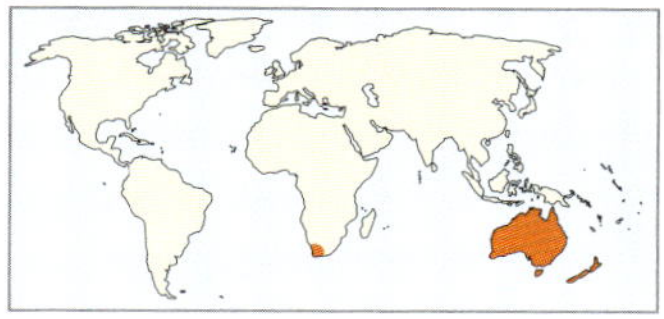

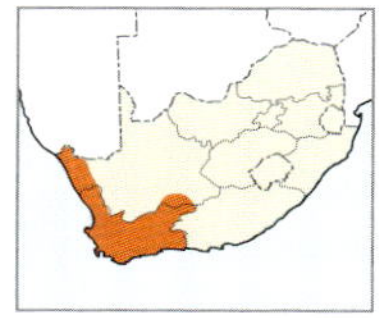

Habitat: On leaves of host plants in agricultural land. **Identification and biology:** Females pear-shaped, 1.0 mm long, velvety black. Legs red and first pair longer than body. Populations develop with onset of winter rains and lower temperatures, which induce egg hatching. Oversummering eggs are produced above 18°C. During spring adults die with oversummering eggs in their bodies for protection. Six-legged larvae emerge from these eggs, followed by 4–5 eight-legged immature stages and eventually adults. Adults live 25–50 days; female/male ratio 3:1. **Introduction and spread:** Origin unknown. First reported from Western Cape in 1908 and now widely distributed there. A pest of vegetables, cereals and pastures. **Impacts:** Feeding causes silvery spots along main leaf vein and eventually discoloration. **Control:** Favour Pigweed (*Arctotheca calendula*), thus weed control can form part of control strategy. An introduced predatory mite, the French anystis mite (*Anystis wallacei*) is an effective biocontrol agent.

A. Weeks

Key reference: Smith Meyer, M.K.P. 1996. Mite pests and their predators on cultivated plants in Southern Africa. Vegetables and berries. *Plant Protection Research Institute Handbook* 6. Pretoria: Agricultural Research Council.

Red palm mite

Raoiella indica

Group: Acari: Tenuipalpidae

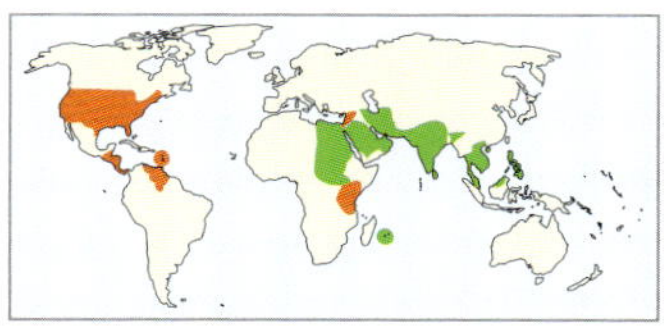

Habitat: On host plants in gardens, orchards and plantations. **Identification and biology:** Female oval, about 0.3 mm long, reddish. Dorsal body hairs long and with broad rounded ends. Legs red, first pair longer than body. Males smaller, tapered posteriorly. Eggs ovoid, red, with slender stalk. Six-legged larvae emerge from eggs, followed by eight-legged nymphs and adults. Development time from egg to adult is 20–28 days. **Introduction and spread:** Native to East Africa and Indian subcontinent. Pest of several ornamental and fruit-producing palm trees, and bananas. Not yet recorded from South Africa, but included because it has spread rapidly since 2000 along east coast of Africa towards South Africa. **Impacts:** Mite infestations are on lower leaf surfaces; heavily infested leaves develop yellow spots and entire leaf turns yellow; eventually flowers and small nuts are aborted. **Control:** Chemicals only means of control but hampered by height of trees. Five predatory mite and insect species feed and develop rapidly on the Red palm mite.

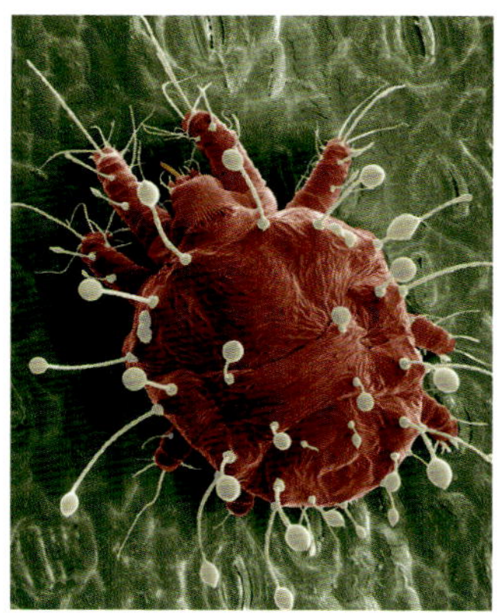

Eric Erbe, Wikimedia Commons

Key reference: Flechtmann, C.H.W. & Etienne, J. 2004. The red palm mite, *Raoiella indica* Hirst, a threat to palms in the Americas (Acari: Prostigmata: Tenuipalpidae). *Systematic and Applied Acarology*, 9:109-110.

Citrus flat mite, Reddish-black mite and Ornamental flat mite

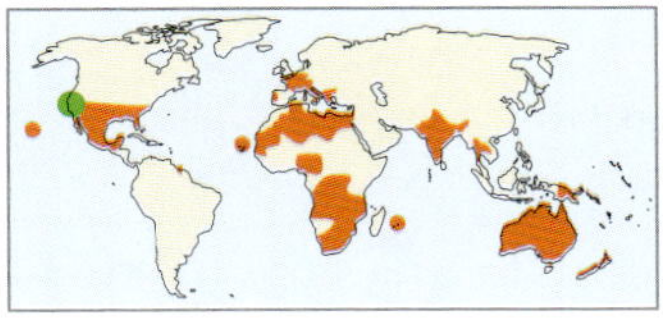

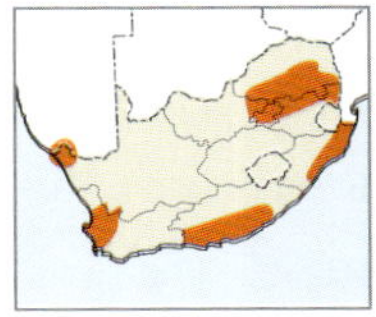

Brevipalpus californicus, B. phoenicis and *B. obovatus*

Group: Acari: Tenuipalpidae

Habitat: On host plants in garden and agricultural settings. **Identification and biology:** These species are all tiny, difficult to see with the naked eye, and very closely related. Females oval and red, with a dark red pattern on middle of body. Body about 0.3 mm long. Unfertilised females produce female offspring and males rare. Bodies of males tapered posteriorly. Duration of egg-laying 14–25 days or longer; approximately one egg produced every 24 hours, usually during daytime. Eggs elliptical, bright red and covered with adhesive substance. Six-legged larvae emerge from eggs, followed by eight-legged nymphs and adults, with resting periods of about four days between each stage. Feed on lower surface of leaves, stems and on leaf petioles. **Introduction and spread:** Mainly pests of citrus, but also feed on cotton, tea, coffee, strawberries, bananas and several ornamental plants, but are of little or no economic importance on these plants. All three species have worldwide distributions and occur in most citrus-growing countries. The first South African records for Ornamental flat mites (type locality France) and Citrus flat mites (type locality Florida; distribution shown in maps) were in 1959, with Reddish-black mites (origin unknown) recorded in 1962 (from tobacco plants). **Impacts:** Citrus flat mite and Reddish-black mite (shown) infestations form bronze and corky, scabby patches on fruit. In South Africa, the fruit most affected is closest to the tree trunk. This results in downgrading of fruit. Ornamental flat mites are not a serious pest in South Africa and can occur together with the other two species. In South America, Citrus flat mites and Reddish-black mites, and probably also Ornamental flat mites, are vectors of the leprosis virus, causing havoc in citrus orchards. **Control:** Chemical control is mainly used. A drastic decrease in numbers of these mites can occur after heavy rain. No successful biological control yet known.

Key references: Bedford, E.C.G., van den Berg, M.A. & de Villiers, E.A (eds). *Citrus pests in the Republic of South Africa*. 2nd ed. Pretoria: Institute for Tropical and Subtropical Crops, Agricultural Research Council.

Smith Meyer, M.K.P. 1981. Mite pests of crops in Southern Africa. *Science Bulletin, Department of Agriculture and Fisheries Republic of South Africa*, No. 397.

Christopher Pooley, Wikimedia Commons

Citrus silver mite

Polyphagotarsonemus latus

Group: Acari: Tarsonemidae

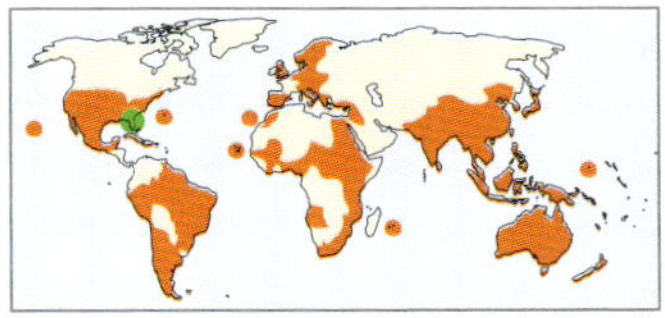

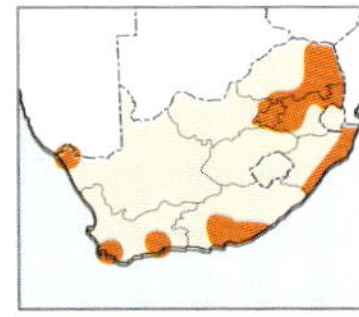

Habitat: On leaves of host plants in agricultural settings. **Identification and biology:** Females small, 0.2–0.3 mm long, broadly oval, translucent, pale yellow, distinct white stripe down middle of back. Males much smaller, but with longer legs. Eggs translucent with longitudinal rows of white knobs. Unmated females produce only males; sons then mate with mothers to produce females. Average fecundity about 40 eggs per mated female; female/male ratio 1:4. Life cycle short. Eggs hatch into six-legged larvae. Adults develop within a sac in the larvae. Males appear first. **Introduction and spread:** Cryptogenic (type locality Washington, DC/Florida), known in Africa since 1890, but also worldwide. Pest of citrus, also damages vegetables, cotton and tea in tropical and subtropical regions. **Impacts:** Feeding damage expressed as curling, crinkling or blistering of foliage; eventually plants drop leaves, stop growing and die. **Control:** Chemical control is the only mechanism to manage these mites; effective predators still unknown.

Christopher Pooley, Wikimedia Commons

Key reference: Smith Meyer, M.K.P. 1996. Mite pests and their predators on cultivated plants in Southern Africa. Vegetables and berries. *Plant Protection Research Institute Handbook 6*. Pretoria: Agricultual Research Council.

Tomato rust mite

Aculops lycopersici

Group: Acari: Eriophyidae

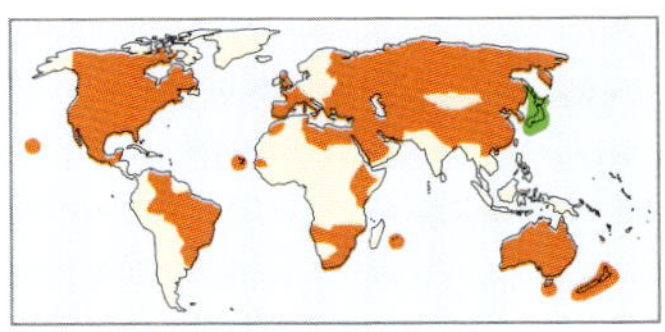

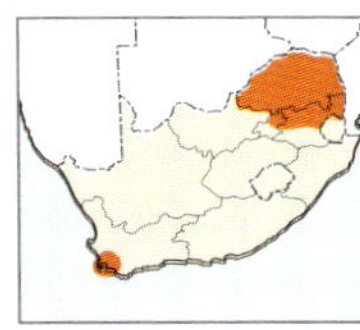

Habitat: On leaves of solanaceous host plants in agricultural and garden settings. **Identification and biology:** Females minute, 0.2 mm long, orange-yellow, bodies torpedo-shaped, with ring-like constrictions and two pairs of legs. Males only differ from females microscopically. Take 6–7 days to develop from egg to adult. Eggs opaline white. All stages four-legged. Females lay about 53 eggs during lifetime. Mainly dispersed by wind. When host begins to die, mites migrate to top of plant, position themselves perpendicular to leaf by means of their anal lobes to be blown to new hosts by wind. Prefer warm temperatures. **Introduction and spread:** Native to Japan. First recorded in South Africa in 1960. Well-known pest of solanaceous plants worldwide. **Impacts:** First signs of infestation appear on lower stems and leaves, which turn brown and wither. Severe mite infestations roughen fruit skin; if not controlled, plants eventually die. **Control:** Chemical control only way to manage these mites.

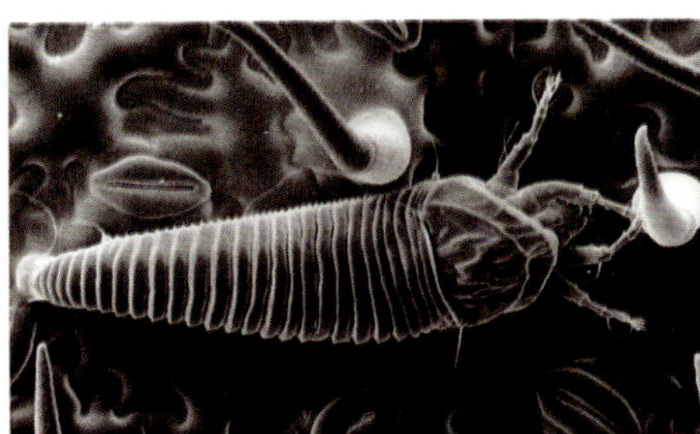

R. Ochoa

Key reference: Smith Meyer, M.K.P. 1996. Mite pests and their predators on cultivated plants in Southern Africa. Vegetables and berries. *Plant Protection Research Institute Handbook 6*. Pretoria: Agricultural Research Council.

Varroa mite

Varroa destructor

Group: Acari: Varroidae

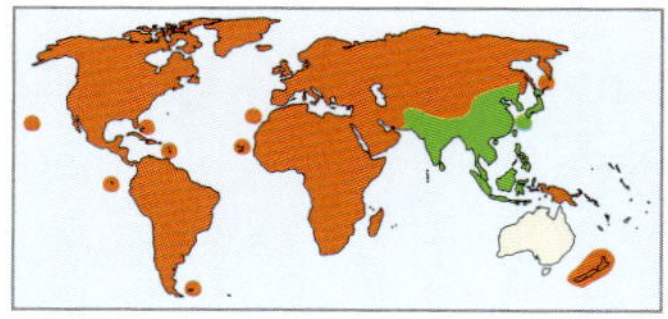

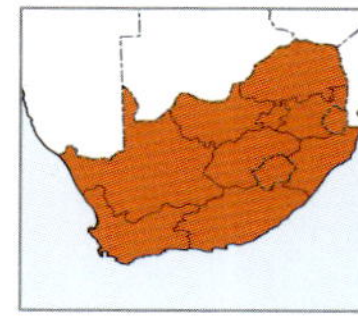

Habitat: Ectoparasite of Western honeybee (*Apis mellifera*) and its various races, found in bee hives. **Identification and biology:** The Varroa mite is the only damaging member of a species complex of 2–5 species. Flattened, fertilised adult females 1–1.8 mm long enter uncapped brood cells in honeybee colonies, feed on the haemolymph of developing larval females (prepupae), and lay 4–6 eggs after brood cells are capped. Young nymphs feed on the haemolymph of the maturing bees. After a week, male mites mature and mate in the sealed cells. Fertilised females are released when the bee emerges, or in the case of weakened or dead brood, when the cells are being cleaned by worker bees. Adult females live on bees. **Introduction and spread:** Originated on the Eastern honeybee (*A. cerana*) in Asia. Discovered in 1997 in the Western Cape of South Africa on the Cape honeybee (*A. m. capensis*). After 1998, spread (probably by commercial beekeepers) from the Western Cape to all parts of the country, where it infests both wild and commercial colonies. **Impacts:** Typically destroys untreated colonies in about two years. In Europe and North America has severely impacted wild honeybee populations, and untreated commercial colonies. Brood are killed and adults deformed, and colonies are weakened and lost. Western honeybee colonies already infected by the fungal disease 'chalkbrood' or viruses (in Europe); or African honeybee (*A. m. scutellata*) colonies invaded by Cape honeybee (in South Africa) are more susceptible to massive Varroa mite infection and collapse. Infected colonies are less effective pollinators, potentially reducing commercial fruit set as well as pollination of indigenous species. Commercial and traditional beekeepers throughout Africa may be affected. **Control:** For commercial colonies, synthetic varroacides (and to a lesser extent natural compounds such as formic acid) reduce mite numbers. Monitoring of varroacide resistance and spread of the mite to neighbouring countries is required, along with restrictions on long-distance transport of colonies. Natural resistance may occur in native populations of honeybees in Africa, which now appear to be tolerant to infection, possibly as a result of high levels of 'hygienic' behaviour.

Key reference: Allsopp, M. 2004. Cape honeybee (*Apis mellifera capensis* Eshscholtz) and varroa mite (*Varroa destructor* Anderson & Trueman) threats to honeybees and beekeeping in Africa. *International Journal of Tropical Insect Science*, 24(1):87-94.

Waugsberg, Wikimedia Commons

Asiatic cattle tick/ Pantropical blue tick

Rhipicephalus microplus

Group: Acari: Ixodidae

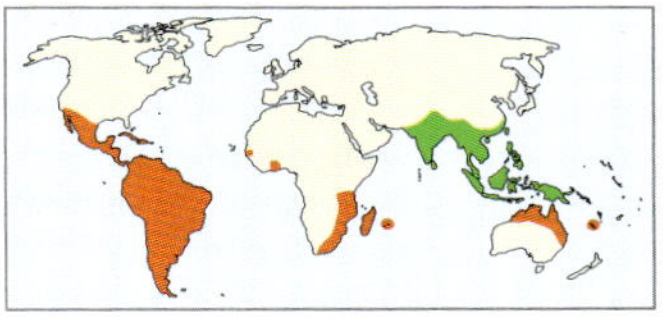

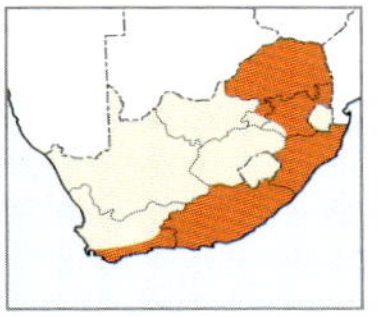

Habitat: Tropical to subtropical zones, parasitic on mammals, especially cattle. **Identification and biology:** A 'hard tick' with brown-grey body, pale cream legs, white sides and front, and a dorsal shield and mouthparts that protrude when viewed from above. Body length increases from 3 mm to 13 mm after engorging. Shield orange-brown in six-legged nymphs. Easily confused with the South African native Blue tick (*Boophilus decoloratus*). Eggs hatch and larvae crawl up vegetation, surviving for 3–6 months without food. Larvae (seed ticks) attach to soft belly skin, moult twice, transforming to nymphs and then adults. Mating occurs on the host, with the female dropping off to lay one egg batch in a crevice before dying. The rapid life cycle is completed in 3–4 weeks. **Introduction and spread:** Endemic to South and Southeast Asia (including associated islands), and subsequently spread to north-eastern Australia, Central and South America, and Madagascar. It was then introduced to West, East and South Africa with cattle imported from Madagascar to offset the 1896 rinderpest outbreak. In South Africa, limited by cold and areas with less than 500 mm annual rainfall. **Impacts:** The most important tick parasite of cattle globally, also occurs on horses, sheep, donkeys, domestic buffalo, goats, pigs, deer and dogs. However, secondary hosts only used where cattle are not present. Non-domesticated hosts include Grey rhebok, Sable, Grey duiker, African buffalo and lion. While heavy infestations can weaken animals, the ticks also transmit the unicellular organisms responsible for the often fatal babesiosis. The unicellular organism, *Babesia digema*, causes African redwater disease, and *B. bovis,* introduced with the Asiatic cattle tick from Madagascar, causes Asiatic redwater disease. The tick also transmits the bacterium *Anaplasma marginale* that causes bovine anaplasmosis. The Asiatic cattle tick displaces the Blue tick (native to South Africa) partially or totally. **Control:** Measures include repeated dipping of cattle with acaricides (acaricide resistance widespread), maintaining pastures free of cattle for 6–9 months, and vaccinations. Ivermectin-laced bait keeps secondary hosts tick-free. Some cattle breeds such as Zebu (*Bos indicus*) are resistant to the tick.

Key reference: Estrada-Pena, A., Bouattour, A., Camicas, J.L., Guglielmone, A., Horak, I., Jongejan, F., Latif, A., Pegram, R. & Walker A.R. 2006. The known distribution and ecological preferences of the tick subgenus *Boophilus* (Acari: Ixodidae) in Africa and Latin America. *Experimental and Applied Acarology*, 38:219-235.

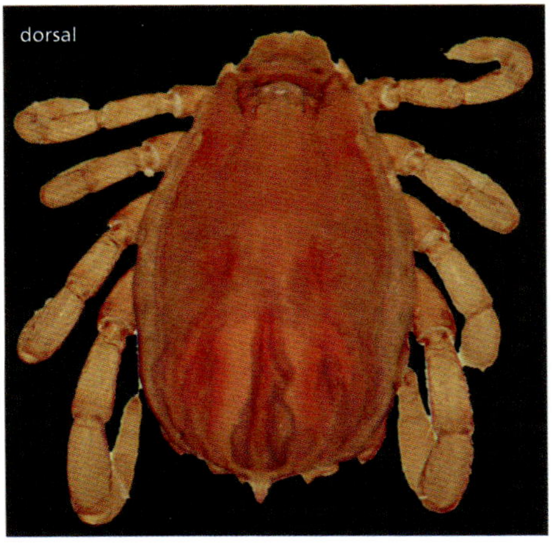

M. Madder

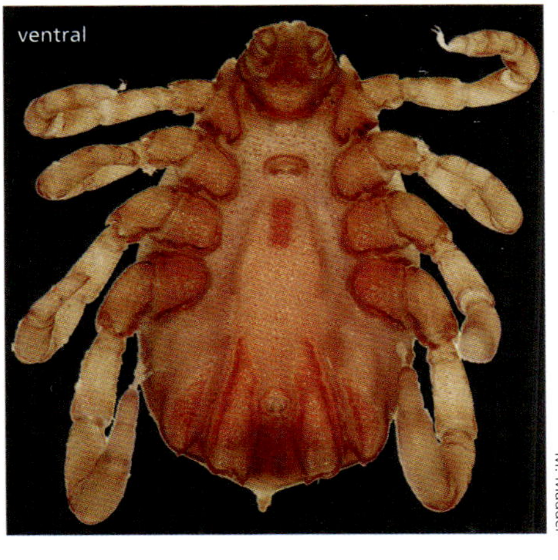

M. Madder

AUTHORS: C. GRIFFITHS

CRUSTACEANS Subphylum Crustacea

Crustaceans are a large, diverse group including crabs, shrimps, woodlice and barnacles. Most are marine, but many occur in freshwater and some on land. All have jointed exoskeletons and have to moult to grow. Body form is very variable, but typically comprises a fused head and thorax, plus segmented abdomen. There are usually two pairs of antennae, the remaining limbs varying greatly in both number and form, and variously adapted for walking, swimming, grasping, filter-feeding, respiration, reproduction and other functions.

Pacific barnacle

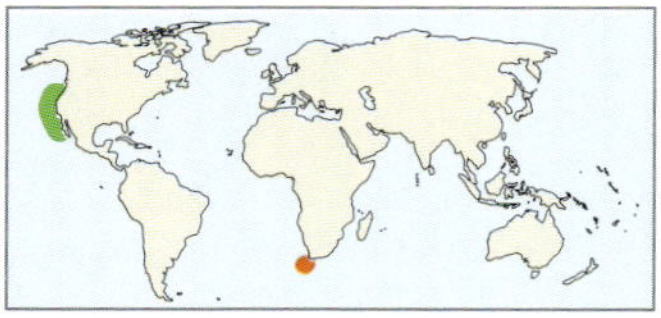

Balanus glandula

Group: Cirripedia: Balanidae

Habitat: Upper levels on intertidal rocky shores. **Identification and biology:** Barnacles are highly modified crustaceans, although the larvae retain a shrimp-like form. Adults are encased in a conical shell of 4–8 hard calcium-carbonate plates and live permanently attached to rocks or similar hard surfaces. When submerged, hairy, tapering legs are extended through an opening at the apex and combed rhythmically through the water to capture food particles. When exposed to air the legs are withdrawn and the shell sealed by four additional shell plates. The Pacific barnacle is by far the most common intertidal barnacle within its range, and is best distinguished from other South African barnacles by having a combination of six shell plates, a calcareous (rather than membranous) base (leaving a white scar on the rock when removed), and an M-shaped opening. Shell shape varies from volcano-shaped with marked radiating ridges in well-spaced individuals, to tall and cylindrical when crowded. Diameter typically 10 mm. **Introduction and spread:** Introduced from Pacific North America before 1992, probably with ballast water, but undetected until 2008. Now the dominant barnacle on exposed rocky shores along much of the west coast, forming a distinct white band in upper intertidal zone. **Impacts:** Dense populations substantially change the shore, monopolising space and eliminating limpets and other large grazers. Invaded shores also support increased number of periwinkles (*Afrolittorina knysnaensis*), which shelter between the barnacles and within their dead shells. Predatory whelks (*Nucella dubia*) also gather to feed on barnacles along the lower edge of the beds. **Control:** Anti-fouling paints and mechanical control can eliminate barnacles from ships, but no known mechanism to control wild populations.

C. Griffiths

Key reference: Laird, M.C. & Griffiths, C.L. 2008. Present distribution and abundance of the introduced barnacle *Balanus glandula* in South Africa. *African Journal of Marine Science*, 30:93-100.

Fish louse

Argulus japonicus

Group: Copepoda: Argulidae

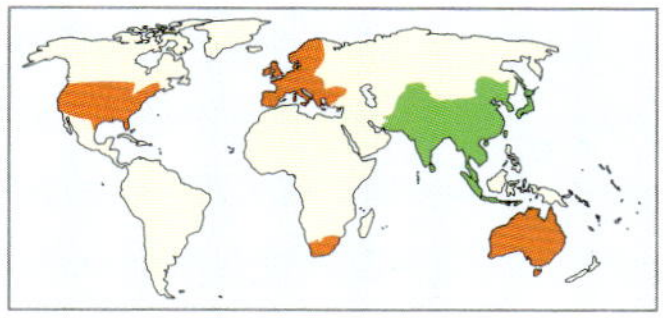

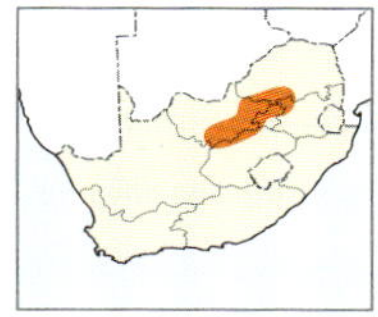

Habitat: External parasite on fish found in freshwater lakes and dams, extending into estuaries. **Identification and biology:** Flattened, transparent crustaceans up to 8 mm long with a rounded, horseshoe-shaped carapace. Attach to skin of fish using sucker-like head appendages, then feed on body fluids of host. Each thoracic segment has a pair of branched swimming legs. After mating, females leave host to deposit up to 400 eggs on a firm substratum. **Introduction and spread:** Accidentally introduced from Asia or Europe before 1983, with Koi carp (p. 52) or other ornamental fish. **Impacts:** Damage caused by low-level infestation not great, but can have serious impact on fish stocks, particularly in fish farms, where high host density facilitates transmission and hundreds of parasites can accumulate on each host, retarding growth and leading to secondary infection by bacteria or fungi, sometimes causing death. **Control:** Chemical control in fish farms achieved using various commercial products, but control in natural water bodies difficult or impossible.

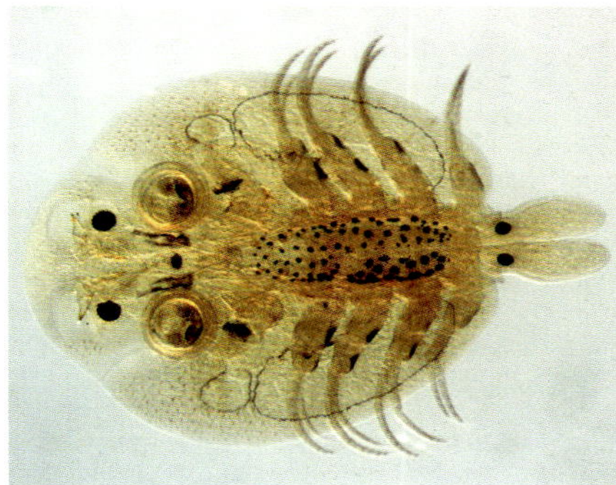

J. van As

Key reference: Shafir, A. & Oldewage, W.H. 1992. Dynamics of a fish ectoparasite population: opportunistic parasitism in *Argulus japonicus* (Branchiura). *Crustaceana*, 62:50-64.

Spinytailed copepod

Acartia spinicauda

Group: Copepoda: Acartiidae

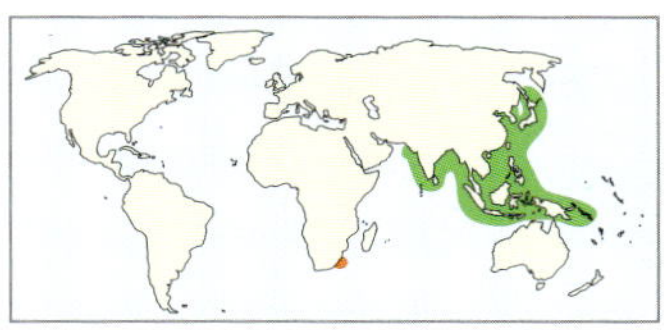

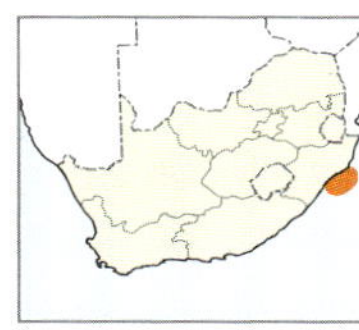

Habitat: Marine, free-swimming in sheltered sites. **Identification and biology:** Tiny, 1–2 mm long, free-swimming crustaceans with tapering cylindrical bodies and long antennae, which project out from the side of the head. Feed on phytoplankton, and form a major link between phytoplankton and fishes. Hard to distinguish from native *Acartia* species without specialised knowledge. **Introduction and spread:** Probably introduced from Asian ports to Richards Bay harbour via ballast water and now common there, reaching 900 per cubic metre of water. **Impacts:** May be important consumers of phytoplankton and food source for indigenous fishes in invaded area. **Control:** Probably impossible, given the free-swimming lifestyle, abundance and rapid rate of reproduction of this species.

C. Griffiths

Key reference: Jerling, H.L. 2008. The zooplankton community of Richards Bay Harbour and adjacent Mhlathuze Estuary, South Africa. *African Journal of Marine Science*, 30:55-62.

Land hopper

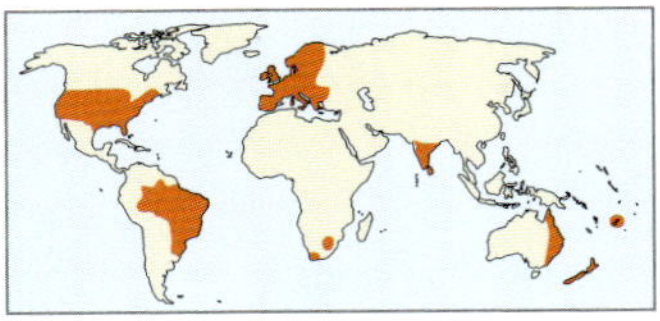

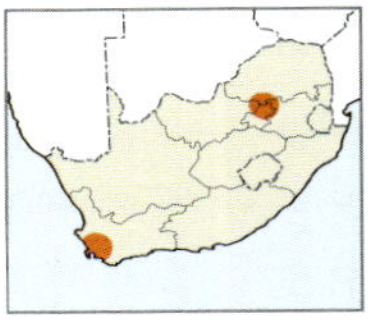

Talitroides topitotum

Group: Amphipoda: Talitridae

Habitat: Terrestrial, in leaf litter, compost heaps and similar moist habitats. **Identification and biology:** Small, laterally flattened, shrimp-like crustaceans known for their ability to jump large distances. Feed on dead organic matter. Females retain eggs in a ventral brood pouch until they hatch into miniature adults. Maximum length 13 mm. Easily confused with similar indigenous species of the genus *Talitriator*, but distinguished by greatly reduced abdominal appendages (pleopods). **Introduction and spread:** A so-called 'tramp species' so widely dispersed around the world that its origin can no longer be determined. Usually transported in soil with potted plants. Reported only from Cape Town and Pretoria although certainly more widespread, but remains undetected because of lack of sampling and similarity to indigenous land hoppers. **Impacts:** Feeds on dead organic matter and not live plants, so unlikely to have significant impacts. **Control:** Not necessary, but can be caught in pit traps.

C. Griffiths

Key reference: Griffiths, C.L. 1999. The terrestrial Amphipoda of South Africa. *Annals of the South African Museum*, 105:345-362.

Beach hopper

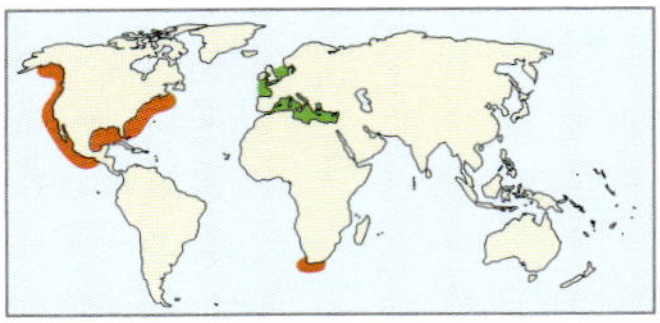

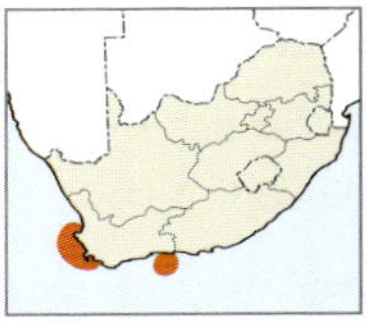

Orchestia gammarella

Group: Amphipoda: Talitridae

Habitat: Estuarine and lagoonal seashores. **Identification and biology:** Small, up to 15 mm long, air-breathing crustaceans commonly associated with, and feeding on, drift seaweeds around the high-water mark on sandy and lagoonal seashores. Distinguished from several indigenous beach hoppers by enlarged, flattened oar-like last pair of walking legs. The introduced South African population was initially described as a new, indigenous species in the related genus *Talorchestia*, an error only corrected decades later. **Introduction and spread:** Almost certainly introduced from Europe with dry ballast at least a century ago. South African distribution restricted to a few present or former harbour sites.

Impacts: Feeds only on dead plant debris, so unlikely to have significant impacts, although may compete with indigenous beach hoppers and form additional food source for shoreline birds. **Control:** Not necessary and no control methods known.

C. Griffiths

Key reference: Griffiths, C.L. 1975. The Amphipoda of Southern Africa Part 5: The Gammaridea and Caprellidea of the Cape Province west of Cape Agulhas. *Annals of the South African Museum*, 67:91-181.

Hitchhiker amphipod

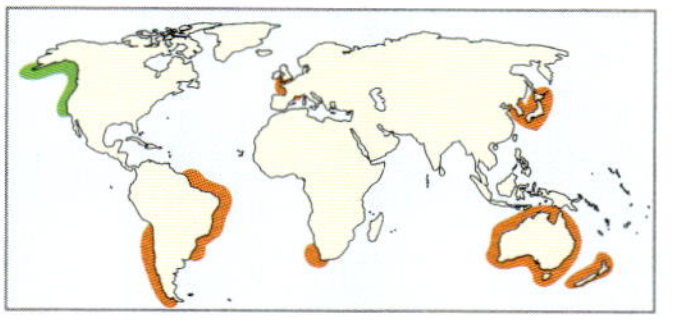

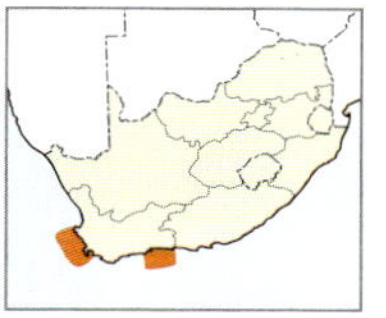

Jassa slatteri

Group: Amphipoda: Ischyroceridae

Habitat: Marine, on seaweeds and other structures, particularly in harbours. **Identification and biology:** Length up to 9 mm. Second pair of antennae enlarged and stout, second pair of nippers of adult male enormously enlarged with an unusual 'thumb' projecting from base of 'hand'. Constructs soft tubes of mud, or crawls around on seaweeds, hydroids and other marine growth. Particularly common on piers, buoys and ship hulls in harbours. South African specimens formerly erroneously identified as *Jassa falcata*, but now reallocated between three other closely related species, all introduced. **Introduction and spread:** Introduced via shipping either directly from Pacific North America or via other invaded regions and particularly common in temperate harbours. **Impacts:** Small and inconspicuous, but occurs in extremely high densities, providing a rich food source for small fish and other predators, so probably has considerable effects on food chains in invaded communities. **Control:** No control method known, probably not feasible given abundance and wide range.

C. Griffiths

Key reference: Conlan, K.E. 1990. Revision of the crustacean amphipod genus *Jassa* Leach (Corophioidea: Ischyroceridae). *Canadian Journal of Zoology*, 68: 2031-2075.

Stout-antenna amphipod

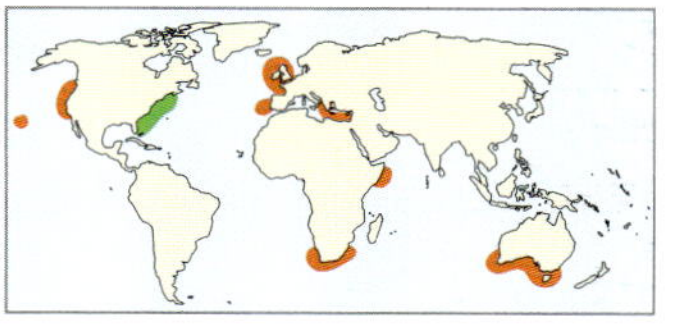

Monocorophium acherusicum

Group: Amphipoda: Coropiidae

Habitat: Marine, on silt-covered surfaces of rocks and piers and in soft substrata, extending into estuaries. **Identification and biology:** Body elongated, up to 5 mm long, cylindrical in cross-section. Second pair of antennae hugely enlarged and swollen in males, a curved tooth on underside of fourth segment. First pair of legs with weak claw, second pair very hairy, but not clawed. Last three abdominal segments fused. Forms tubes in sandy or muddy habitats. **Introduction and spread:** Originated in western North Atlantic and first recorded in South Africa in 1915; abundant in South African harbours. **Impacts:** Can occur in densities of thousands per square metre and readily consumed by fish and other predators, so probably affects food chains, otherwise a minor contributor to fouling fauna. **Control:** No known control method.

C. Griffiths

Key reference: Bousfield, E.L. 1973. *Shallow water* Gammaridean Amphipoda *of New England*. New York: Cornell University Press.

Hermit amphipod

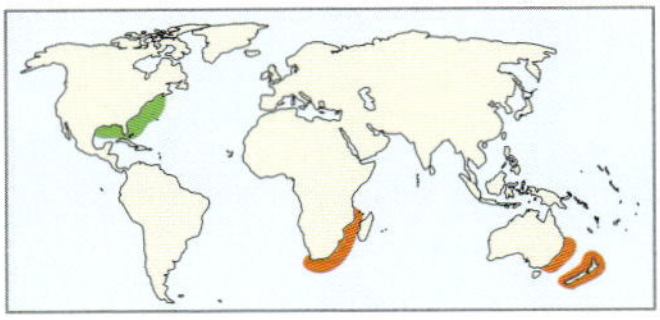

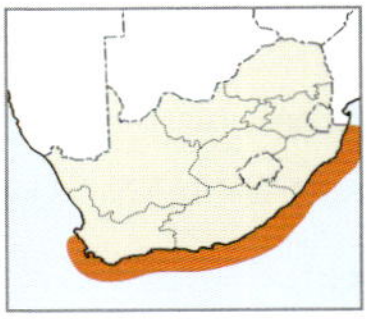

Cerapus tubularis

Group: Amphipoda: Ischyroceridae

Habitat: On mud or sand, carrying a portable case, much like a hermit crab. **Identification and biology:** Elongated, up to 5 mm long, cylindrical amphipods that live concealed in mud or sand tubes, juveniles often occupying smaller tubes branching off those of adults. Second claw of males has enlarged and toothed fifth segment, such that sixth and seventh segments together form the 'finger' of the claw. Last two pairs of abdominal appendages greatly reduced. **Introduction and spread:** Probably introduced from Atlantic North America via shipping or ballast water and now widespread throughout region. **Impacts:** Minimal, although an important component of the fouling fauna and probably commonly eaten by fish and other small predators. **Control:** None attempted.

C. Griffiths

Key reference: Bousfield, E.L. 1973. *Shallow water* Gammaridean Amphipoda *of New England*. New York: Cornell University Press.

Wood-boring amphipod

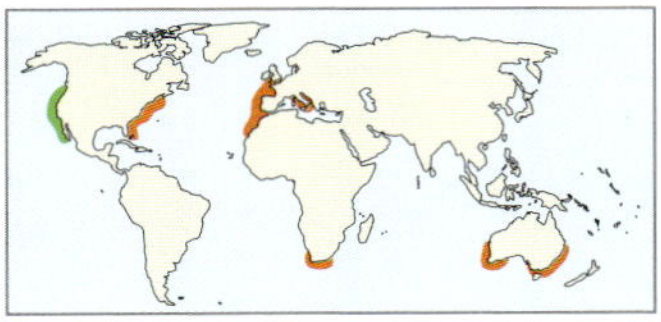

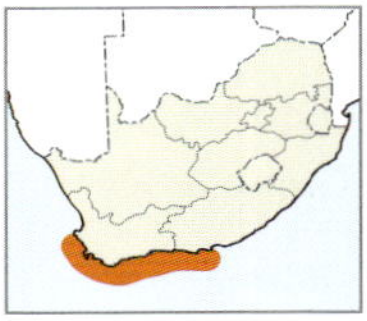

Chelura terebrans

Group: Amphipoda: Cheluridae

Habitat: Marine, enlarging and living within shallow burrows in pilings, stakes and other wooden structures previously excavated by wood-boring isopods known as Gribbles (*Limnoria* species, p. 165). **Identification and biology:** Length 4–6 mm. Very distinctive amphipods with a rounded, cylindrical body and head, short antennae with club-like segments, and a large dorsal spike projecting from the abdomen. Abdominal segments fused and their appendages (uropods) greatly modified, lobed and enlarged, projecting well beyond end of body. Also easily identified by unusual wood-burrowing habits. **Introduction and spread:** Almost certainly introduced via wooden vessels, either directly from its native Pacific North America or secondarily via invaded sites in Europe. **Impacts:** Bore into wooden structures, but appear only to enlarge and expand tunnels made by Gribbles, rather than creating their own tunnels. **Control:** Specific control measures against *Chelura* are generally not necessary, as they are dependent on previous infestation of habitat by Gribbles.

C. Griffiths

Key reference: Bousfield, E.L. 1973. *Shallow water* Gammaridean Amphipoda *of New England*. New York: Cornell University Press.

Gribble

Limnoria tripunctata

Group: Isopoda: Limnoriidae

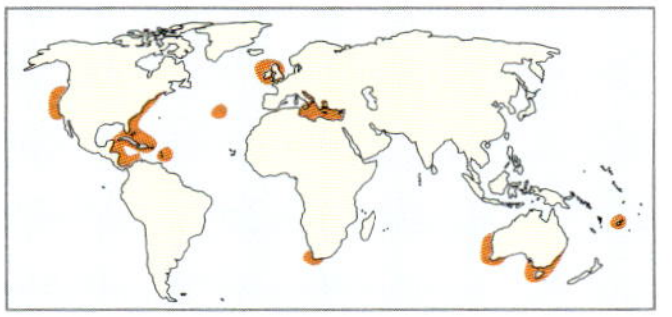

Habitat: Marine, boring into wood. **Identification and biology:** Tiny (3–4 mm), white, elongated-oval crustaceans with very short antennae. Best identified by their unusual habit of excavating burrows in submerged wood. There are several species, this one identified by three rounded tubercles on top of the last body segment (pleotelson). Often have two dark spots on pleotelson caused by another introduced species, the ciliate parasite *Mirofolliculina limnoriae*. **Introduction and spread:** Widely dispersed via wooden ships. First reported in Table Bay in only 2008, but certainly a much earlier introduction; origins obscure, probably Indo-Pacific. **Impacts:** Despite their small size a serious pest, responsible for destruction of marine structures like wharfs and pier pilings. Infected beams eroded into a characteristic hourglass shape. **Control:** Numerous control measures proposed, including various chemicals, use of ultrasound, etc., but tolerant to creosote, the most common treatment against marine wood-borers, and hence able to damage treated wood extensively.

C. Griffiths

Key reference: Johnson, M.W. & Menzies, R.J. 1956. The migratory habits of the marine gribble *Limnoria tripunctata* Menzies in San Diego Harbor, California. *Biological Bulletin*, 110:54-68.

Horned isopod

Dynamene bidentata

Group: Isopoda: Sphaeromidae

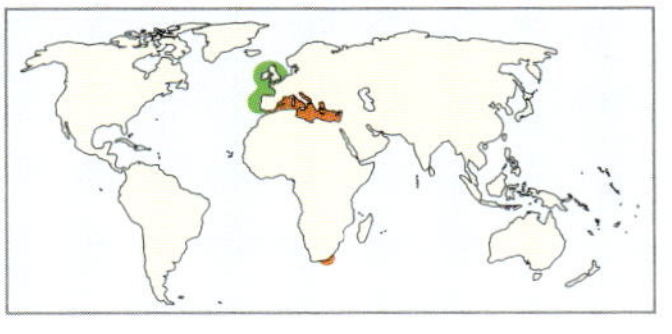

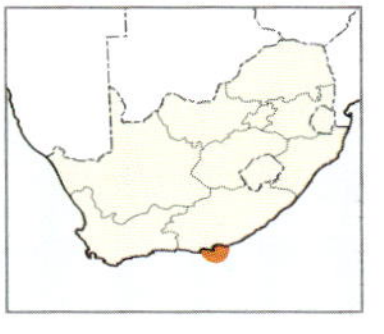

Habitat: Marine, on rocky shores, among barnacles and algae. **Identification and biology:** Small, up to 7 mm long, flattened isopods with a rounded head and anterior section. Strongly sexually dimorphic, male having a distinctive pair of dorsal processes or 'horns' that project backwards from the sixth thoracic segment over the abdomen, also last pair of abdominal appendages (uropods) have a greatly enlarged outer branch projecting well beyond the end of body. Females lack both these features and are smooth-bodied. **Introduction and spread:** Only known record from Port Elizabeth harbour, to which the species would have been introduced from Europe either as a fouling organism, or via ballast water. **Impacts:** Minimal at this stage. **Control:** No control method known or attempted.

C. Griffiths

Key reference: Holdich, D.M. 1968. Reproduction, growth and bionomics of *Dynamene bidentata* (Crustacea: Isopoda). *Journal of Zoology*, 156:137-153.

Pill isopod

Sphaeroma walkeri

Group: Isopoda: Sphaeromidae

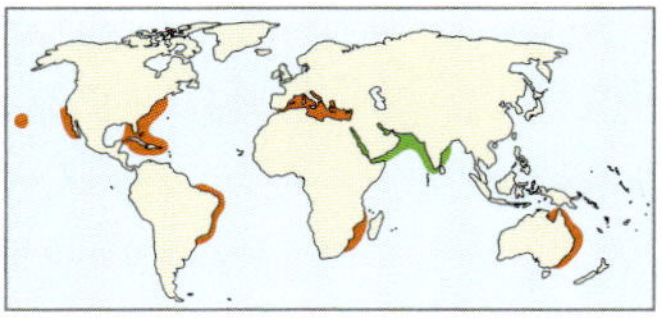

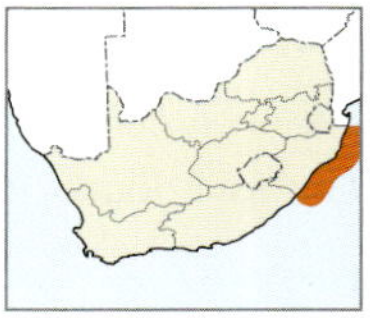

Habitat: Marine, rocky shores, estuaries and harbours. **Identification and biology:** Reaches 10 mm in length and best identified by the ability to roll into a ball, from which the generic name is derived, as well as the unusual serrate outer branch of the last abdominal appendage (uropod). The pleotelson has several longitudinal rows of tubercles. Lives in fouling communities and on rocky shores and is an omnivore. **Introduction and spread:** One of the most widely introduced of all marine species. Probably transported from ports in the tropical Indian Ocean to Durban harbour via fouling on vessels or more recently via ballast water. **Impacts:** Probably insignificant, although related burrowing species cause significant damage to submerged wooden structures and prop roots of mangroves. **Control:** None known or proposed.

C. Griffiths

Key reference: Carlton, J.T. & Iverson, E.W. 1981. Biogeography and natural history of *Sphaeroma walkeri* Stebbing (Crustacea: Isopoda) and its introduction into San Diego Bay, California. *Journal of Natural History*, 15:31-48.

Sponge isopod

Paracerceis sculpta

Group: Isopoda: Sphaeromatidae

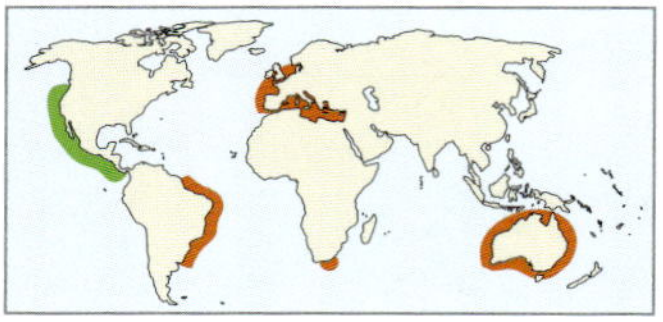

Habitat: Marine, on both rocky shores and sand to depths of about 70 m, often associated with sponges. **Identification and biology:** Body rounded and flattened, reaching about 8 mm length. Abdomen hairy, with various granular knobs. Last abdominal segment (pleotelson) split by a wide median toothed notch. Outer branch of last pair of abdominal appendages (uropods) projecting into a long point. Herbivorous, feeding on coralline algae in its native range. Juveniles can be carnivorous. **Introduction and spread:** Widely dispersed by shipping and likely secondarily introduced to Port Elizabeth harbour via Europe or Australia rather than from its native Pacific North America. **Impacts:** Probably minimal, although fed upon by many fish species. **Control:** None attempted or proposed.

C. Griffiths

Key reference: Espinosa-Perez, M.C. & Hendrickx, M.E. 2002. The genus *Paracerceis* Hansen, 1905 (Isopoda, Sphaeromatidae) in the Eastern Tropical Pacific with a description of a new species. *Crustaceana*, 74:1169-1187.

Smooth woodlouse

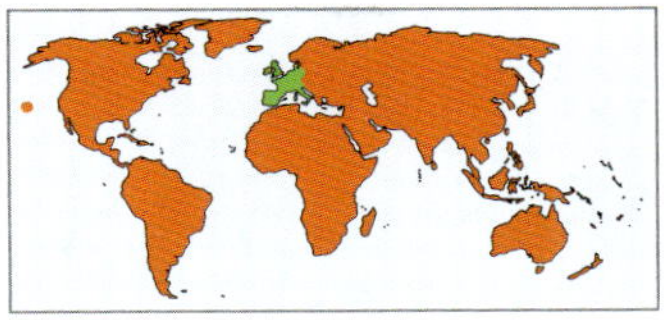

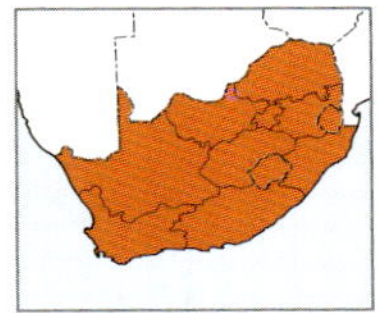

Porcellio laevis

Group: Isopoda: Porcellionidae

Habitat: Terrestrial, leaf litter, under logs and rocks, in compost heaps and similar damp, protected habitats. **Identification and biology:** Broadly oval, flattened, mottled brown woodlouse with smooth surface, length up to 15 mm. Head with distinct side-lobes, set back into first thoracic segments and bearing long, elbowed second antennae. Abdomen not distinctly narrower that thorax, sides of each segment forming backward-facing points. Terminal uropods flattened, with spear-shaped outer branches projecting well beyond end of body. Prefers damp habitats and mostly nocturnal. Feeds on leaf litter and other dead organic matter. Eggs brooded in a pouch under female's body and young resemble miniature adults. Grows to 20 mm. **Introduction and spread:** Easily transported with logs, pot plants and similar goods and has been widely dispersed from its native Europe. First recorded from Cape Town in 1932 and almost certainly widely distributed in other urban areas, but distribution inadequately studied. **Impacts:** Useful decomposers of organic matter and dominant component of fauna in compost heaps in Cape Town gardens. Harmless to humans. **Control:** Not required.

C. Griffiths

Key reference: Barnard K.H. 1932. Contributions to the crustacean fauna of South Africa. No. 11. Terrestrial. Isopoda. *Annals of the South African Museum*, 30:179-388.

Rough woodlouse

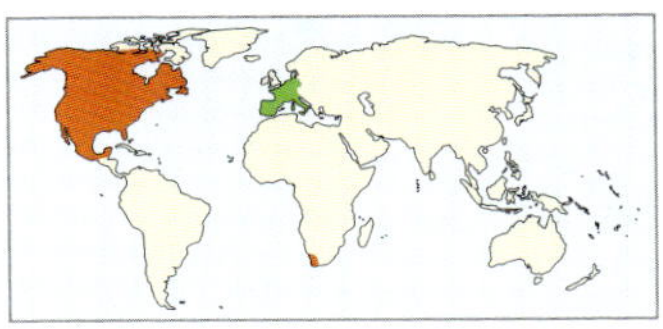

Porcellio scaber

Group: Isopoda: Porcellionidae

Habitat: Terrestrial, leaf litter, under logs and rocks, compost heaps and similar damp, protected habitats. **Identification and biology:** Length up to 15 mm. Similar to smooth woodlouse, above, but readily distinguished by the 2–3 transverse rows of small rounded tubercles across the head and each thoracic segment. Usually mottled grey, although brown, yellow and orange-tinged individuals also found. Habits similar to smooth woodlouse, but less common in Cape Town and typically in somewhat drier habitats. **Introduction and spread:** Probably introduced from Europe with timber or other produce and first reported from Cape Town in 1885. Almost certainly widespread in towns and cities across South Africa, but distribution not adequately surveyed. **Impacts:** Useful decomposers of organic matter. Harmless to humans. **Control:** Not required.

C. Griffiths

Key reference: Barnard, K.H. 1932. Contributions to the crustacean fauna of South Africa. No. 11. Terrestrial. Isopoda. *Annals of the South African Museum*, 30:179-388.

Pill bug

Armadillidium vulgare

Group: Isopoda: Armadillidiidae

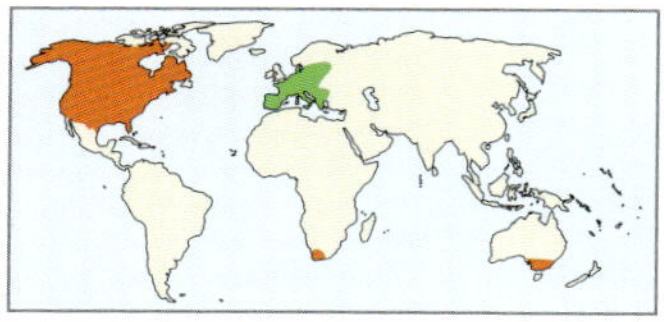

Habitat: Under stones and logs or in leaf litter, including in drier habitats. **Identification and biology:** Smooth, glossy, hard-bodied woodlouse with a regular elongated-oval outline, fat, humped body and relatively short antennae. Colour brown to red with paler flashes. Reaches 18 mm. Terminal uropods short and quadrangular and do not project beyond end of body. Best recognised by the ability to roll into a tight closed ball when disturbed. Able to tolerate drier conditions than most other woodlice, hence found in grassland, as well as forest and urban habitats. Feeds on decaying organic matter, but also lichens and algae from tree bark and walls. **Introduction and spread:** Native to Europe. First South African records from Cape Town in 1943. Probably widespread, insufficient survey data available to assess current distribution. **Impacts:** Decomposer of dead organic matter, so unlikely to have negative impacts. Harmless to humans. **Control:** Not necessary.

Key reference: Barnard, K.H. 1932 Contributions to the crustacean fauna of South Africa. No. 11. Terrestrial. Isopoda. *Annals of the South African Museum*, 30:179-388.

C. Griffiths

Fast woodlouse

Philoscia muscorum

Group: Isopoda: Oniscidae

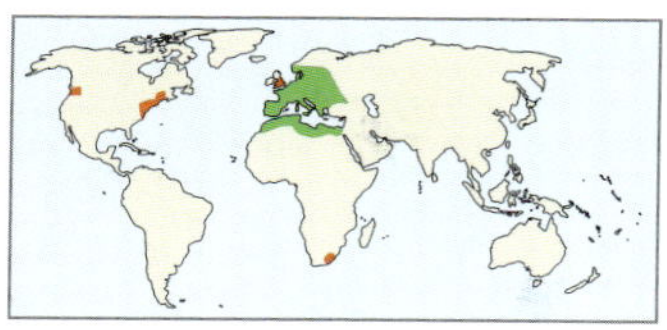

Habitat: Decomposing plant matter, under logs, in gardens and other moist habitats. **Identification and biology:** Relatively slender, almost parallel-sided woodlouse reaching 11 mm length, abdomen abruptly narrower than thorax. Faster moving that the species above and does not roll into a ball when disturbed. Usually has a darker stripe along back. Head lacks lateral lobes and uropods have the inner branch about half length of outer branch. **Introduction and spread:** Of European origin, recorded from central KwaZulu-Natal but almost certainly more widespread, as no systematic surveys have been undertaken. **Impacts:** A detritus feeder, so unlikely to have significant impacts. **Control:** No control known or required.

Key reference: Barnard, K.H. 1932. Contributions to the crustacean fauna of South Africa. No. 11. Terrestrial. Isopoda. *Annals of the South African Museum*, 30:179-388.

C. Griffiths

Marron, freshwater crayfish

Cherax tenuimanus

Group: Decapoda: Parastacidae

Habitat: Freshwater, bottom-dwelling in dams and rivers. **Identification and biology:** A large, robust freshwater crayfish with strong equal-sized chelae or claws and series of marked ridges along dorsal surface of carapace. Can attain 40 cm length and 2.5 kg mass. Mainly a detritus feeder, consuming dead plant debris, but also eats live water plants. Reaches sexual maturity in third year and then spawns annually during spring. Eggs attached to abdominal appendages (pleopods) of female and undergo preliminary development there, before being released as free-swimming juveniles. **Introduction and spread:** Deliberately introduced from Australia since the 1980s and stocked in a number of aquaculture facilities in the Western Cape. Previously also in Eastern Cape, where it escaped into the Buffalo River, but apparently failed to establish a self-sustaining population there. As with other freshwater crayfish, can live for several days out of water and walk short distances overland, making them hard to contain in captivity. Two other freshwater crayfish are also known to have been introduced to South Africa and to have escaped into the wild. Importation of the North American Red swamp crayfish (*Procambarus clarkii*), which is a destructive burrower, is not permitted, but at least one illegally imported population has escaped and established in the Crocodile River near Dullstroom. Similarly, a population of Australian redclaw (*Cherax quadricarinatus*) from an abandoned aquaculture facility in Swaziland have escaped and established in the Sand River Dam, on a tributary of the Crocodile River, from where they are spreading into irrigation canals in the region. Yet another Australian species, the Yabby (*C. destructor*), is also held in captive aquaculture facilities in the Free State, but does not appear to have established wild populations. **Impacts:** May compete with and displace indigenous crabs and shrimps. Because of their large size and high meat yield, also a potential candidate for aquaculture. **Control:** Can be trapped, but complete elimination of escaped populations difficult.

Key reference: De Moor, I. 2002. Potential impacts of alien freshwater crayfish in South Africa. *African Journal of Aquatic Science*, 27:125-139.

User:OpenCage, Wikimedia Commons

European shore crab

Carcinus maenas

Group: Decapoda: Portunidae

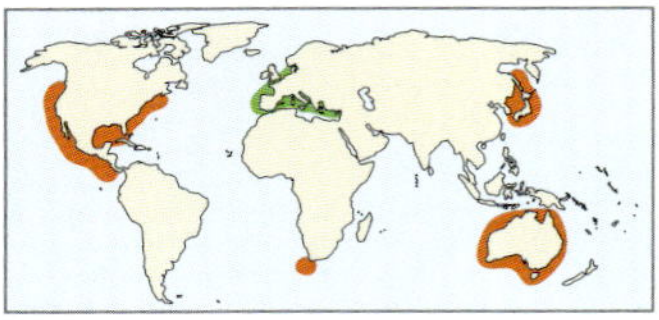

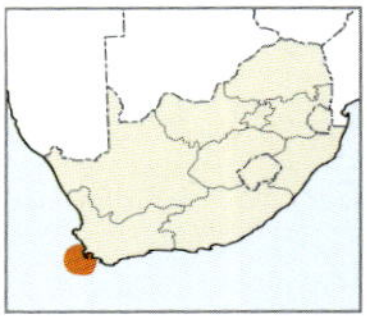

Habitat: Marine, on sheltered rocky shores and in harbours, its native range also extends into estuaries. **Identification and biology:** Robust and aggressive crabs with strong claws, mottled green or sometimes red on upperside, with semicircular row of white dots across carapace; yellow or orange on underside. Front margin of carapace with three teeth between eyes and five prominent triangular teeth along either margin. Voracious predators, often aggressive when handled and consuming a wide range of prey, especially shellfish such as mussels and clams, but also carrion, can also be cannibalistic. Reach 100 mm carapace width. **Introduction and spread:** Introduced from European shores to Table Bay docks where they were first reported in 1983; now very common in that harbour, where they have largely eliminated mussels and other shellfish. A few small colonies have established in sheltered sites in the vicinity, notably in Hout Bay harbour, but appear to be incapable of colonising the open, wave-swept shores typical of most of the South African coastline. There is, however, concern that they will sooner or later make the jump to Saldanha Bay, where there are large areas of suitable habitat and abundant shellfish resources. **Impacts:** An important pest of shellfisheries in North America and other sites to which they have been introduced. Pose a severe threat if introduced to the Saldanha Bay-Langebaan Lagoon system, which is both the centre of the marine aquaculture industry in South Africa and site of a major marine national park. **Control:** Can readily be caught in traps, but a major effort would be required to eliminate the extensive population in Table Bay docks manually. Biocontrol using the Castrating barnacle (*Sacculina carcini*) has been proposed, but this method not yet sufficiently developed to be practical.

Key reference: Hampton, S. & Griffiths, C.L. 2007. Why *Carcinus maenas* cannot get a grip on South Africa's wave exposed coastline. *African Journal of Marine Science*, 29:123-126.

C. Griffiths

Black-fingered crab

Xantho incisus

Group: Decapoda: Xanthidae

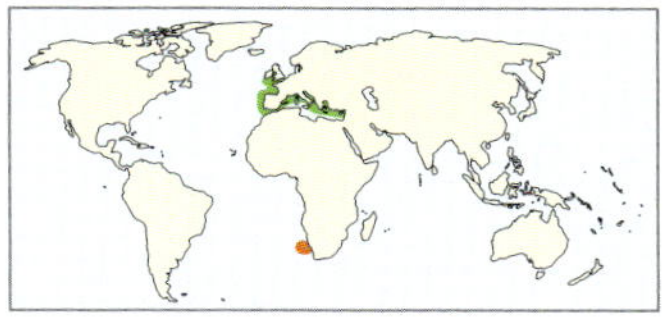

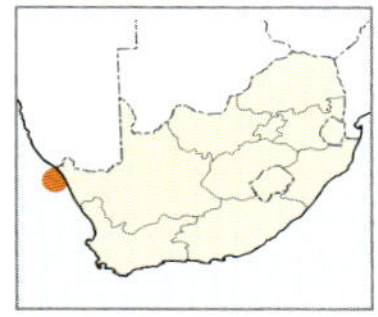

Habitat: Marine, rocky shores and shallow reefs. **Identification and biology:** A heavily built crab reaching about 40 mm across the carapace. Easily identified from the few other coastal crabs on the west coast by the smooth surface texture, lack of teeth on sides of carapace and huge, black-tipped claws. Brown when alive (photo shows dried specimen). **Introduction and spread:** Known only from the oyster farm at Kleinsee on the west coast, to where it was almost certainly introduced with oyster spat imported from Europe. **Impacts:** Not significant at present, given its restricted range, but a powerful predator that could be problematic if it spreads into the open sea. **Control:** Could be eliminated by trapping while the population remains confined to the oyster dam.

C. Griffiths

Key reference: Haupt, T.M., Griffiths, C.L., Robinson, T.B. & Tonin, A.F.G. 2010. Oysters as vectors of marine aliens, with notes on four introduced species associated with oyster farming in South Africa. *African Zoology*, 45:52-62.

Pea crab

Pinnixa occidentalis

Group: Decapoda: Pinnotheridae

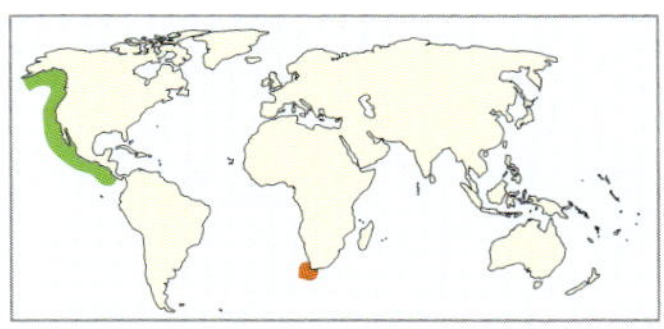

Habitat: Marine, in shallow sand and mud, probably associated with burrows of Sand prawns (*Callianassa* species). **Identification and biology:** Small crabs with a smooth outline and broadly rectangular body only about 20 mm wide. Eye-stalks very short. Claws of male greatly enlarged with strongly curved finger and held above head. In native range often found in burrows of echiurans (spoon worms), but in South Africa may occupy burrows of Sand prawns, which are abundant burrow-forming species in invaded sites. **Introduction and spread:** A Pacific North American crab. Locally known only from Saldanha Bay, where it was first collected in 2004. Has increased greatly in abundance since then and is now spreading into Langebaan Lagoon. **Impacts:** Only recently detected and no studies have been done on its impact. **Control:** None proposed to date.

C. Griffiths

Key reference: Zmarzly, D.L. 1992. Taxonomic review of pea crabs of the genus *Pinnixa* (Decapoda: Brachyura: Pinnotheridae) occurring on the Californian shelf, with descriptions of two new species. *Journal of Crustacean Biology*, 12:677-713.

ROUNDWORMS Phylum Nematoda

Nematodes, also known as roundworms or eelworms, are thin, rather featureless worms with a tough external cuticle. Free-living forms abound in marine, freshwater and terrestrial habitats, while numerous other species parasitise plants and animals, some being important agricultural pests. Easily transported with hosts or soil and many probably introduced, but as identification requires specialist knowledge, only two are detailed here.

Potato cyst nematode

Globodera rostochiensis

Group: Heteroderidae

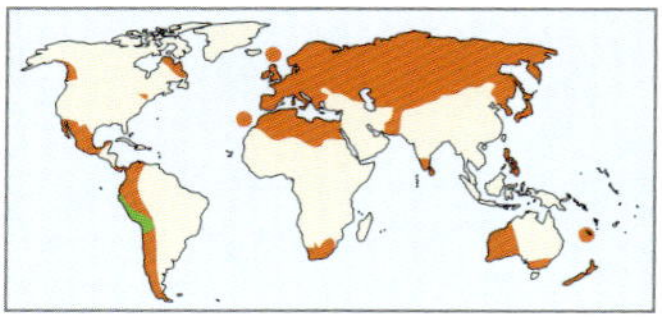

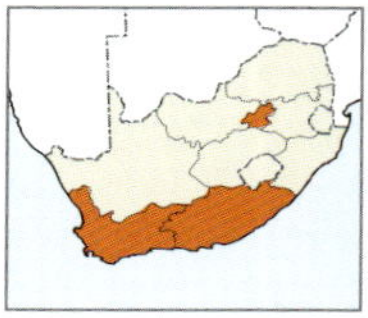

Habitat: Roots of potato, brinjal and tomato plants. **Identification and biology:** Adult females about 0.5 mm long and become swollen and form hard, round yellow cysts on roots of affected plants; eggs are protected in cysts. Males 1 mm long and remain worm-like and mobile. Second-stage larvae constitute the mobile infection stage; later development occurs within roots of host. At maturity, body lies mostly outside the root, only head-end remaining embedded. **Introduction and spread:** Native to Bolivia. Introduced to Europe in 19th century, from where it spread worldwide. **Impacts:** Most important nematode parasite of potatoes worldwide. Nematode feeding blocks transport of water and nutrients; infected plants become small, with sparse, yellow foliage and low yields. **Control:** Treated with soil nematicides, but resistant to control, as eggs protected in tough cysts. Crop rotation, or planting resistant cultivars, greatly reduces infection rates.

Key reference: Keetch, D.P. & Heyns, J. (eds). 1982. Nematology in southern Africa. *Science Bulletin 400*. Department of Agriculture and Fisheries, Republic of South Africa.

Ulrich Zunke, Mactode Publications

Woodwasp nematode

Beddingia siricidicola

Group: Beddingiidae

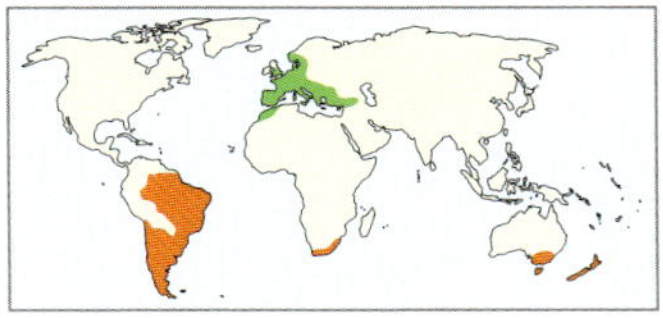

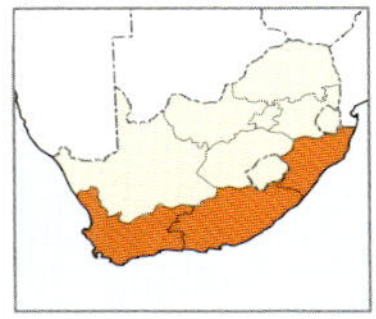

Habitat: Spend part of life cycle within pine trees, feeding on fungus, then ingested by woodwasp larvae, finally sterilising adult female wasps. **Identification and biology:** Length 1–2 mm. Best identified by their extraordinary life cycle. Adult nematodes feed on the fungus (*Amylosterum areolatum*), which is itself introduced and spread by the Sirex woodwasp (p. 141). Nematodes are subsequently ingested by feeding wasp larvae and reproduce within the wasp pupae, finally entering and killing the eggs of developing females. When nematode-infected female wasps emerge and lay eggs in new trees, they are thus distributing nematodes, rather than their own eggs. **Introduction and spread:** Native to Europe. Deliberately introduced as a biocontrol agent from 2004. **Impacts:** Most important control agent for Sirex woodwasps, which are serious pests of pine trees. **Control:** Beneficial, therefore no control desirable.

Key reference: Hurley, B.P., Slippers, B. & Wingfield, M.J. 2007. A comparison of control results for the alien invasive woodwasp, *Sirex noctilio*, in the southern hemisphere. *Agricultural and Forest Entomology*, 9:159-171.

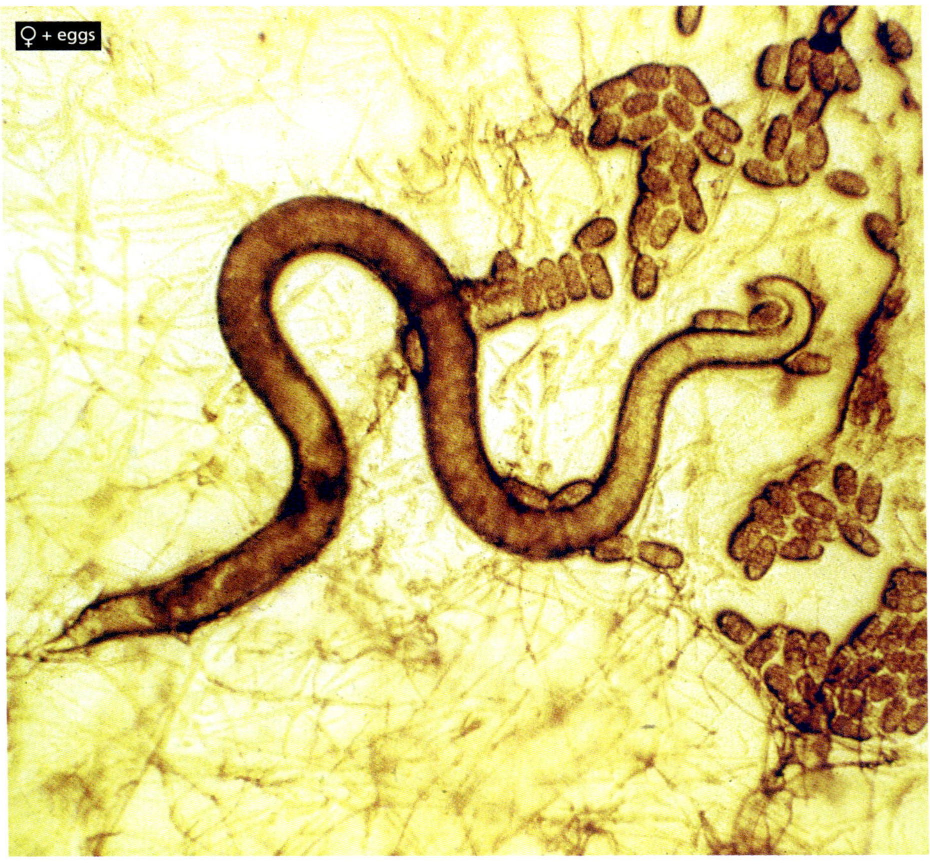

R. Bedding

AUTHORS: D. PLISKO & C. GRIFFITHS

SEGMENTED WORMS Phylum Annelida

Segmented worms comprise three distinct groups. **Leeches** (class Hirudinea) are predatory or blood-sucking forms with 34 body segments and a prominent sucker at either end, but none are known to have been introduced to South Africa. **Oligochaetes** (class Oligochaeta) are smooth-bodied with short setae and comprise two subgroups: (1) The megadrile oligochaetes or 'earthworms' are terrestrial, preferring moist habitats and burrowing in soil or mud. They are hermaphrodite and usually reproduce sexually, although parthenogenetic reproduction and self-fertilisation are common. Fertilised eggs are laid in protective cocoons secreted by a swollen glandular body section. Fifty species have been introduced to the region. (2) The microdrile oligochaetes include terrestrial, freshwater and semi-aquatic species, but little is known about invasive species in South Africa and they are not included here. **Polychaetes** or bristle worms (class Polychaeta) are a diverse group of mostly marine species that include both mobile predators and tube-dwelling scavengers and filter-feeders. Most have distinctive and often elaborate head appendages and 'hairy' segmental appendages. Fertilisation is external and eggs develop into tiny swimming larvae. Only some of the numerous introduced species are detailed below.

Composting earthworm

Eisenia fetida

Group: Oligochaeta: Lumbricidae

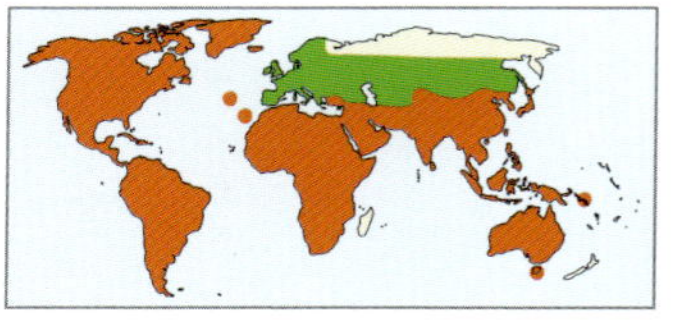

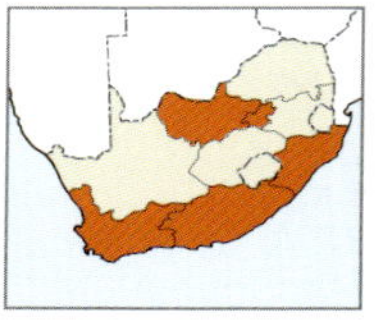

Habitat: Terrestrial, composting.
Identification and biology: Body length 35–130 mm, colour variable, brownish to red, sometimes with alternating bands of red-brown on segments, yellow band between segments; eight setae per segment, closely paired, arranged in four regular rows; male pores with large glandular swelling on segment 15; clitellum saddle-shaped, from segments 24 or 25 to 32. Lives in compost heaps, manure, decaying vegetation and soil high in organic matter and moisture. Cannot survive in arable fields, in dry natural habitats or soil low in organic matter, and when introduced to garden soil usually dies after utilising available organic matter. Commonly used in 'worm farms'. Reproduces sexually, although parthenogenetic reproduction and self-fertilisation common. **Introduction and spread:** Origin unknown, supposedly in Caucasus (Eurasia). Now cosmopolitan, following spread by humans for worm farming. **Impacts:** Known in South Africa for centuries, common around human habitation and propagated by worm farming. Used worldwide in various laboratories for experimental study. **Control:** No control necessary, but as worm cultures may contain many unknown soil organisms, the distribution of these worms and the compost in which they live should be controlled.

C. Griffiths

Key reference: Plisko, J.D. 2010. Megadrile earthworm taxa introduced to South African soils (Oligochaeta: Acanthodrilidae, Eudrilidae, Glossoscolecidae, Lumbricidae, Megscolecidae, Ocnerodrilidae). *African Invertebrates*, 51:289-312.

Pink soil earthworm

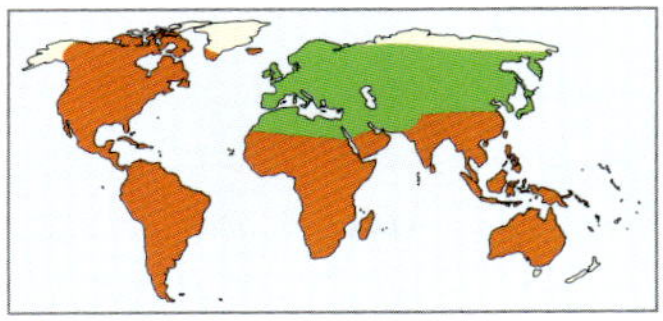

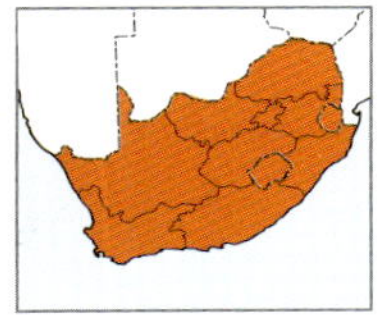

Aporrectodea rosea

Group: Oligochaeta: Lumbricidae

Habitat: Terrestrial, a burrow-forming, soil-eater. **Identification and biology:** Body length 20–120 mm, colour pale with reddish tint. Eight setae per segment, closely paired, arranged in four regular rows; male pores on segment 15, with glandular crescents protruding into neighbouring segments. Clitellum saddle-shaped, from segments 24, 25 or 26 to 32 or 33. Reproduction sexual, although frequent parthenogenesis creates different morphs, complicating identification. Casts discarded in the soil and on the surface. **Introduction and spread:** Native to temperate Europe and Asia, now cosmopolitan due to human introduction. **Impacts:** Occurs in pastures, cultivated fields, plantations and forests, introduced with imported plants and soil samples. Prefers moist habitats; in dry soil under high temperature often undergoes seasonal aestivation, coiling up and remaining in a small cavity moistened with excreted mucous. **Control:** Widely distributed and not especially invasive, the distribution of these worms and the compost in which they live should be controlled.

B. Muller

Key reference: Plisko, J.D. 2010. Megadrile earthworm taxa introduced to South African soils (Oligochaeta: Acanthodrilidae, Eudrilidae, Glossoscolecidae, Lumbricidae, Megscolecidae, Ocnerodrilidae). *African Invertebrates*, 51:289-312.

Bark-eating earthworm

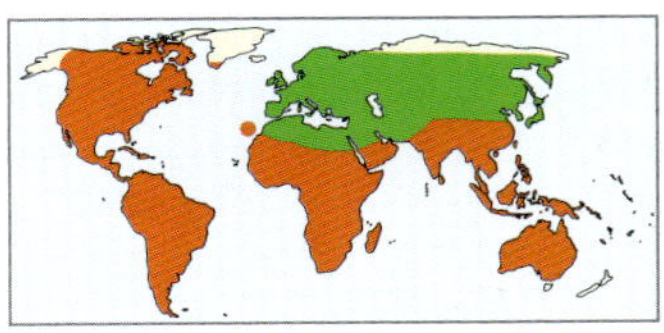

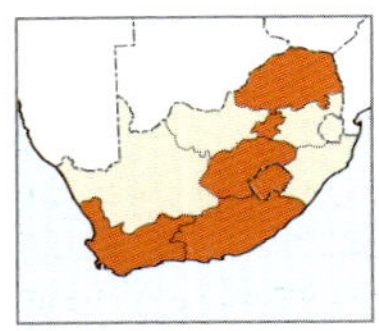

Dendrodrilus rubidus

Group: Oligochaeta: Lumbricidae

Habitat: Terrestrial, burrow-forming, bark and litter-eater. **Identification and biology:** Small worm, body length 30–60 mm, colour red-violet, darker dorsally; eight setae per segment, moderately paired, ventrally wider, laterally closer; male pores on segment 15, glandular, extended onto two neighbouring segments. Clitellum saddle-shaped, on segments 26 or 27 to 31 or 32. Reproduces sexually, but parthenogenetic forms common, often identified as subspecies. **Introduction and spread:** Native to Europe and northern Asia, but introduced widely by humans, becoming cosmopolitan. Widely distributed in South Africa in protected areas, forests, bushveld, grasslands, riverbanks and cultivated land, from sea level to high altitudes. Occurs in large populations, being active nearly year-round in our native environment in litter, fallen logs and between plant roots. **Impacts:** A great eater of organic matter and might compete for food usually utilised by native invertebrates. **Control:** The distribution of these worms and the soil in which they occur should be controlled.

Key reference: Plisko, J.D. 2010. Megadrile earthworm taxa introduced to South African soils (Oligochaeta: Acanthodrilidae, Eudrilidae, Glossoscolecidae, Lumbricidae, Megscolecidae, Ocnerodrilidae). *African Invertebrates*, 51:289-312.

B. Muller

Quincunx earthworm

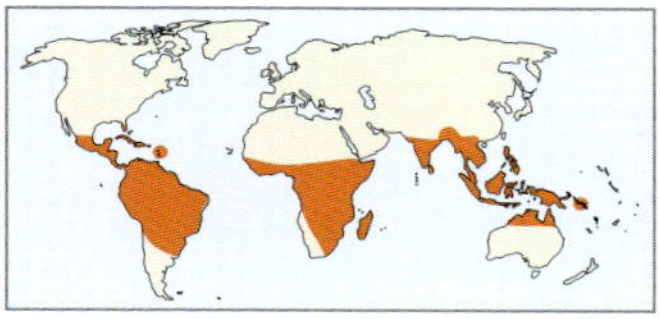

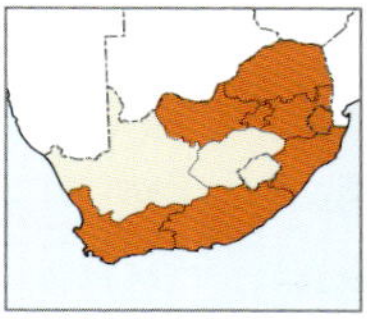

Pontoscolex corethrurus

Group: Oligochaeta: Glosscoscolecidae

Habitat: Terrestrial, burrow-forming, composting. **Identification and biology:** Moderate size, length 45–120 mm, colour greyish-white, eight setae per segment, evenly paired anteriorly, irregular in posterior part. Clitellum saddle-shaped, on segments 15–23. Always parthenogenetic. **Introduction and spread:** Exact origin unknown, possibly Amazonian Brazil. Distributed by humans worldwide. Abundant in tropical and subtropical habitats, in greenhouses and worm composting farms. Recently used to enhance soil fertility in experimental fields where climatic conditions permit. **Impacts:** Common in north-eastern South Africa, always in large populations. May be used increasingly in worm composting production and shows high potential for agricultural application in the improvement of arable land, and thus crop improvement. Intentionally transported to many African countries. **Control:** Distribution of these worms and the compost in which they live should be controlled.

C. Griffiths

Key reference: Plisko, J.D. 2001. Notes on the occurrence of the introduced *Ponthoscolex corethrurus* (Müller, 1857) in South Africa (Oligochaeta: Glossoscolecidae). *African Invertebrates*, 42:323-334.

Snake earthworm

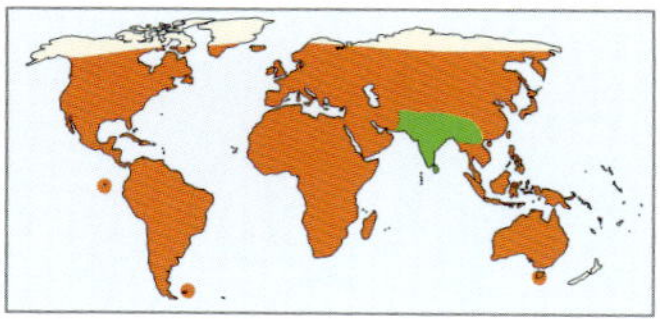

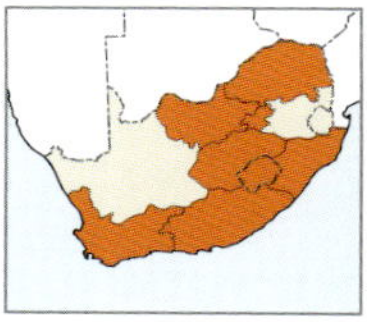

Amynthas diffringens (species-group)

Group: Oligochaeta: Megascolecidae

Habitat: Terrestrial. **Identification and biology:** Variable size, body length 50–170 mm; clearly segmented; reddish-purple, violet to brownish dorsally, lighter ventrally; setae numerous, in a circle on each segment, variable number on segments and on individuals; usually around 46. Male pores on segment 18, each pore a circular to transversely elliptical disc. Clitellum pale, on segments 14–16. Sexual and parthenogenetic reproduction both common. Thrashes when exposed. **Introduction and spread:** Origin probably China; distributed worldwide by humans. Common, restricted only by climatic conditions. In cooler zones frequent in greenhouses and worm composting farms. **Impacts:** Forms large populations in sugar plantations and natural biotopes with suitable moist and organic matter. Also compost heaps and farms, although often mistaken for other earthworms. **Control:** Distribution of these worms and the compost in which they live should be controlled.

C. Griffiths

Key reference: Sims, R.W. & Easton, E.G. 1972. A numerical revision of the earthworm genus *Pheretima* auct. (Megascolecidae: Oligochaeta) with the recognition of new genera and an appendix on the earthworms collected by the Royal Society North Borneo Expedition. *Biological Journal of the Linnean Society*, 4:169-268.

Estuarine tube-worm

Ficopomatus enigmaticus

Group: Polychaeta: Serpulidae

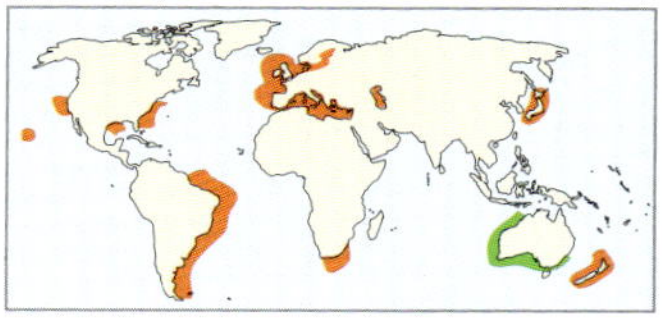

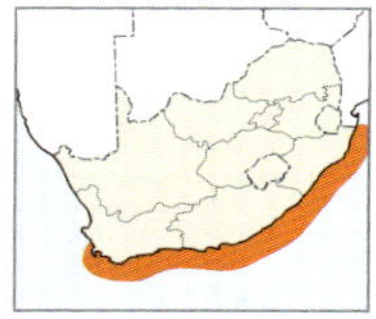

Habitat: Low-salinity sections of estuaries, attached to rocks, jetties, boats and any other hard surface. **Identification and biology:** Forms extensive (50 cm diameter), rounded, reef-like colonies of narrow, intertwined, white, calcareous tubes. The opening of each tube is flared like a trumpet, other rims down the length of each tube mark the positions of earlier openings. When feeding, the tiny worms (25 mm long) extend a crown of mottled brown, feather-like tentacles from the aperture of the tube and use these to filter fine food particles from the water column. When disturbed they retract rapidly and the openings to the tubes are sealed by a cone-like operculum. **Introduction and spread:** Accidentally introduced to South Africa prior to 1951, either from their native Australia or secondarily via Europe, probably with dry ballast or shipping, and now found in estuaries from Cape Town eastwards. Easily transported between sites on the hulls of boats, or with buoys, anchors and other objects. **Impacts:** Forms large colonies that foul the hulls of boats, constrict pipes and channels, coat ropes and buoys, etc. The calcareous tubes are sharp and can injure and inconvenience paddlers and swimmers. Dense populations filter formidable volumes of water, but the ecological effects of this, which include clarifying the water, converting suspended material into bottom mud and competing with indigenous filter feeders, have not been studied adequately. Reefs also support a host of smaller species, including other aliens. **Control:** It is fairly easy to remove large colonies mechanically, but as the species has mobile larvae and a fast growth rate, keeping structures clear is likely to require continuous and expensive effort.

Key reference: Davies B.R., Stuart, V. & de Villiers, M. 1989. The filtration activity of a serpulid polychaete population *Ficopomatus enigmaticus* (Fauvel) and its effects on water quality in a coastal marina. *Estuarine, Coastal and Shelf Science*, 29:613-620.

C. Griffiths

C. Griffiths

Spiral fanworm

Neodexiospira brasiliensis

Group: Polychaeta: Serpulidae

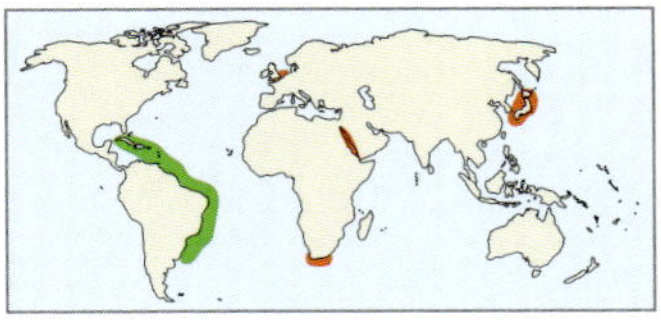

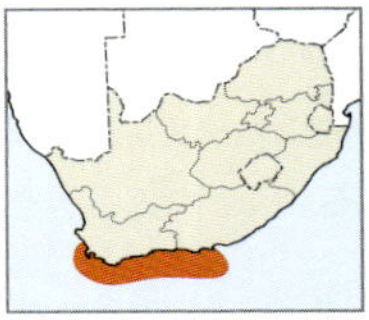

Habitat: Marine, rocky shores, usually on seaweeds. **Identification and biology:** Tiny red worms enclosed in coiled white, calcareous shell 2–3 mm in diameter. Shell coils anti-clockwise (another introduced polychaete worm, *Janua pagenstecheri*, also has an anti-clockwise shell, but with three longitudinal ridges running around the spiral; several related indigenous species coil clockwise). Common on seaweeds in the intertidal zone and on shallow structures. When feeding, worms extend red, feathery tentacles from the opening to filter-feed, giving the colony a red blush. When disturbed, tentacles withdraw rapidly and shell openings are sealed with a stalked cap or operculum. **Introduction and spread:** Probably introduced as a fouling species with shipping. Native range eastern coastline of South America. **Impacts:** A common fouling species, but impact minimal due to its very small size. **Control:** None known or proposed, but easily scraped off fouled surfaces.

C. Griffiths

Key reference: Day, J.H. 1967. *A Monograph on the Polychaeta of South Africa. Part 2. Sedentaria.* Trustees of the British Museum (Natural History). London, pp. 459-878.

Black coral worm

Dodecaceria fewkesi

Group: Polychaeta: Cirratulidae

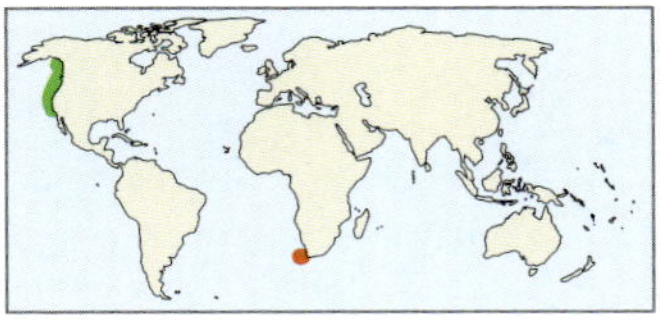

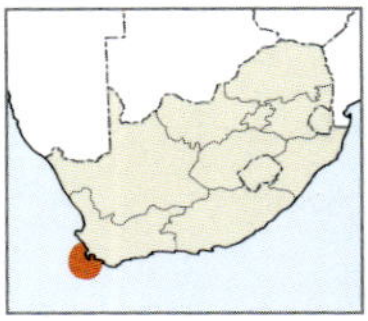

Habitat: Marine, on hard substrata like harbour wharfs. **Identification and biology:** Colonies form hard, grey, hemispherical reefs about 100 mm across comprised of hard, calcareous tubes on harbour walls, wharfs and rocks. Each colony houses hundreds of 10–20 mm long jet-black worms, each with a long pair of feeding tentacles, behind which lie 4–5 pairs of elongated tubular gills. Colonies are derived asexually from single individuals, hence retaining the sex of the founder. **Introduction and spread:** Probably accidentally introduced from Pacific USA by shipping, first recorded in Table Bay Harbour in 2007. This remains the only known location, although few other sites have been searched for this species. **Impacts:** Forms hard coral-like colonies that are difficult to remove, but to date no serious impacts noted. **Control:** Can be removed mechanically, but no control programme initiated or proposed.

C. Griffiths

Key reference: Blake J.A. 1996. Family Cirratulidae. In: Blake, J.A., Hilbig, B. & Scott, P.H. (eds). *Taxonomic atlas of the Santa Maria Basin and Western Santa Barbara Channel*, Vol 6. Annelida, Part 3. Santa Barbara Museum of Natural History, pp. 263-384.

Mudworm

Polydora hoplura

Group: Polychaeta: Spionidae

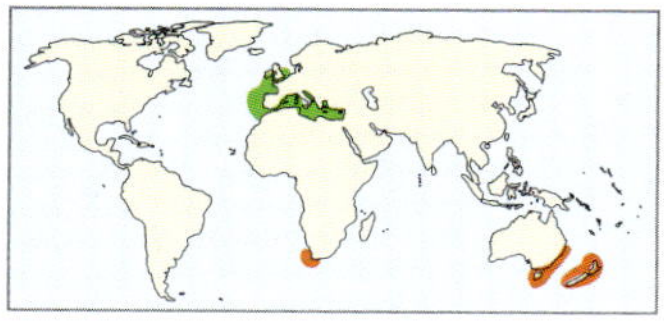

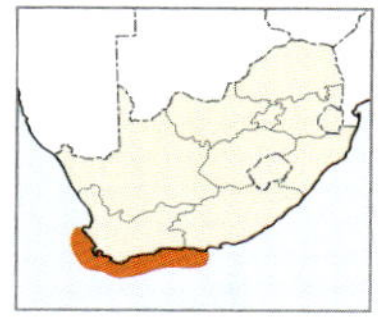

Habitat: Marine, build muddy tubes and burrow into shells of molluscs, particularly oysters, forming distinctive mud-filled blisters. Also infect farmed abalone. **Identification and biology:** Length about 40 mm. Head keeled, bearing single pair of long, grooved palps that are waved from end of tube, catching small particles of detritus. Fifth body segment swollen and bearing lateral spines to grip tube. Gills only present posterior to segment five. Last segment expanded into a distinctively saucer-shaped disc. **Introduction and spread:** One of several introduced species in this genus. Probably originally imported from Europe as fouling organism, also transported with live shellfish. **Impacts:** Well-known pest of cultured shellfish, reducing condition and increasing mortality of infected individuals, also reduces value of infected oysters, therefore incurring significant economic costs. **Control:** Infected oysters can be treated by placing them in fresh water for 12 hours, or in heated water (70°C) for 40 seconds. Abalone are more sensitive, but can be treated by air-drying, or chemically.

Key reference: Simon, C.A., Ludford, A. & Wynne, S. 2006. Spionid polychaetes infesting cultured abalone *Haliotis midae* in South Africa. *African Journal of Marine Science*, 28:167-171.

C. Griffiths

Shell worm

Boccardia proboscidea

Group: Polychaeta: Spionidae

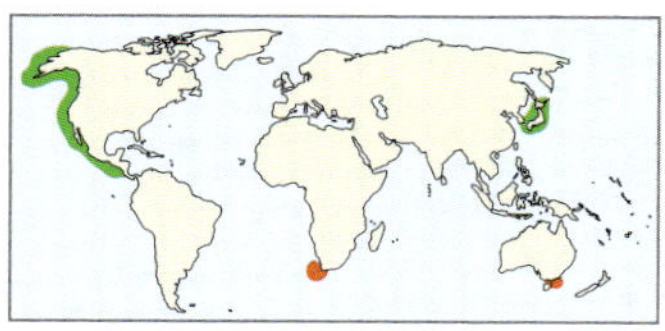

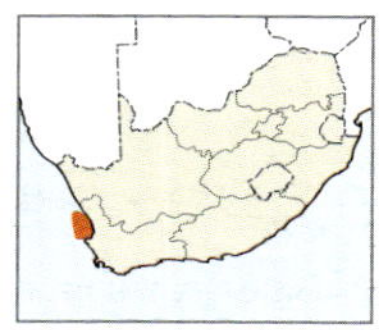

Habitat: Tube-dwelling worms found in shallow, sand-lined burrows secreted in grooves on the surfaces of oysters, abalone and other shellfish. **Identification and biology:** Length about 20 mm. Body yellow when alive, red blood vessels conspicuous, head with pair of elongated feeding tentacles and black eyes, gills on segments 2–4 then segment six onwards; tail rounded disc with dorsal notch. **Introduction and spread:** Native to the Pacific coast of North America and Japan; in South Africa previously known only from abalone and oyster farms, to which it was apparently introduced with Californian abalone in the 1980s. Recently recorded in Saldanha Bay outside of aquaculture facilities. **Impacts:** Reduces the condition of cultured abalone, increasing production costs. **Control:** Can be treated by soaking infected shellfish in fresh or heated water.

Key reference: Simon, C.A., Thornhill, D.J., Oyarzun, F. & Halanych, K.M. 2009. Genetic similarity between *Boccardia proboscidea* from Western North America and cultured abalone, *Haliotis midae*, in South Africa. *Aquaculture*, 294:18-24.

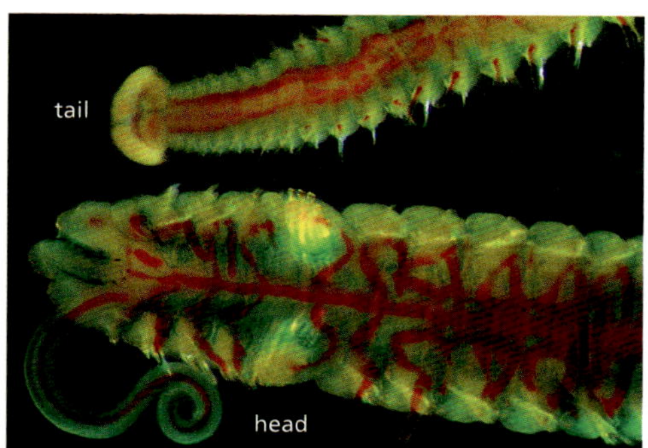

C. Simon

Pileworm

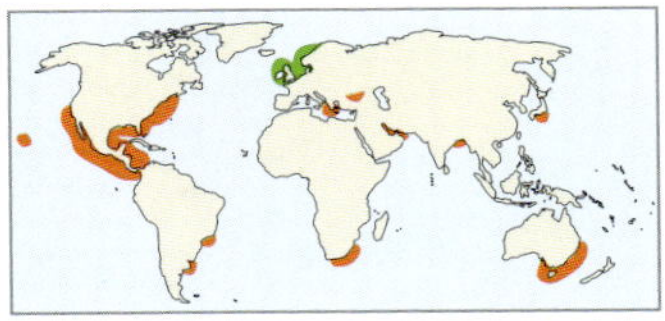

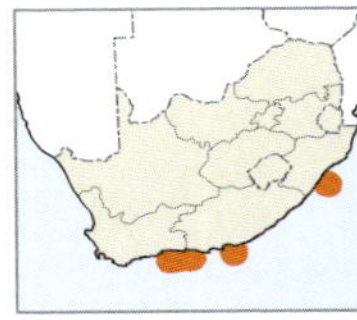

Neanthes succinea

Group: Polychaeta: Nereidae

Habitat: In sand and mud and among mussels and other fouling organisms, particularly common in muddy estuaries. **Identification and biology:** Mobile, reddish-brown worm up to 80 mm long. Head with four large eyes, four pairs of short tentacles, one pair of short antennae and one long pair of feeding palps. Eversible proboscis bears a pair of strong, black jaws and numerous small tooth-like dots (paragnaths). Posterior limbs have greatly enlarged upper lobes that distinguish this from other similar nereid worms. Opportunistic species that can feed on detritus, graze on plant material or consume small invertebrates and that can tolerate a wide range of salinities and temperatures. **Introduction and spread:** Introduced from Europe by shipping. **Impacts:** Can be common and therefore an important food for predatory fish and birds. **Control**: No known control method attempted.

C. Griffiths

Key reference: Day, J.H. 1967. *A Monograph on the Polychaeta of South Africa. Part 1. Errantia.* Trustees of the British Museum (Natural History). London, pp. 1-458.

Pink spindle worm

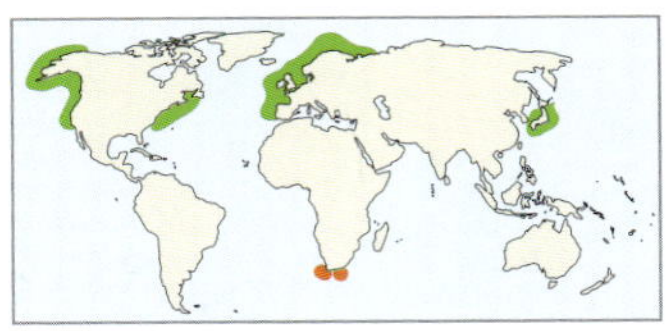

Travisia forbesii

Group: Polychaeta: Opheliidae

Habitat: Among other organisms that attach to buoys, jetties and similar substrates. **Identification and biology:** Length up to 30 mm. Unusual fat, pink and spindle-shaped, almost maggot-like, body with few appendages. Head conical, gills cylindrical and tapering. Small, conical swellings above and below the posterior limbs. Live head-downwards in mud and sand, feeding on organic material within the sediment. **Introduction and spread:** Native to North Atlantic, Arctic and also Japan. In South Africa known only from harbours in Table Bay and Port Elizabeth, to which it was probably introduced with ballast water, or as a fouling organism. **Impacts:** Given its feeding habits unlikely to have substantial impacts, other than as an additional food resource for predators. **Control:** No known method and not deemed necessary at this stage.

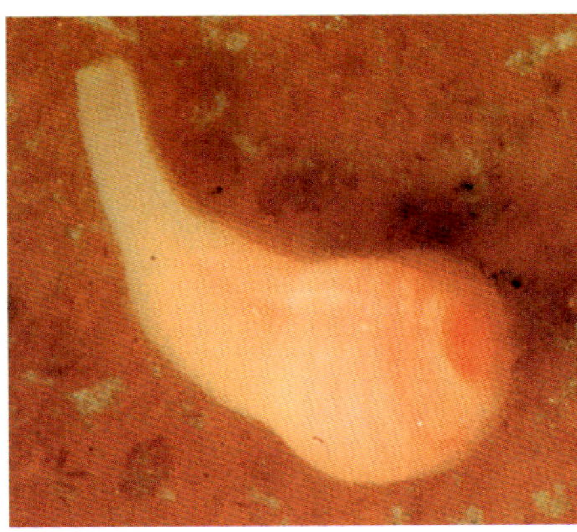

C. Griffiths

Key reference: Day, J.H. 1967. *A Monograph on the Polychaeta of South Africa. Part 2. Sedentaria.* Trustees of the British Museum (Natural History). London, pp. 459-878.

AUTHORS: D. HERBERT, C. GRIFFITHS & C. APPLETON

MOLLUSCS

Phylum Mollusca

Molluscs are one of the largest invertebrate groups, with over 5 000 species in South Africa. They are found in terrestrial, freshwater and marine habitats and have unsegmented bodies, divided into a head, a visceral mass that contains the body organs, and a muscular foot. Most secrete calcium carbonate shells, but some have secondarily lost the shell. The major groups are the **bivalves** (mussels, clams and oysters) that have paired shells and enlarged sieve-like gills to filter food particles from the water; the **gastropods**, a large and diverse group including snails and shell-less slugs; and the **cephalopods** (octopi and squids).

Mediterranean mussel

Mytilus galloprovincialis

Group: Bivalvia: Mytilidae

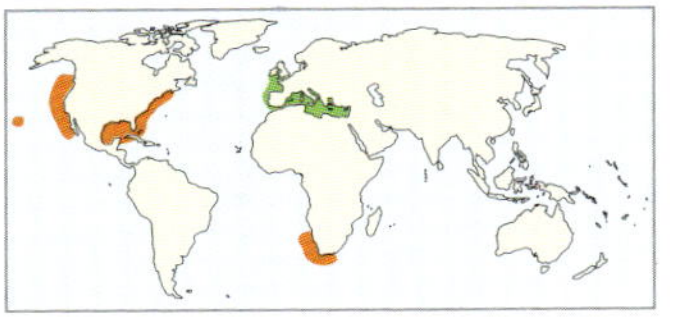

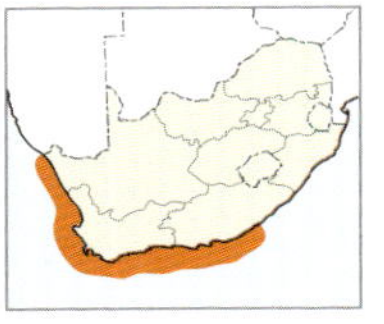

Habitat: Marine, on rocky shores and other hard surfaces. **Identification and biology:** Large (typically 50 mm but up to 120 mm long), smooth-shelled black to blue mussels forming dense beds. Easily confused with indigenous Black mussel (*Choromytilus meridionalis*), but fatter in cross-section, usually have eroded shells and occur higher on the shore, away from sand. Attach firmly to rocks using strong byssal threads secreted by a mobile foot. Feed by pumping water through enlarged sieve-like gills. Sexes separate and gonads extend throughout body, which is cream in males and pink in females. Prolific broadcast spawners, releasing millions of eggs and sperm at each spawning. Fertilised eggs develop into free-swimming larvae that disperse widely. Grow rapidly, reaching 70 mm length in first year at favourable sites. **Introduction and spread:** Accidentally introduced from Europe to south-western Cape in about 1979, almost certainly by shipping. Quickly spread northwards into Namibia and more slowly to Eastern Cape. Now abundant along entire west coast, while on south coast form mixed beds with indigenous Brown mussel (*Perna perna*). **Impacts:** Form dense, often multilayered beds that monopolise space on intertidal rocks. Displace indigenous mussels and large limpets, but smaller species flourish within the beds. Thus invaded shores have increased biomass and more species. Eagerly consumed by predators, including whelks, starfish and birds, notably African black oystercatchers (*Haematopus moquini*), which have benefited from enhanced food availability. Popular seafood and form basis of significant aquaculture industry. **Control:** Probably impossible, given wide range, prolific reproductive habits and widely dispersing larvae.

Key reference: Robinson, T.A., Griffiths, C.L., McQuaid, C.D. & Rius, M. 2005. Marine alien species of South Africa – status and impacts. *African Journal of Marine Science*, 27:297-306.

C. Griffiths

Bisexual mussel

Semimytilus algosus

Group: Bivalia: Mytilidae

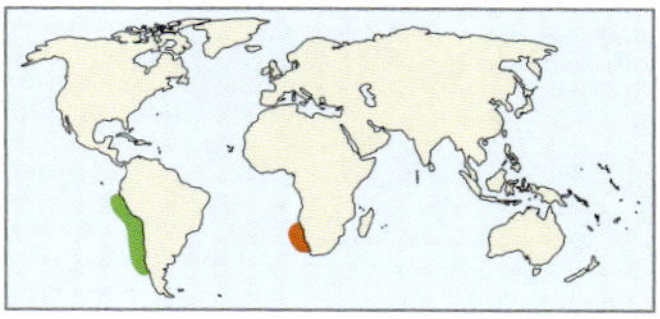

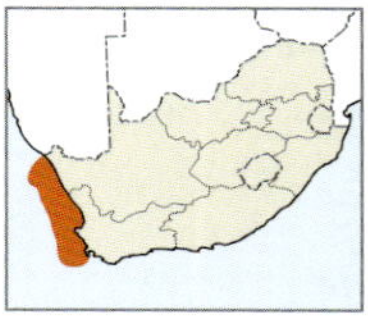

Habitat: Marine, on rocky shores. **Identification and biology:** Small, elongated, relatively fat and smooth brown mussel, shells tinged with green. Grows to only 50 mm length and abundant intertidally on rocky shores. Most specimens are hermaphroditic, having a male gonad on one side and female one on the other. **Introduction and spread:** Long known from Namibia where it was presumed to have been introduced from Chile by shipping. First recorded from South Africa only in 2010; found along west coast in mixed beds with another alien mussel, the Mediterranean mussel (p. 181). **Impacts:** Beds provide habitat for a variety of other smaller species, while the mussels themselves enhance food availability for predators. **Control:** None attempted or proposed and probably impossible.

Key reference: Branch, G.M., Griffiths, C.L., Branch, M.L. & Beckley, L.E. 2010. *Two oceans: A guide to the marine life of Southern Africa*. Cape Town: Struik.

C. Griffiths

Japanese oyster

Crassostrea gigas

Group: Bivalvia: Ostreidae

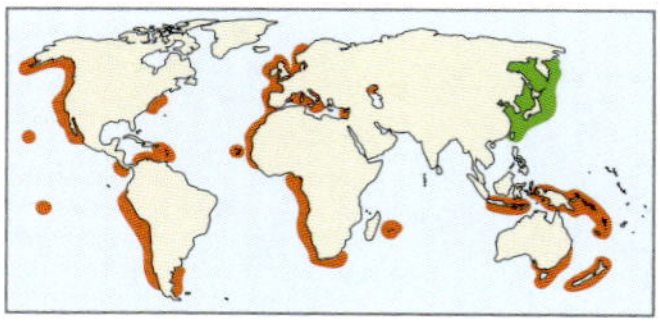

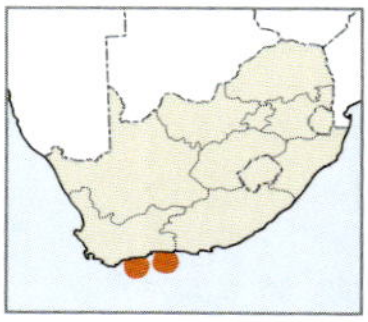

Habitat: Marine, attached to rocks in estuaries. **Identification and biology:** Typically elongated, about twice as long as broad; adult length 100 mm. Two shells unequal; lower shell cup-shaped, permanently attached to rock; upper shell flat with thin, concentric, wavy external flanges. Colour white, with dark blotches, interior matt-white. An important seafood species widely spread by aquaculture. Filter fine particles from water using sieve-like gills. Broadcast spawner with planktonic larvae. **Introduction and spread:** Native to Japan, introduced to South Africa via Europe or Chile for aquaculture purposes, but until recently not thought to present a threat as an invasive species. First wild populations reported in 2005 from three estuaries in southern Cape. **Impacts:** May compete with indigenous shellfish for space and food. **Control:** Can be controlled manually by harvesting adults. Currently not considered problematic, as readily controlled by collectors.

Key reference: Robinson, T.B., Griffiths, C.L., Tonin, A., Bloomer, P. & Hare, M.P. 2005. Naturalised populations of oysters *Crassostrea gigas* along the South African coast: distribution, abundance and population structure. *Journal of Shellfish Research*, 24:443-450.

C. Griffiths

European oyster

Ostrea edulis

Group: Bivalvia: Ostreidae

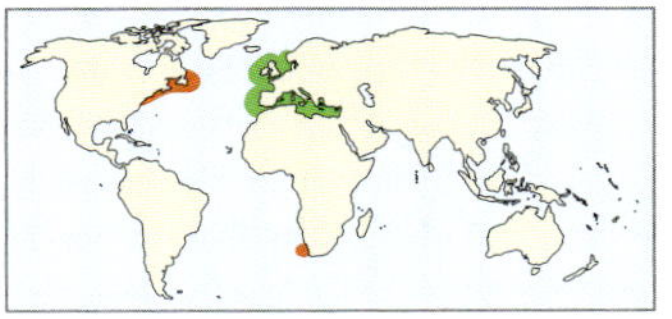

Habitat: Marine, attached to rocks in estuaries. **Identification and biology:** Length typically 80 mm. Lower valve cup-shaped and upper one flat, usually pinkish in colour, best distinguished from the Japanese oyster (p. 182) by its round shell. **Introduction and spread:** Deliberately introduced to Knysna from Europe for aquaculture purposes from 1946, but until recently was thought to have become locally extinct. A wild, breeding population was recently discovered living in oyster dams at Alexander Bay on the west coast. Relict population probably also present in other historical oyster-culture sites, such as Knysna. **Impacts:** Currently not considered to be problematic. **Control:** No control considered necessary at this stage.

C. Griffiths

Key reference: Haupt, T.M., Griffiths, C.L., Robinson, T.B. & Tonin, A.F.G. 2010. Oysters as vectors of marine aliens, with notes on four introduced species associated with oyster farming in South Africa. *African Zoology*, 45:52-62.

Shipworm

Teredo navalis

Group: Bivalvia: Teredinidae

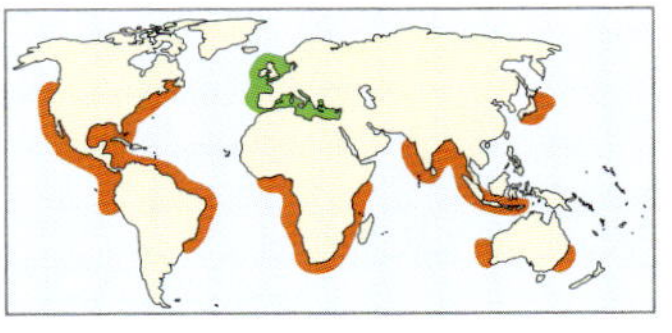

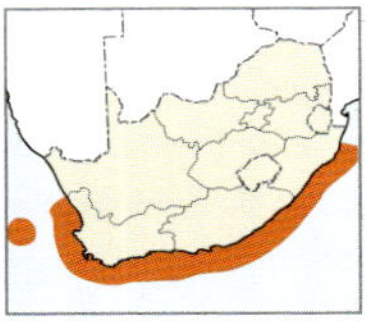

Habitat: Bores into submerged wood, including wharfs, pier-pilings and wooden ships. **Identification and biology:** Despite common name not a worm, but highly specialised, elongated, worm-like bivalve mollusc that has adopted boring habits. Length up to 200 mm. Paired shells no longer enclose body but are rounded, ridged cutting tools covering only anterior end, rocked back and forth to bore into wood and extend burrow, which is lined with calcareous material. Short, paired posterior siphons extend from open end of burrow, which is closed by additional shell-like structures termed pallets – the shape of which constitute an important identification feature. Feeds by filtering food particles from water and digesting wood shavings scraped from burrow. **Introduction and spread:** Early European introduction with wooden ships, widespread, already causing major damage to South African jetties and breakwaters by 1800s. **Impacts:** Very important pest, responsible for many shipwrecks and destruction of wooden structures. **Control:** Numerous methods proposed, including painting wood with creosote, periodic removal of structures from the sea and sun-baking, charring of wood, poisoning, etc.

C. Griffiths

Key reference: Kilburn, R. & Rippey, E. 1982. *Sea shells of Southern Africa*. Johannesburg: Macmillan South Africa.

Lagoon snail

Littorina saxatilis

Group: Gastropoda: Littorinidae

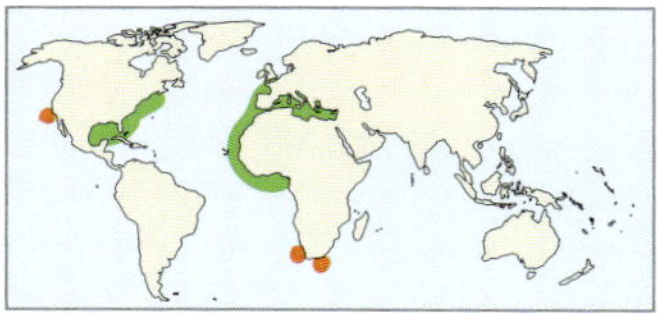

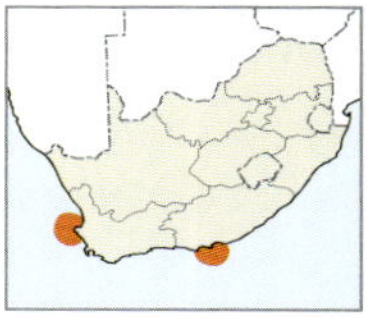

Habitat: Upper intertidal zone on estuarine and lagoonal shores. **Identification and biology:** Small, dark, conical snail, typically 5 mm long and either smooth, or with faint spiral sculpture. Operculum rounded and transparent. Air-breathes and found crawling on stems of Cord grass (*Spartina maritima*). Feeds on microscopic encrusting algae. Sexes separate and fertilisation internal, developing embryos retained in brood pouch of female and hatch into crawling young. **Introduction and spread:** Probably accidentally introduced from North Atlantic with dry ballast and recorded at just two sites in South Africa – Langebaan Lagoon and Knysna Estuary. **Impacts:** Prefers sheltered sites, so not expected to spread onto open coast. Unlikely to have significant ecological impacts due to its small size and specialised habitat. **Control:** Can be gathered manually, but no control measures proposed or in use.

Key reference: Hughes, R.N. 1979. South African populations of *Littorina rudis. Zoological Journal of the Linnean Society*, 65:119-126.

C. Griffiths

Blanford's whelk

Thais blanfordi

Group: Gastropoda: Muricidae

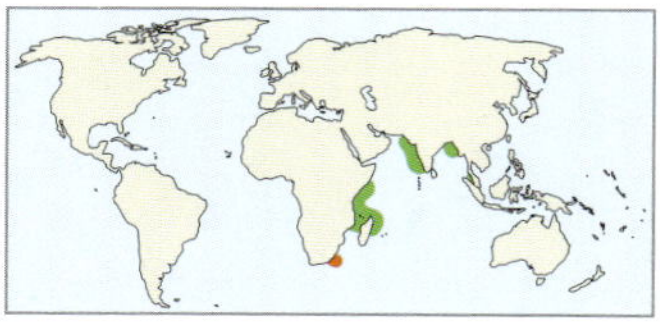

Habitat: Found on rocky shores in Durban Bay. **Identification and biology:** Shell ridged, typically 30 mm long, with 8–10 distinct knobs on each body whorl and pale pink interior. **Introduction and spread:** An Indo-Pacific species, but appears to have been collected from South Africa only after 1950, which, given its known association with shipping, suggests it is an introduction to the region. Also introduced to Durban Bay is the related Tissot's rock shell (*Thais tissoti*), which has a more strongly ridged shell with a brown interior and a similar history, originating in the western Indian Ocean but only recorded in South Africa since 1950, suggesting that its southward movement has been facilitated by shipping. **Impacts:** None known to date. **Control:** None attempted, but could be hand picked.

Key reference: Kilburn, R. & Rippey, E. 1982. *Sea shells of Southern Africa*. Johannesburg: Macmillan South Africa.

D. Herbert

Columbian nudibranch

Catriona columbiana

Group: Nudibrachia: Tergipedidae

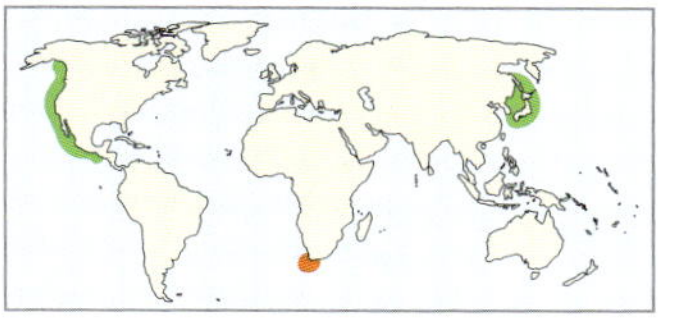

Habitat: Marine: in harbours, associated with hydroids of the genus *Tubularia*. **Identification and biology:** No shell, body length 10 mm, off-white, a pair of tentacles on anterior corners of head, followed by a longer pair of smooth rhinophores. Dorsal surface covered with bunches of light pink cerata that contain branches of digestive gland and serve to increase surface area for respiration. Feeds exclusively on hydroids of the genus *Tubularia*. **Introduction and spread:** Introduced from North Pacific to Table Bay docks prior to 1972, probably as a fouling species on the hulls of ships, not yet recorded outside the harbour area. **Impacts:** No significant impacts recorded or expected, given specialised feeding habits. **Control:** No control attempted or required.

T. Gosliner

Key reference: Gosliner, T.M. & Griffiths, R.J. 1981. Description and revision of some South African Aeolidacean Nudibranchia (Mollusca, Gastropoda). *Annals of the South African Museum*, 84:105-150.

Mystery snail/Apple snail

Pomacea diffusa

Group: Gastropoda: Ampullariidae

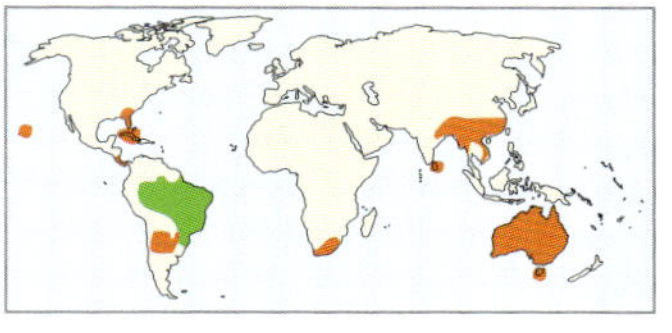

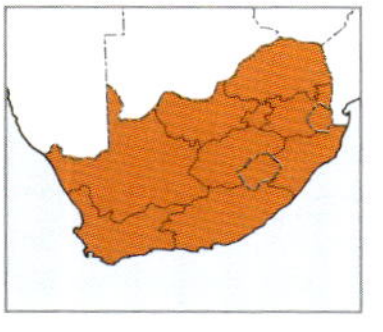

Habitat: Freshwater. **Identification and biology:** Operculum present, shell coiling to left, smooth and spherical. Wild-type grey with dark spiral bands, but golden-yellow variety popular with aquarists. Shell length up to 42 mm. Bottom dwellers, but lay eggs in conspicuous pink clusters out of water on any solid surface. Juveniles drop into the water after hatching. **Introduction and spread:** Native to Brazil but common in aquaria and ornamental ponds in South Africa. No evidence that this species has become invasive in South Africa, but it has done so in Southeast Asia and Hawaii, particularly in rice paddies, so should be considered potentially invasive here. **Impacts:** None known, but *Pomacea* species cause considerable economic losses to rice growers in Southeast Asia. **Control:** Has been controlled in rice paddies by hand picking and using molluscicide pellets.

C. Appleton

Key reference: Hayes, K.A., Joshi, R.C., Thiengo, S.C. & Cowie, R.H. 2008. Out of South America: multiple origins of non-native apple snails in Asia. *Diversity and Distributions*, 14:701-712.

Quilted melania

Tarebia granifera

Group: Gastropoda: Thiaridae

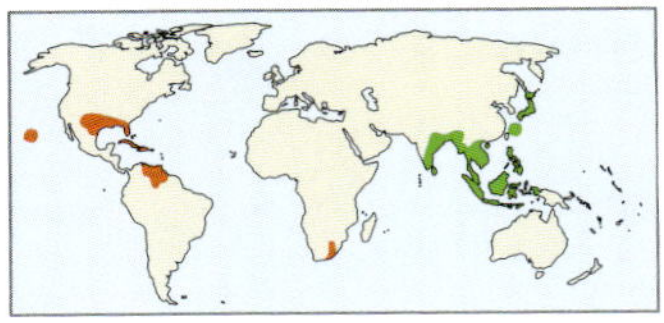

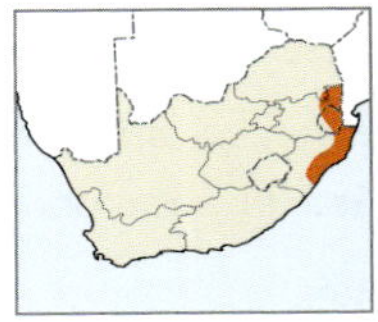

Habitat: Freshwater and estuarine. **Identification and biology:** Shell elongated, up to 29.5 mm long, conical, with conspicuous tubercles, coiling to right, operculum present, colour pale yellow/brown with early whorls dark brown; sometimes uniformly dark brown. Males scarce, females parthenogenetic and ovoviviparous. Juveniles develop to shelled stage in an internal brood pouch that contains as many as 27 embryos at any time. Shelled embryos born at 1–2 mm length and at a rate of two per day. Bottom-dwelling and detritus-feeding. Usually abundant (>1 000 m^{-2}), with highest recorded density of 21 000 m^{-2} in Nseleni River, KwaZulu-Natal. **Introduction and spread:** Native to Southeast Asia. First collected in northern KwaZulu-Natal in 1999, but probably introduced some years earlier via aquarium trade. Currently found over most of KwaZulu-Natal below altitude of 300 m, although there are a few records from higher altitudes. Also known from Swaziland and the Mpumalanga Lowveld and will probably continue spreading. Occurs in all types of flowing and standing water bodies, both natural and artificial. Also found in estuaries, where tolerant of salinities up to 20%. Probably distributed by birds, which have been found to feed on juveniles and void them later, most crushed but some intact. **Impacts:** Not known, but the high densities attained must negatively affect diversity of benthic invertebrates of colonised rivers and lakes. May be displacing indigenous benthic snails, notably the Red-rimmed melania (*Melanoides tuberculata*) and *Thiara amarula*. Quilted melania have damaged machinery at several industrial complexes in the Richards Bay area of northern KwaZulu-Natal after being pumped out of reservoirs into the water supply of industrial plants. **Control:** Probably impossible in natural water bodies because of its parthenogenicity and high reproductive rate. Control may be possible in artificial reservoirs and ponds, which could be drained, cleaned and the walls and bottom painted with a liquid molluscicide. Even if adult snails are killed, juveniles in brood pouch are not affected.

Key reference: Appleton, C.C., Forbes, A.T. & Demetriades, N.T. 2009. The occurrence, bionomics and potential impacts on the invasive freshwater snail *Tarebia granifera* (Lamarck, 1822) in South Africa. *Zoologische Mededelingen (Leiden)*, 83:525-536.

H. Madsen

Reticulate pond snail

Lymnaea columella

Group: Gastropoda: Lymnaeidae

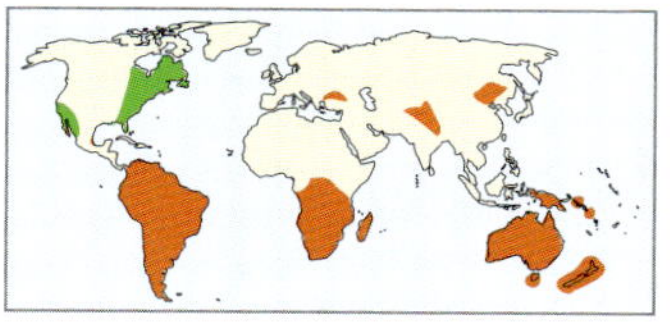

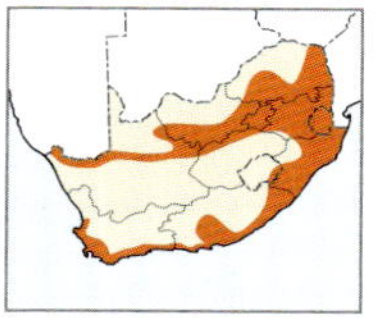

Habitat: Freshwater. **Identification and biology:** Pulmonate (air-breathing) snail without an operculum. Shell up to 16 mm long, coiled to right, dark brown and with diagnostic reticulate sculpture. Hermaphroditic. Partially amphibious, often on damp mud on riverbanks, but eggs laid in jelly-like masses underwater. As with the other pulmonate snails discussed below, eggs hatch into miniature snails after 2–3 weeks, depending on temperature. **Introduction and spread:** Native to North America. First recorded from South Africa in Somerset West in 1942, but now invasive in river systems and artificial water bodies over a wide area. **Impacts:** None recorded, but anecdotal observations suggest it may be displacing indigenous pond snail (*Lymnaea natalensis*). Serves as intermediate host for Common liver fluke (*Fasciola hepatica*) of domestic stock in Australasia and South America and probably also South Africa. **Control:** Wide distribution and amphibious habit preclude any attempts at control. Molluscicides might be effective in artificial habitats.

D. Herbert

Key reference: de Kock, K.N., Joubert, P.H. & Pretorius, S.J. 1989. Geographical distribution and habitat preferences of the invader freshwater snail species *Lymnaea columella* in South Africa. *Onderstepoort Journal of Veterinary Research*, 56:271-275.

Rust-coloured pond snail

Radix rubiginosa

Group: Gastropoda: Lymnaeidae

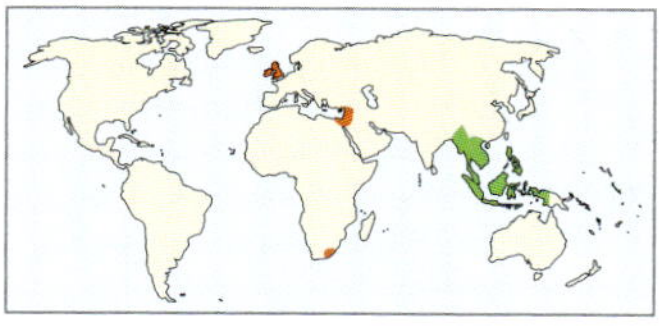

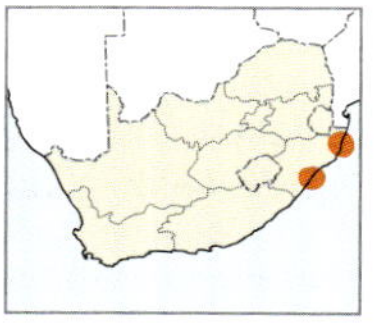

Habitat: Freshwater. **Identification and biology:** Pulmonate snail without operculum. Shell coils to right, smooth, uniformly pale to reddish-brown with outer margin of body whorl more-or-less straight. Size 25 x 13 mm. Hermaphroditic. **Introduction and spread:** Native to Southeast Asia. Discovered in 2006 in artificial ponds in northern KwaZulu-Natal, where it breeds prolifically. Collected in garden ponds in Durban in 2008. Not known to be invasive, but should be monitored. **Impacts:** Unknown, but serves as intermediate host for at least 13 trematode species in Southeast Asia, including Giant liver fluke (*Fasciola gigantica*). **Control:** Impossible in natural water bodies, but molluscicides may be successful in artificial habitats. Along with others, introduced via the aquarium trade. Careful inspection of imported aquarium supplies by trained personnel needed at ports of entry.

H. Madsen

Key reference: Chitramvong, Y.P., Upatham, E.S. & Sukhapanth, N. 1981. Effect of some physico-chemical factors on the survival of *Bythinia siamensis, Radix rubiginosa* and *Indoplanorbis exustus*. *Malacological Review*, 14:43-48.

Chinese ram's-horn snail

Gyraulus chinensis

Group: Gastropoda: Planorbidae

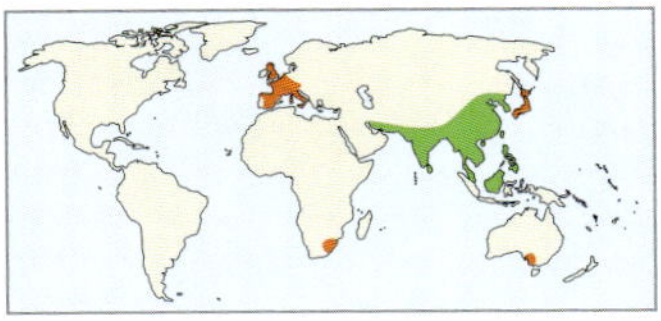

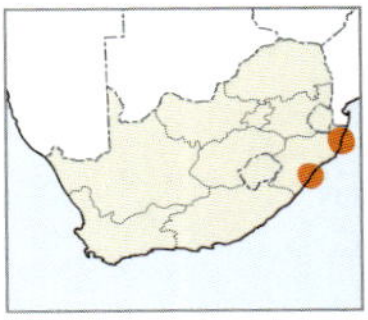

Habitat: Freshwater. **Identification and biology:** Pulmonate snail without operculum. Shell disc-like, 6 mm diameter, coils to left, brown with fine growth lines. Whorls rounded. Head, tentacles and mantle roof have diagnostic dark markings. Hermaphroditic. **Introduction and spread:** Native to Southeast Asia. Discovered in 2007 in artificial ponds in northern KwaZulu-Natal and subsequently in Durban, but not known to have spread. **Impacts:** Unknown, but should be monitored because invasive in Europe and must be considered potentially invasive here. Serves as intermediate host of Giant intestinal fluke (*Fasciolopsis buski*) in Asia. **Control:** Probably introduced via aquarium trade. More thorough inspection of imported aquarium supplies by trained personnel needed at ports of entry.

H. Madsen

Key reference: Brown, D.S., Grácio, M.A.A. & Meier-Brook, C. 1998. The Asian freshwater snail *Gyraulus chinensis* (Dunker, 1848) (Planorbidae) in West Africa and Europe. *Journal of African Zoology*, 112:203-213.

Dury's ram's-horn snail

Helisoma duryi

Group: Gastropoda: Planorbidae

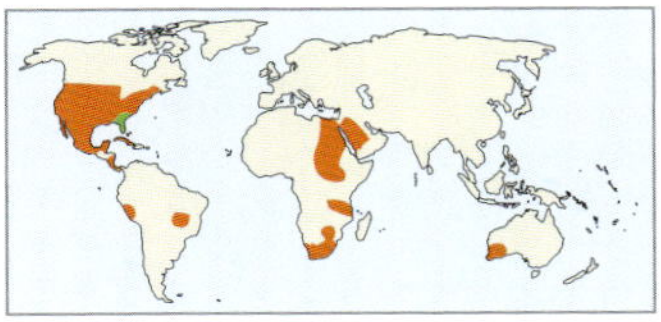

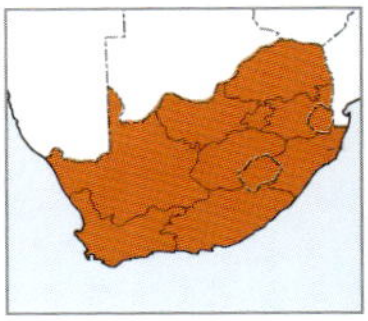

Habitat: Freshwater. **Identification and biology:** Pulmonate snail without operculum. Shell disc-shaped, 22 mm diameter, coils to left, smooth and brown. Hermaphroditic, but unusual in that preputial organ on male genitalia also serves as a holdfast, acting like a suction cup to attach one partner to neck of its mate and vice versa during copulation. Soft parts may be orange coloured. Easily confused with indigenous genus *Biomphalaria,* but distinguished by its greater height and more angular shoulders to whorls. **Introduction and spread:** Native to USA, first reported from South Africa in 1964, now widespread in aquaria, artificial ponds, reservoirs and occasional streams and small rivers. Not invasive, at least partly because of its inability to self-fertilise. **Impacts:** Shown experimentally to interfere mechanically with embryonic development of indigenous snails inside egg masses and subsequent growth of hatchlings. **Control:** Unnecessary.

D. Herbert

Key reference: Appleton, C.C. 2003. Alien and invasive freshwater Gastropoda in South Africa. *African Journal of Aquatic Science*, 28:69-81.

Sharp-spired bladder snail

Physa acuta

Group: Gastropoda: Physidae

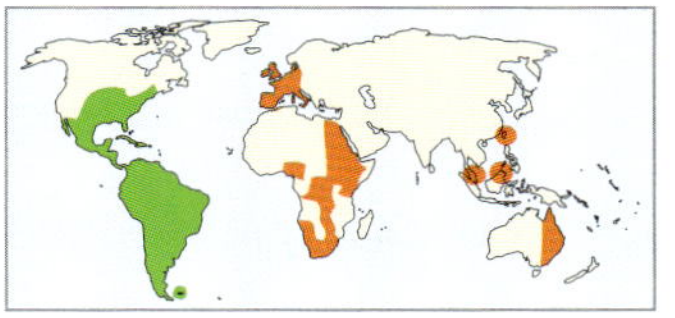

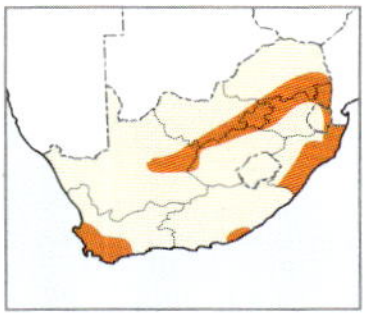

Habitat: Freshwater. **Identification and biology:** Pulmonate snail without operculum. Shell ovate, 12 mm long, coiled to left, smooth and sometimes shiny. Spire sharply pointed. Mantle margin with finger-like processes. Hermaphroditic. Physids (see also Slender bladder snail, p. 190) possess several attributes that give them an advantage over indigenous pulmonate snails. Among these is a unique set of muscles that allows them to swing their shells from side to side to prevent predators gaining purchase. They move more quickly than other snails, their locomotion is more directional, and they tolerate faster current speeds, up to 0.6 m/sec. Particularly tolerant of mild organic pollution, earning it the nickname 'sewage snail'. A short-lived, adaptable species that can switch between a single breeding season per year and up to seven generations per year. Hence can survive in both permanent and seasonal habitats, as well as those subject to repeated disturbances. This is probably because it responds to disturbances more rapidly in terms of breeding than indigenous species. Frequently lays eggs on shells of other snails; a useful behaviour because when these snails follow receding water levels, as often happens in rivers, the eggs will not dry out. Also a more versatile feeder than most indigenous snails. Its radular teeth are more numerous, finer and with longer and sharper cusps, enabling it to scrape the substratum clean, piercing food as it does so. **Introduction and spread:** Native to South America and some Caribbean islands. First collected in South Africa in Umsinduzi River, Pietermaritzburg, in 1954. Has since become invasive and spread through most of the country's major river systems and artificial water bodies. **Impacts:** Often becomes the dominant snail in colonised rivers after floods. Its habit of denuding feeding surfaces may be detrimental to other grazing invertebrates, especially when present in large numbers. **Control:** Not attempted, but could be controlled in artificial habitats with molluscicides.

Key reference: Appleton, C.C. 2003. Alien and invasive freshwater Gastropoda in South Africa. *African Journal of Aquatic Science*, 28:69-81.

H. Madsen

Slender bladder snail

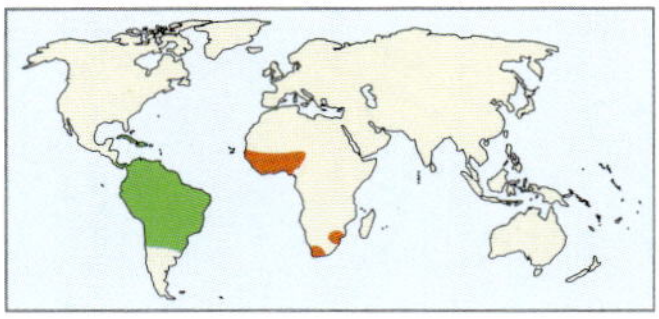

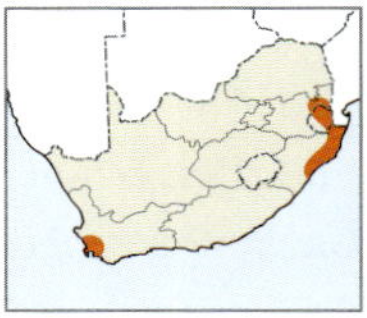

Aplexa marmorata

Group: Gastropoda: Physidae

Habitat: Freshwater. **Identification and biology:** Pulmonate snail without operculum. Shell slender and shiny with pointed spire, length 15 mm, coiled to left. Mantle margin with finger-like processes. Hermaphroditic. Breeding cycle less flexible than that of the Sharp-spired bladder snail (p. 189). This may explain why it has not colonised rivers, which in South Africa are prone to disturbances and variable flow. **Introduction and spread:** Native to South America. First recorded in KwaZulu-Natal in 1986, but since spread over most of the province's lowlands, northwards into Mpumalanga and also in Cape Town. Found in standing water, such as swamps, ponds, canals and reservoirs, but not rivers. **Impacts:** None known. **Control:** Not attempted, but could be controlled in artificial habitats using molluscicides.

N. Miranda

Key reference: Appleton, C.C. & Dana, P. 2005. Re-examination of *Physa mosambiquensis* Clessin, 1886 and its relationship with other Aplexinae (Pulmonata: Physidae) reported from Africa. *African Invertebrates*, 46:71-83.

Vermiculate snail

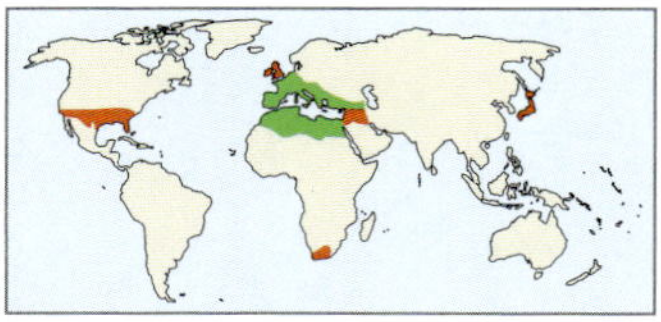

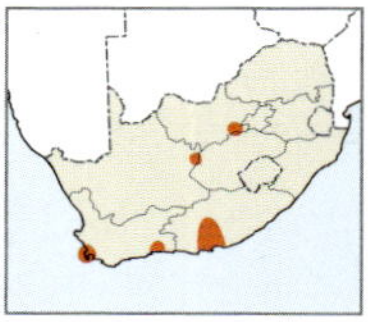

Eobania vermiculata

Group: Gastropoda: Helicidae

Habitat: Terrestrial. **Identification and biology:** Shell diameter up to 33 mm with rather low spire; aperture set at oblique angle, lip strongly flaring and noticeably thickened basally; umbilical region covered with pale callus deposit. Colour generally pale, some strongly marked with bold chocolate-brown spiral bands, others almost uniformly pale brown; nearly all with numerous, fine, whitish, worm-like (vermiculate) markings. Generalised herbivore, locally abundant in urban habitats. **Introduction and spread:** Native to Mediterranean region, introduced to South Africa mid-1980s. Primarily in Port Elizabeth area, but spreading. Also in Cape Town, probably as result of second introduction. **Impacts:** Local pest in gardens, with potential to become significant horticultural problem. **Control:** Regular hand picking and removal of hiding places, combined with use of standard molluscicide baits, can provide some control.

D. Herbert

Key references: Herbert, D.G. 2010. The introduced terrestrial Mollusca of South Africa. *SANBI Biodiversity Series* 15. Pretoria: South African National Biodiversity Institute.

Herbert, D.G. & Sirgel, W.F. 2001. The recent introduction of two potentially pestiferous alien snails in South Africa and the outcomes of differing pest management practices: an eradication and a colonisation. *South African Journal of Science*, 97:301-304.

•European brown garden snail

Cornu aspersum

Group: Gastropoda: Helicidae

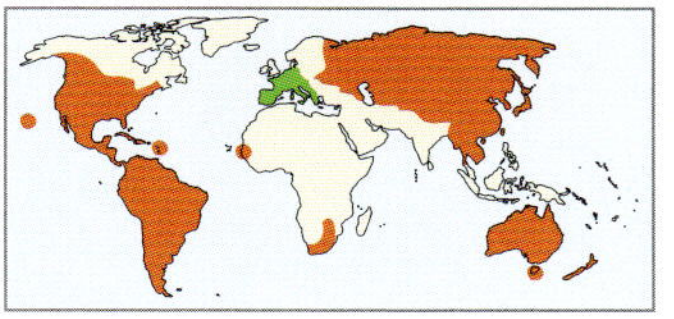

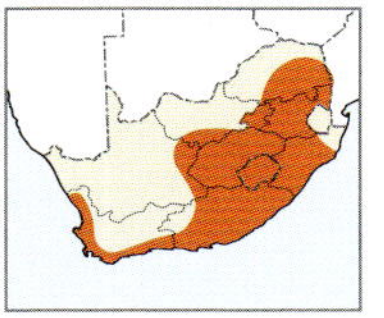

Habitat: Terrestrial. **Identification and biology:** Shell of moderate size (diameter 25–40 mm), but usually thin, comprising 4.5–5 evenly rounded whorls; aperture relatively large, more or less circular and set at an oblique angle, its lip strongly flaring in adults (not so in juveniles); basal region lacking an umbilicus. Colour pattern variable, generally mottled brown with up to five broad spiral bands in darker brown, sometimes more or less uniformly brown; flaring lip white. A herbivore, readily consuming a wide variety of green plant material; largely associated with humans in South Africa, occurring in a range of transformed habitats, particularly in domestic gardens, sheltering among vegetation and in shaded positions, often climbing walls. Does not appear to have spread widely into natural habitats, but may sometimes occur in coastal scrubland or fynbos adjacent to human habitation. In regions where there is limited calcium in the soil the shell may be extremely brittle. **Introduction and spread:** Native to western Europe and the Mediterranean, now widely dispersed around the globe by human agency. The first South African record dates from 1855, but the species was almost certainly introduced to Cape Town well before this; now very widespread, occurring wherever there is human habitation, even in very remote areas, from sea level to over 2 000 m. Also recorded in Lesotho and Zimbabwe. **Impacts:** Undoubtedly the most widespread alien snail in South Africa and a significant pest in gardens and crops, particularly vineyards and orchards. An edible species often served as 'escargot' and commercially farmed for such purposes in Europe. **Control:** Populations can be kept at bay by application of metaldehyde. Regular hand picking after dark and removal of hiding places can also be effective in domestic gardens, and in the eastern parts of the country, the indigenous cannibal snails (*Natalina* species), which will readily consume this pest, can be useful allies.

Key reference: Herbert, D.G. 2010 The introduced terrestrial Mollusca of South Africa. *SANBI Biodiversity Series* 15. Pretoria: South African National Biodiversity Institute.

D. Herbert

Dune snail/ White garden snail

Theba pisana

Group: Gastropoda: Helicidae

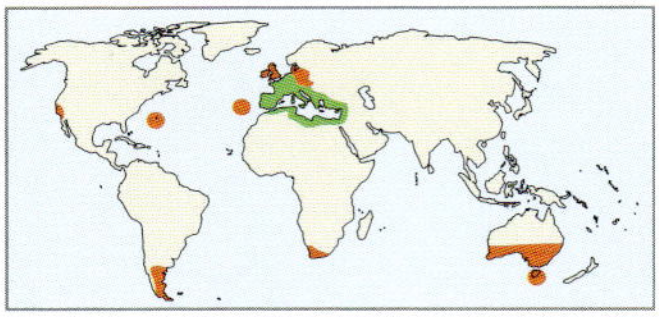

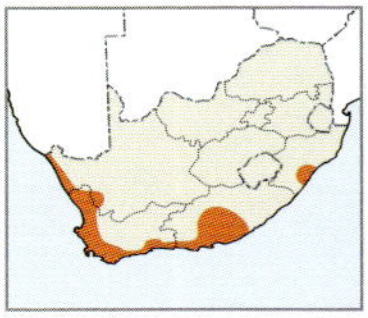

Habitat: Terrestrial. **Identification and biology:** Shell relatively small, diameter seldom exceeding 20 mm, pale, with rather low spire; whorls rounded, usually with slight angle level with suture (more prominent in juveniles); sculpture of microscopic growth-lines and incised spiral lines; base with small umbilicus, partly concealed by reflected inner lip of aperture; aperture lip not flaring outward in adults, but interior frequently with a slightly thickened ridge set back from edge. General colour dirty white to pale ginger, usually with numerous uneven and broken dark brown spiral lines, sometimes uniformly pale; inner lip frequently tinged with pink; apex frequently dark purplish-brown. Herbivorous, bush-dwelling species from coastal Mediterranean, hence thriving in similar habitats in South Africa and Australia. Can occur at very high population densities and typically climbs upwards to escape high summer temperatures at ground level, adorning shrubs, bushes, fence posts and telegraph poles in vast numbers. Reproductive cycle annual or biennial, depending on conditions. **Introduction and spread:** Originally around the Mediterranean; introduced to Cape Town before 1881; now widespread from Orange River mouth to East London, particularly near coast; also recorded from, but not established in, Durban and Maputo and in Namibia. **Impacts:** Agricultural and garden pest reportedly damaging a range of plants, including vegetables, cereals, ornamental flowers and shrubs, deciduous and citrus fruit trees, berries and vines. May also act as vector for plant pathogens and animal parasites, render pasture crops unpalatable to livestock and contaminate cereal crops at harvest time. Has invaded natural coastal ecosystems, such as West Coast National Park, to an alarming degree and its impacts on these habitats are cause for concern. In Australia, may out-compete native snail species. **Control:** Metaldehyde will help reduce populations in cropland and gardens, but this is an ongoing battle. Little hope of effective control in natural ecosystems and although biocontrol options (e.g. sarcophagid flies) have been investigated, none has yet proved effective.

C. Griffiths

C. Griffiths

Key references: Herbert, D.G. 2010 The introduced terrestrial Mollusca of South Africa. *SANBI Biodiversity Series* 15. Pretoria: South African National Biodiversity Institute.

Odendaal, L.J., Haupt, T.M. & Griffiths, C.L. 2008. The alien invasive snail *Theba pisana* in the West Coast National Park: is there cause for concern? *Koedoe*, 50:93-98.

Small conical snail

Cochlicella barbara

Group: Gastropoda: Cochlicellidae

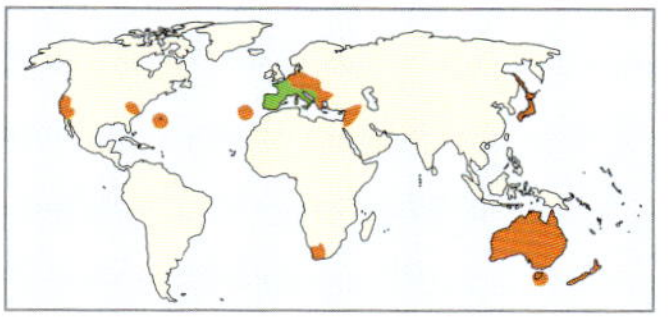

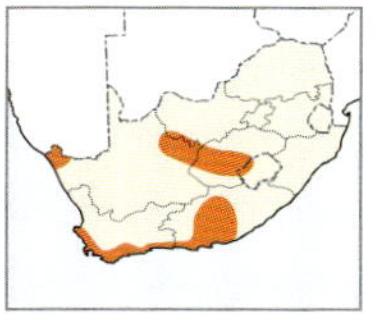

Habitat: Terrestrial. **Identification and biology:** Small snail with an elongated-conical shell reaching 12 mm length, comprising 7–8 convex whorls; surface dull, smooth except for some irregular axial riblets just beneath suture line; umbilicus open, narrow and partially concealed by reflected inner lip of aperture. Coloration variable, ranging from almost uniform dirty-white to pale translucent buff mottled with white-and-brown blotches and flecks, also frequently with a broad, darker brown spiral band at mid-whorl. Favours areas with a relatively dry, Mediterranean climate, especially near the coast. Common in gardens, cultivated fields and wastelands, where it can occur in large numbers under favourable conditions. Generally of cryptic habits and may thrive largely unseen, beneath the surface of lawns; rarely climbs vegetation. Easily translocated with plants, building materials and household goods. **Introduction and spread:** Native to the Mediterranean region, introduced to Cape Town prior to 1909. Now widely distributed along the Cape south coast, from False Bay to Port Alfred. Also recorded from isolated localities on the Atlantic coast as far north as the South Africa/Namibia border, and from scattered inland localities in the Cape. Lack of records from KwaZulu-Natal suggests that the warm, summer-rainfall climate of the eastern seaboard is not favourable for this species. **Impacts:** Herbivorous and although small, may cause appreciable damage when population densities are high (over 1 000 snails per m^2 recorded in New Zealand pasture). Evidently a pest of wheat and canola crops in the Bredasdorp area, Western Cape. **Control:** Metaldehyde will help reduce populations in both cropland and gardens, but eradication is all but impossible.

Key reference: Herbert, D.G. 2010. The introduced terrestrial Mollusca of South Africa. *SANBI Biodiversity Series* 15. Pretoria: South African National Biodiversity Institute.

C. Griffiths

Orchid snail

Zonitoides arboreus

Group: Gastropoda: Gastrodontidae

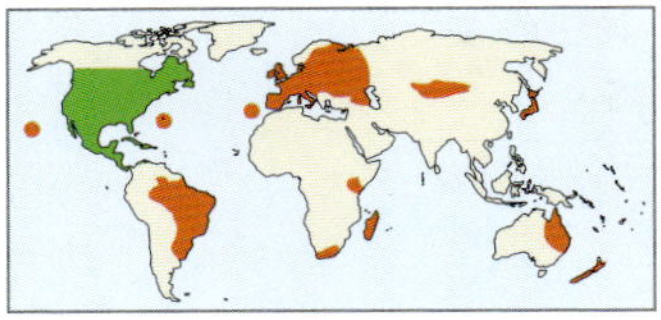

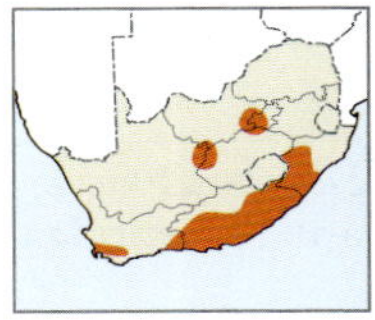

Habitat: Terrestrial. **Identification and biology:** Shell very small, round and low-spired, diameter <6.0 mm, comprising 4.5–5 rather narrowly expanding whorls; superficially smooth and glossy, but with close-set growth lines; umbilicus wide and deep. Colour translucent, pale honey-brown, darker when alive; living animal mostly dark blue-grey. Tolerant of transformed habitats, including gardens and forestry plantations; under logs, flowerpots and groundcover plants, in leaf-litter under trees and behind flaking bark.

D. Herbert

Introduction and spread: Native to North America, but widely introduced; first recorded in South Africa in 1898 in Port Elizabeth and now known from widely scattered localities countrywide; mostly in association with human habitation and transformed habitats, spreading into adjacent natural habitats. Also found in Namibia. **Impacts:** Common greenhouse pest in many countries and a particular problem for orchid growers; also potential root pest of sugarcane. **Control:** Regular inspection and treatment of infected plants with standard molluscicides.

Key reference: Herbert, D.G. 2010. The introduced terrestrial Mollusca of South Africa. *SANBI Biodiversity Series* 15. Pretoria: South African National Biodiversity Institute.

Draparnaud's glass snail

Oxychilus draparnaudi

Group: Gastropoda: Oxychilidae

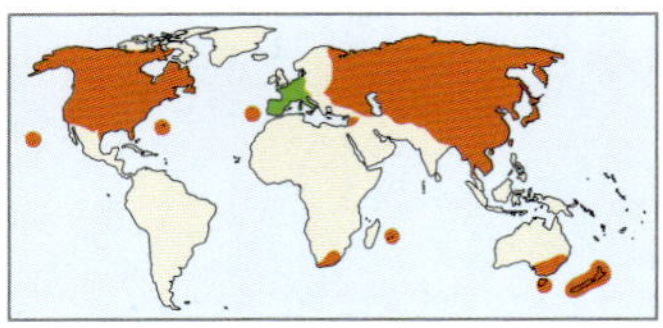

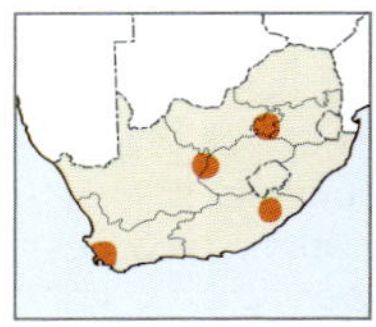

Habitat: Terrestrial. **Identification and biology:** Shell disc-like, diameter up to 16 mm, smooth and somewhat glossy; base with moderately wide umbilicus. Slightly translucent, pale horn-brown, paler and rather milky on base. Living animal blue-grey. Occurs primarily in suburban gardens and nurseries, but also adjacent indigenous habitats in Western Cape. At least partly carnivorous, rarely feeding on green plants. The indigenous Dwarf cannibal snail (*Nata vernicosa*) is similar, but less flattened and has fine radial ribs below suture line. **Introduction and spread:** Native to western Europe; introduced to Cape Town before 1908 and mostly confined to south-western Cape, with isolated records elsewhere. **Impacts:** Insignificant horticultural pest on account of largely carnivorous diet, but abundant in indigenous Afromontane forest on Cape Peninsula, where it may negatively impact indigenous snail fauna. **Control:** Difficult once established.

D. Herbert

Key reference: Herbert, D.G. 2010. The introduced terrestrial Mollusca of South Africa. *SANBI Biodiversity Series* 15. Pretoria: South African National Biodiversity Institute.

Marsh, Brown field and Grey field slugs

Deroceras laeve, D. panormitanum and *D. reticulatum*

Group: Gastropoda: Agriolimacidae

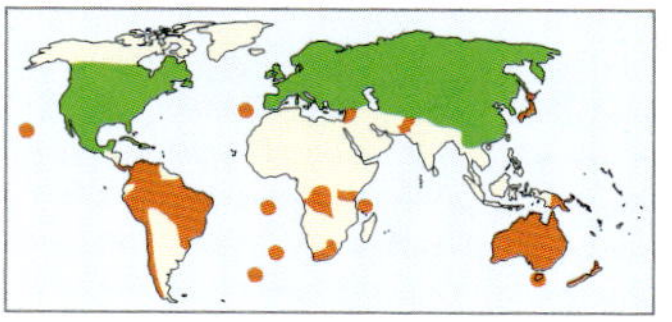

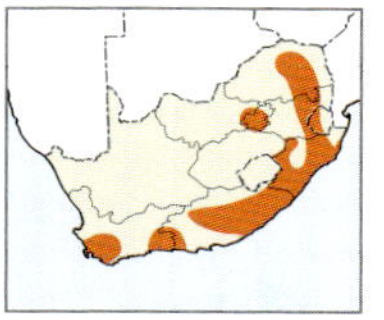

Habitat: Terrestrial. **Identification and biology:** Small, rather nondescript, fawn to dark brown slugs that resemble banded garden slugs (*Lehmannia* species, p. 196), but are generally smaller, mostly less than 45 mm in length, and lack distinct dark longitudinal bands. Mantle shield with fine concentric ridges resembling fingerprint, the central point of which lies to right of mid-line, just above lung opening (pneumostome). Pneumostome on right, behind mid-point of mantle shield; posterior portion of tail with short, indistinct keel; tail tip steeply sloping; foot sole longitudinally tripartate. Three *Deroceras* species are recorded from South Africa, but species discrimination based on external features can be frustratingly difficult; accurate identification requires dissection of distal genital tract. The Marsh slug (*D. laeve*) and Brown field slug (*D. panormitanum*, top image) are small, dark brown slugs commonly found in the vegetable patch. The third species, the Grey field slug (*D. reticulatum*, bottom image), is usually slightly larger and paler, speckled with dark grey, and likewise all too common in gardens. Though commonly associated with humans, field slugs are not confined to transformed habitats and have spread into natural habitats, even in remote areas. They are able to reproduce by self-fertilisation (perhaps even by parthenogenesis), enhancing their capacity to spread and invade new territory.

Introduction and spread: All three species native to Europe and Holarctic, but have now been widely dispersed by humans. Marsh slug and Grey field slug first recorded in South Africa in 1898, and Brown field slug in 1964. Have been recorded in almost all provinces and are likely to continue to spread into natural habitats in wetter regions. (Maps apply to all species.) **Impacts:** Herbivorous and can be serious pests in a wide range of agricultural and horticultural crops, as well as in domestic gardens. Particularly problematic in market garden vegetables and this association with human food crops has undoubtedly contributed to their wide passive dispersal. **Control:** Metaldehyde will help reduce populations in cropland and gardens.

Key reference: Herbert, D.G. 2010 The introduced terrestrial Mollusca of South Africa. *SANBI Biodiversity Series* 15. Pretoria: South African National Biodiversity Institute.

D. Herbert

D. Herbert

Striped and Three-banded garden slugs

Lehmannia nyctelia and *L. valentiana*

Group: Gastropoda: Limacidae

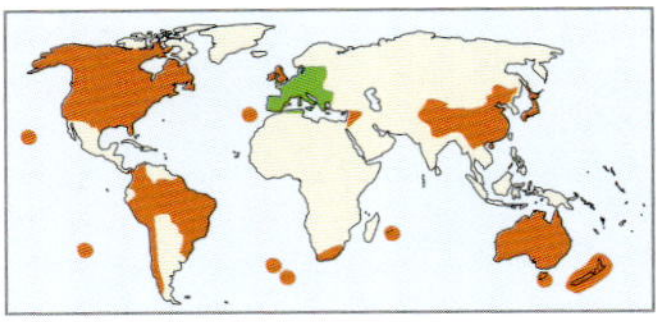

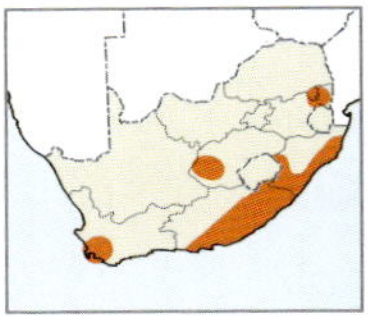

Habitat: Terrestrial. **Identification and biology:** Medium-sized, adult length 40–60 mm, grey-buff to brown, somewhat translucent, usually with a pair of dark, dorso-lateral longitudinal bands, one on each side of mid-line, sometimes broken up into spots; mantle shield similarly marked and often with a third band along mid-line; concentric ridges on mantle shield centred about the mid-line (compare with *Deroceras* species, p. 195). Pneumostome on right, behind mid-point of mantle shield; posterior keel short and not particularly obvious, tail tip gently sloping; foot sole longitudinally tripartate. Two species recorded from South Africa, the Striped garden slug (*Lehmannia nyctelia*) and Three-banded garden slug (*L. valentiana*, image below), but indistinguishable externally and dissection of distal genital tract needed for accurate identification. Both herbivorous and generally associated with humans, but the Striped garden slug is more common and widespread in region – being one of most commonly encountered slugs in South African gardens. Live in sheltering microhabitats, such as beneath stones and logs, among plant debris and in greenhouses. **Introduction and spread:** Both species are native to Europe and first South African records date from 1939 and 1962 respectively. Now occur widely in southern and eastern parts of the country, even in remote rural areas, but seem largely absent from the Highveld. There has been some spread into indigenous habitats, particularly where these abut agricultural land or human settlements. (Maps apply to both species.) **Impacts:** Both species identified as greenhouse pests in other temperate regions, damaging orchid flowers and ornamental plants. Almost certainly pestiferous in gardens and in the horticulture sector in South Africa. **Control:** Good horticultural hygiene and application of molluscicides, such as metaldehyde, will be necessary if damage is considered significant.

Key reference: Herbert, D.G. 2010. The introduced terrestrial Mollusca of South Africa. *SANBI Biodiversity Series* 15. Pretoria: South African National Biodiversity Institute.

D. Herbert

Yellow garden slug

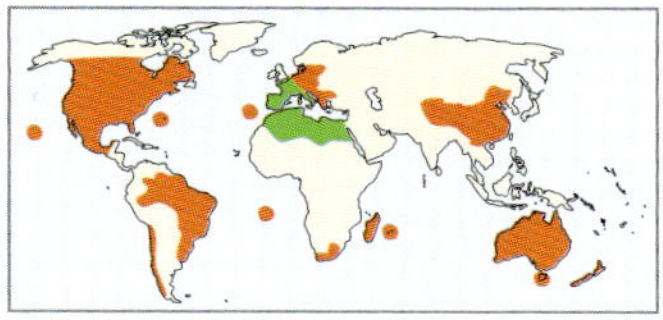

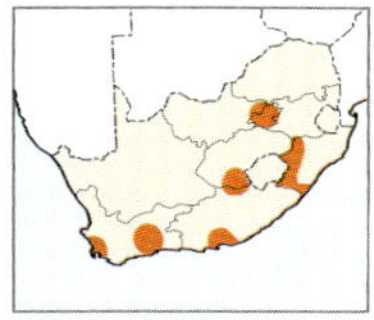

Limacus flavus

Group: Gastropoda: Limacidae

Habitat: Terrestrial. **Identification and biology:** Moderately large, striking slug, extended length 75–100 mm, typically with yellowish ground colour overlain with greyish or greenish mottling, appearing grey-green with yellow spots; mantle shield similarly patterned; tentacles bluish; no longitudinal bands on body or mantle shield. Generally associated with humans, in gardens, outbuildings, cellars and kitchens, in sheltering microhabitats, such as under logs, pots, discarded carpets, sacks and timber. Omnivorous, active chiefly at night, feeds on decaying vegetation, lichens and fungi, also domestic refuse and pet food. **Introduction and spread:** Probably from Mediterranean region, but now widely dispersed. Introduced to South Africa prior to 1898 and currently in most southern and eastern provinces, and Gauteng. **Impacts:** Damages ornamental plants and market garden vegetables, also stored fruit and vegetables. **Control:** Can generally be controlled by hand picking, combined with removal of food sources and hiding places.

D. Herbert

Key reference: Herbert, D.G. 2010. The introduced terrestrial Mollusca of South Africa. *SANBI Biodiversity Series* 15. Pretoria: South African National Biodiversity Institute.

Black keeled slug

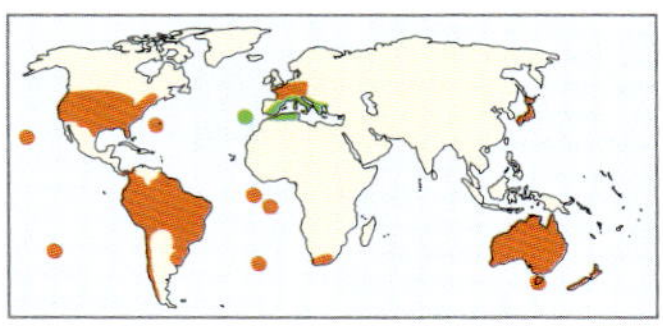

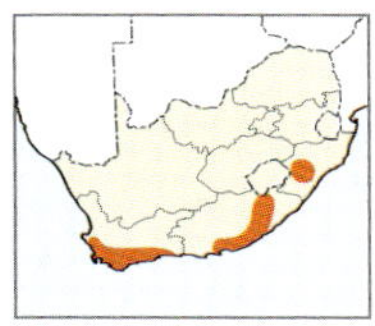

Milax gagates

Group: Gastropoda: Milacidae

Habitat: Terrestrial. **Identification and biology:** Dark grey to black slug, length up to 60 mm, with a prominent keel running from tail tip to hind end of mantle shield. **Introduction and spread:** Introduced to South Africa from Europe prior to 1873, now widespread in Western Cape, Eastern Cape and to a lesser extent KwaZulu-Natal. **Impacts:** Feeds on seeds and soft plant material and a common pest of fruit, root and tuber crops, sunflower, maize, cereals, pasture and newly sown crops in general. Reported to be an episodic pest of canola, barley, wheat and lucerne in the southern Cape. **Control:** Application of molluscicides may be necessary where there is a risk of crop damage.

D. Herbert

Key reference: Herbert, D.G. 2010. The introduced terrestrial Mollusca of South Africa. *SANBI Biodiversity Series* 15. Pretoria: South African National Biodiversity Institute.

Giant garden slug

Limax maximus

Group: Gastropoda: Limacidae

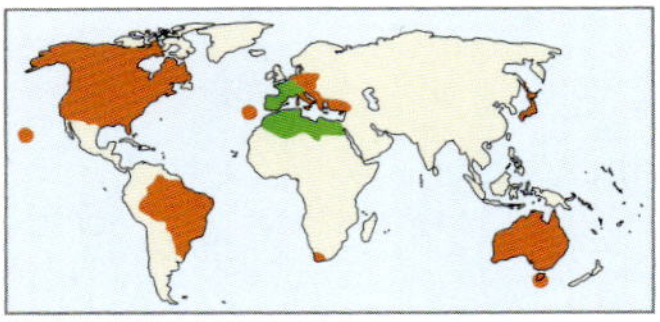

Habitat: Terrestrial. **Identification and biology:** Very large and distinctively marked slug up to 200 mm long; usually pale fawn-brown to grey, with two or three longitudinal rows of bold, almost black, spots or blotches on each side of body, sometimes coalescing into bands; mantle shield spotted or marbled with darker coloration, but not banded. Head and tentacles pinkish to reddish-brown. Omnivorous, active primarily at night. **Introduction and spread:** Native to Europe. First South African record from Newlands, Cape Town, in 1898, but not seen again until 2006, when it was rediscovered in the same forests above Newlands and Kirstenbosch. Widespread in Australia. **Impacts:** Rarely cited as a significant pest. **Control:** Probably not warranted.

D. Herbert

Key reference: Herbert, D.G. 2010. The introduced terrestrial Mollusca of South Africa. *SANBI Biodiversity Series* 15. Pretoria: South African National Biodiversity Institute.

Hedgehog slug

Arion intermedius

Group: Gastropoda: Arionidae

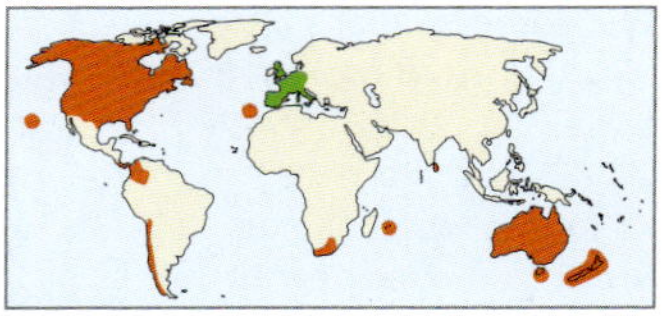

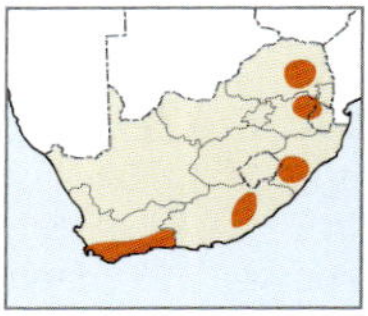

Habitat: Terrestrial. **Identification and biology:** Small slug with extended length up to 20 mm, yellowish-white to greyish-brown, sometimes with indistinct longitudinal band on each side. Mantle shield finely granular, lacking pattern of concentric ridges and pneumostome on right, in front of mid-point of shield (compare with *Deroceras* species, p. 195). Mucus yellow. When slug contracts, dorsal skin granules become exaggerated, giving its back a somewhat prickly texture. Consumes vegetable matter, and may be important decomposers. **Introduction and spread:** Native to western Europe. Introduced to South Africa prior to 1898 and now quite widely distributed; found mostly in natural habitats adjacent to urban areas, rather than in gardens. **Impacts:** Pest status is uncertain, but probably rarely a serious problem. **Control:** Probably not warranted.

D. Herbert

Key reference: Herbert, D.G. 2010. The introduced terrestrial Mollusca of South Africa. *SANBI Biodiversity Series* 15. Pretoria: South African National Biodiversity Institute.

LAMP SHELLS Phylum Brachiopoda

A small group of filter-feeding animals, representing the remnants of what was once an important phylum, with thousands of species in the fossil record. Body covered by paired calcium carbonate shells, superficially resembling those of bivalve molluscs, although the two shells enclose the body dorsally and ventrally rather than side to side, as is the case in bivalves. Also the filter-feeding organ is a large, feather-like lophophore, rather than the paired gills found in bivalves. The lower shell is usually the larger and is pierced posteriorly by a hole, through which a short stalk, the pedicel, protrudes to attach the animal to the substratum. The sexes are separate and free-swimming larvae are released into the water.

Disc lamp shell

Discinisca tenuis

Group: Discinidae

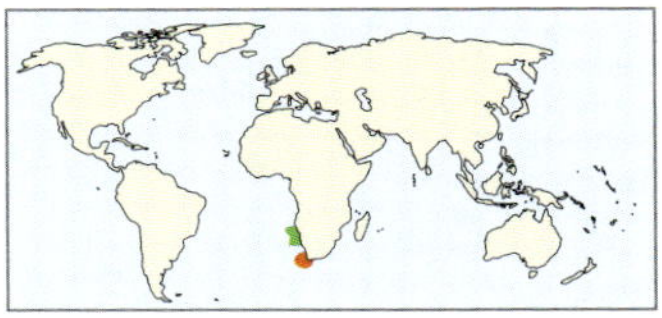

Habitat: Marine, firmly attached to hard surfaces, including the shells of molluscs. **Identification and biology:** An unusual brachiopod in that the shells are flat, round, transparent horny discs up to 20 mm across, that show concentric growth lines and may be fringed by long, straight 'hairs' (particularly evident in juveniles). In their natural range they mainly attach to one another to form large clumps, but introduced populations are generally attached singly, or in small groups, to oysters or other shellfish. **Introduction and spread:** Endemic to Namibia and first recorded in South Africa only in 2008 on oysters being cultured in Saldanha Bay. Currently appear to be restricted to oyster farms. **Impacts:** Presently minimal, although may reduce the visual appeal of oysters for market. Reach very high densities in their home range and could become significant fouling species if they propagate in the natural environment outside of the oyster farms, which is likely to happen, given their broadcast spawning habits. **Control:** Hand picking, or soaking in fresh or hot water are possible methods for killing animals attached to commercially marketed oysters. Control in the natural environment is unlikely to be possible.

Key reference: Haupt, T.M., Griffiths, C.L., Robinson, T.B. & Tonin, A.F.G. 2010. Oysters as vectors of marine aliens, with notes on four introduced species associated with oyster farming in South Africa. *African Journal of Zoology*, 45:52-62.

C. Griffiths

MOSS OR LACE ANIMALS Phylum Bryozoa

Colonial organisms consisting of large numbers of tiny individuals, called zooids, each enclosed in a tiny coffin-like casing made of chitin or lime and with a crown of filter-feeding tentacles that are extended from an opening on the case. Growth form can be very variable and includes flat or encrusting species, upright branching ones that often resemble seaweeds, and stony ones that are easily confused with corals. Bryozoans are common elements of the fouling fauna on ships; as a result many species have been widely distributed around the world. The species detailed below represent the more obvious introductions to South Africa. Some other species have now been so widely dispersed that their origins and natural distribution will only be determined after detailed genetic research has been done.

Red-rust bryozoan

Watersipora subtorquata

Group: Cheilostomata: Watersiporidae

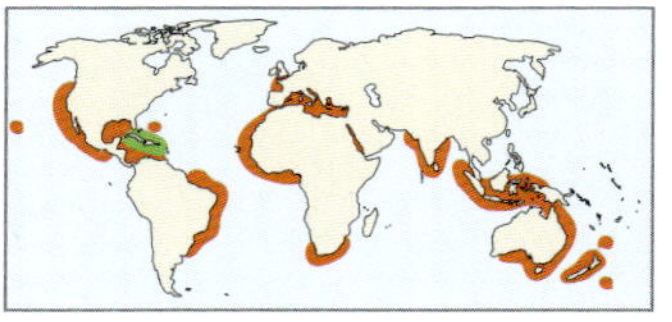

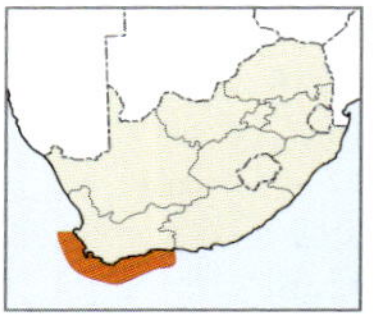

Habitat: Attached to hard surfaces, including rocks and the shells of molluscs, most commonly found on the undersurfaces of rocks intertidally and in shallow water. **Identification and biology:** An encrusting colonial species composed of numerous tiny individuals or zooids arranged in concentric circles to form a flat, plate-like colony up to 40 mm across. Colour a characteristic orange-brown with black dots. If observed closely the tiny zooids are visible, each enclosed in an urn-shaped case with an oval opening at the one end, through which a circle of feeding tentacles can be extended. The tentacles withdraw rapidly when disturbed. Filter feeds on small particles in the water. **Introduction and spread:** Originally Caribbean, but now widely distributed and probably secondarily introduced to South Africa on the hulls of ships. Now one of the most common bryozoans found on south-western Cape shores. **Impacts:** Probably minimal, but has not been studied. **Control:** No control method known or attempted.

Key reference: Florence, W.K., Hayward, P.J. & Gibbons, M.J. 2007. Taxonomy of shallow-water Bryozoa from the west coast of South Africa. *African Natural History*, 3:1-58.

C. Griffiths

Blue dentate moss animal

Bugula dentata

Group: Cheilostomata: Bugulidae

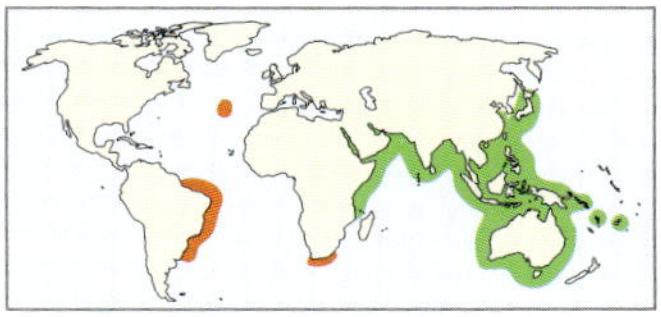

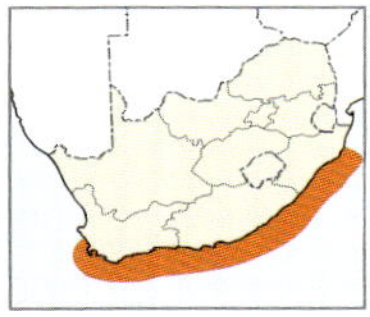

Habitat: Marine, attached to hard surfaces like ship hulls, wharfs and rocks, usually hanging from vertical surfaces. **Identification and biology:** Erect colonial form, tiny zooids linked in double rows to form a grey to deep-blue bushy colony up to 50 mm tall and easily mistaken for a seaweed. When observed closely individual zooids are visible, each enclosed in an oval case with an aperture and several prominent spines at the distal end. Specialised protective individuals, termed avicularia and resembling bird heads in shape, project from the sides of the colony. **Introduction and spread:** Indo-Pacific in origin, but now widely dispersed by shipping. An early introduction to the south and east coasts of Africa, first reported in 1852. **Impacts:** Common and minor nuisance as a fouling species. The bushy colonies provide habitat for many small invertebrates, but being attached filter-feeders they are unlikely to have other significant ecological or economic impacts. **Control:** No control method known or proposed.

G. Branch

Key reference: Florence, W.K., Hayward, P.J. & Gibbons, M.J. 2007. Taxonomy of shallow-water Bryozoa from the west coast of South Africa. *African Natural History*, 3:1-58.

Purple dentate moss animal

Bugula neritina

Group: Cheilostomata: Bugulidae

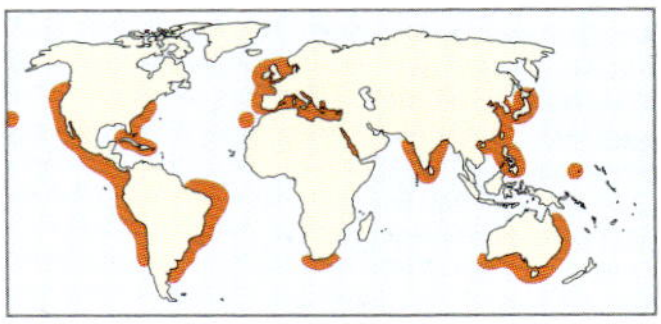

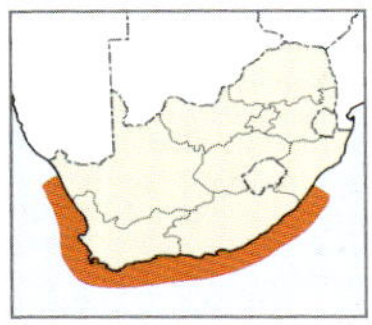

Habitat: Marine, attached to hard surfaces and seaweeds, especially in harbours. **Identification and biology:** A common fouling species, predominantly found growing on seaweeds. Forms delicate, sparsely branching reddish-purple colonies about 30 mm tall. Zooids are long and slender, arranged in double rows, distally formed into sharp lateral points, and lack both terminal spines and protective avicularia. **Introduction and spread:** Probably European in origin and an early introduction, but first published record 1944. Occurs throughout the region, mainly in harbours. **Impacts:** This species is the source of the anti-carcinogenic compound bryostatin. Ecological impact probably minimal, but a minor fouling species. **Control:** No control method known or proposed.

C. Griffiths

Key reference: Florence, W.K., Hayward, P.J. & Gibbons, M.J. 2007. Taxonomy of shallow-water Bryozoa from the west coast of South Africa. *African Natural History*, 3:1-58.

AUTHORS: C. GRIFFITHS & K. CHRISTISON

FLATWORMS Phylum Platyhelminthes

Free-living flatworms have flattened bodies and a muscular feeding organ or pharynx on the underside, leading into a sac-like gut. They glide along on a bed of fine 'hairs' or cilia. Parasitic forms are also abundant and are found attached to the body surface or gills, or in the guts, of vertebrates. Some parasitic species have lost their gut entirely, absorbing food directly through the body surface. Many have complex life cycles, involving one or more intermediate hosts. Both freshwater and terrestrial flatworms have been introduced to South Africa.

Spade-headed planarian

Bipalium kewense

Group: Tricladida: Geoplanidae

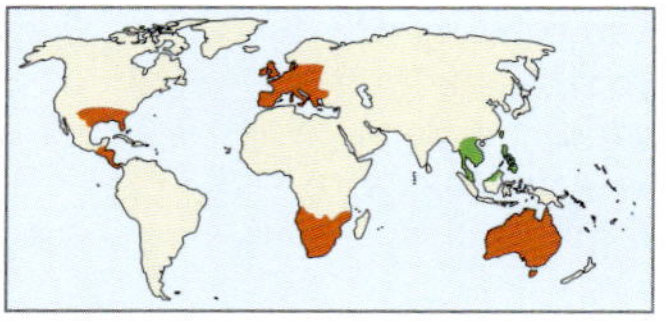

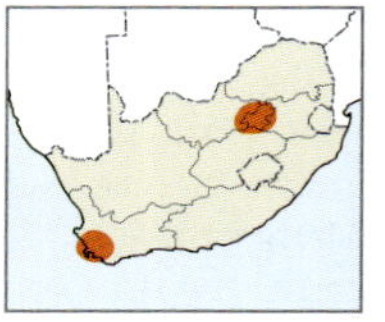

Habitat: Terrestrial, in dark, moist areas, particularly under rocks and logs and in leaf litter. May emerge onto soil surface after rain. **Identification and biology:** Elongated, flexible, flat-bodied, glossy and moist-bodied free-living worms, growing up to 30 cm long when extended. Easily recognised by unusual diamond- or spade-shaped head; several prominent dark stripes running lengthwise along body; and pale underside. Glides along smoothly, propelled by a bed of tiny cilia that cover ventral surface. A common body opening, serving as both mouth and anus, situated halfway along the underside and equipped with a muscular pharynx to grasp prey. Mostly active at night, feeding on earthworms and slugs; may also be cannibalistic. Reproduction occurs by fragmentation at posterior end, short sections of body sticking to substratum as parent pulls away. Fragments are immediately mobile and develop a head within about 10 days. Sexual reproduction also takes place, eggs being laid in red cocoons and hatching in about three weeks. **Introduction and spread:** Thought to be native to Indo-China, but now widely introduced, especially to tropical areas; probably distributed with potted plants. Photographed in Cape Town, but almost certainly occurs in moist areas throughout region, although distribution poorly studied. **Impacts:** Voracious predators of earthworms and can eradicate populations in worm farms. Prey detected by chemoreception and subdued by slimy secretion. Pharynx then extruded through mouth and into prey. Can survive many weeks without food, gradually shrinking in size as own body resources are used. Harmless to humans. **Control:** No known method of control.

Key reference: Windsor, L. 2008. A revision of the cosmopolitan land planarian *Bipalium kewense* Moseley, 1878 (Turbellaria: Tricladidae: Terricola). *Zoological Journal of the Linnean Society*, 79:61-100.

Paul J. Morris, Wikimedia Commons

Asian tapeworm

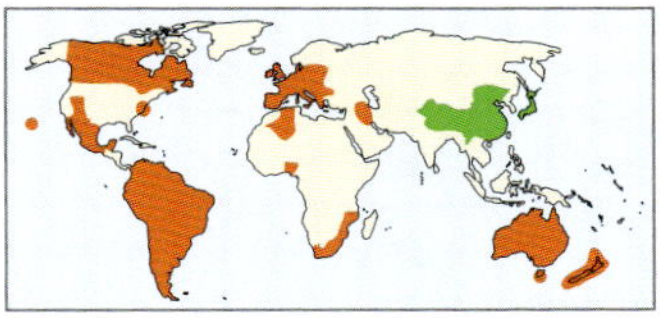

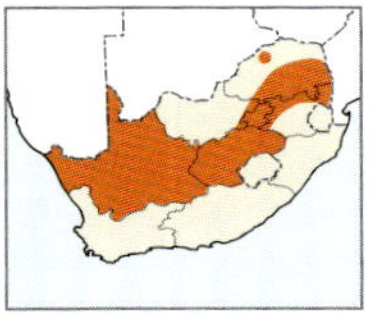

Bothriocephalus acheilognathi

Group: Cestoda: Bothriocephalidae

Habitat: Freshwater; parasite in gut of freshwater fishes. **Identification and biology:** Segmented parasitic tapeworm up to 20 cm long, found in intestines of carp and other cyprinid fishes. Elongated head or scolex with two deep lateral grooves and prominent terminal disc; body shows well-developed segmentation. Adult hermaphroditic, fertilised eggs released into host's gut cavity, emerging into surrounding water with faeces. There they develop into free-swimming coracidium larvae that are ingested by a copepod host, where development continues. Infected copepods in turn consumed by fish, completing life cycle. **Introduction and spread:** Native to China, but probably introduced from West Germany with Grass carp (p. 51) imported to Fisheries Research Station in Marble Hall in 1975. **Impacts:** Known to cause severe mortalities in Common carp (p. 52) in Europe. Infestations of over 1 000 worms in a single fish recorded in South Africa. **Control:** Only possible in controlled conditions (fish ponds).

J. van As

Key reference: De Moor, I.J. & Bruton, M.N. 1988. *Atlas of alien and translocated indigenous aquatic animals in Southern Africa*. South African National Scientific Programmes Report 144.

Fish skin fluke

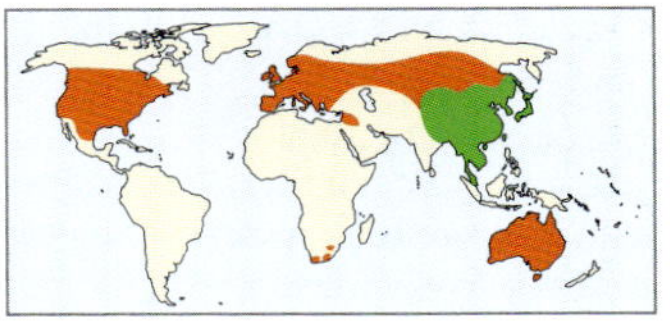

Gyrodactylus kherulensis

Group: Monogenea: Gyrodactylidae

Habitat: Freshwater, parasitic, attached to skin and fins of fishes. **Identification and biology:** Tiny spindle-shaped parasitic flatworms less than 1 mm long. Live on skin and fins of fishes, attaching to host by means of large, specialised posterior organ (haptor) armed with marginal hooks, anchors (hamuli) and supporting bars. Life history unusual in that they display both viviparity and 'polyembryonism' – juvenile worms developing inside mothers already carrying embryos of the next generation within their uteri. **Introduction and spread:** Confirmed introductions from Asia with alien fishes such as koi and carp, with high likelihood of establishment on feral populations of these fishes locally. Can potentially also infect indigenous fish species, although this has not yet been reported. **Impacts:** Skin lesions, caused by parasite attachment and feeding, compromise the osmotic integrity of host epidermis and render host susceptible to secondary infections. **Control:** Many different chemical treatments or 'dips' proposed and are detailed in specialised aquaculture literature.

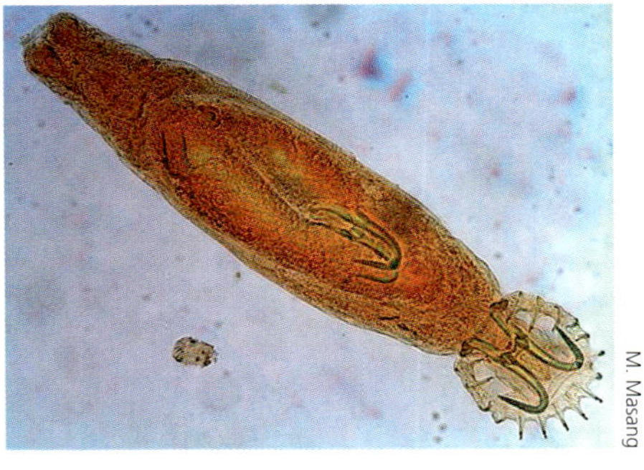

M. Masang

Key reference: Bakke, T.A., Cable, J. & Harris, P.D. 2007. The biology of gyrodactylid monogeneans: The 'Russian-doll killers'. *Advances in Parasitology*, 64:161-378.

Eel gill fluke

Pseudodactylogyrus anguillae

Group: Monogenea: Dactylogyridae

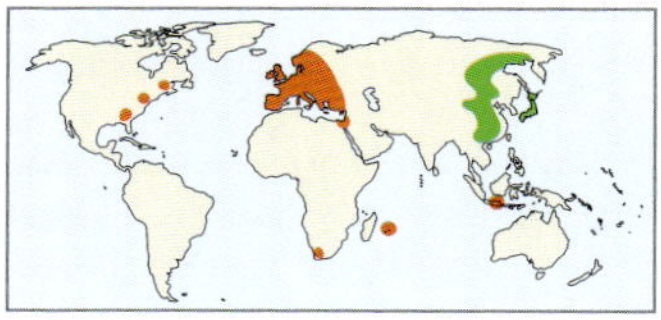

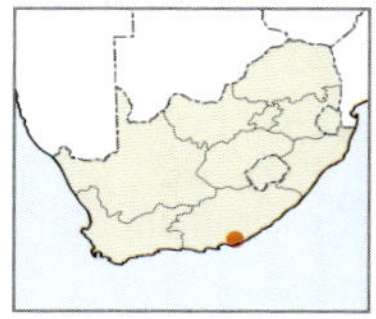

Habitat: Freshwater, parasitic, attached to gills of freshwater eels.

Identification and biology: Small parasitic flatworms less than 2 mm in length. Live on the gills of freshwater eels, attaching themselves to the host gill tissue by means of a posterior attachment organ (the haptor). The haptor consists of two large, ventrally directed anchors (hamuli) and 14 marginal hooklets distributed along the outer membrane of the haptor. Parasites have a direct life cycle, whereby eggs are released into the environment (up to 29 eggs produced per worm per 24 hours). Egg development and embryonation are temperature-dependent; larvae take about six days to hatch at temperatures between 20°C and 24°C. A free-swimming ciliated larva (onchomiracidium) hatches from the egg and actively seeks out a new host, where it will attach and mature to complete the life cycle. **Introduction and spread:** Native range East Asia, from where it was introduced to Europe and North America. Global distribution of this species has often been explained by the active introduction of infected eels for aquaculture purposes, however, no such evidence is currently available to account for its recent discovery on South African eels, as there is no documented evidence of exotic eels being introduced to this region, although unregulated introductions may have occurred. **Impacts:** Infections by large numbers of these flukes usually result in excessive mucus production by the gills and the destruction of gill structures. Severe haemorrhaging and extensive gill epithelial overgrowth have been attributed to attachment and feeding of these parasites. **Control:** Many different chemical treatments or 'dips' have been proposed and are detailed in the specialised aquaculture literature.

Key references: Buchmann, K., Mellergaard, S. & Køie, M. 1987. *Pseudodactylogyrus* infections in eel: a review. *Diseases of Aquatic Organisms*, 3:51-57.

Christison, K.W. & Baker, G.C. 2007. First record of *Pseudodactylogyrus anguillae* (Yin & Sproston, 1948) (Monogenea) from South Africa. *African Zoology*, 42:279-285.

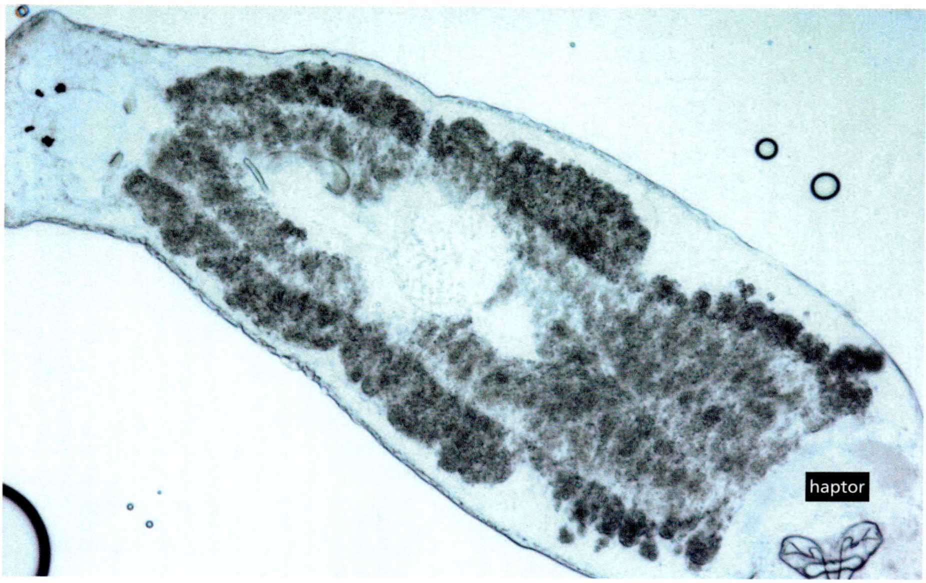

K. Buchmann

ANEMONES, JELLYFISH, HYDROIDS Phylum Cnidaria

The Cnidaria include the sea anemones, corals, jellyfish and hydroids and have a simple, circular, sac-like body, which can either take the form of an upright tentacle-bearing polyp, which is often colonial, or of a swimming medusa. The body has a single opening that serves both as the mouth and anus. The most characteristic feature is the possession of stinging cells used to capture prey. Two anemones, a freshwater jellyfish and several hydroids are recognised as introduced to the region.

Plumose anemone

Meridium senile

Group: Actiniaria: Metridiidae

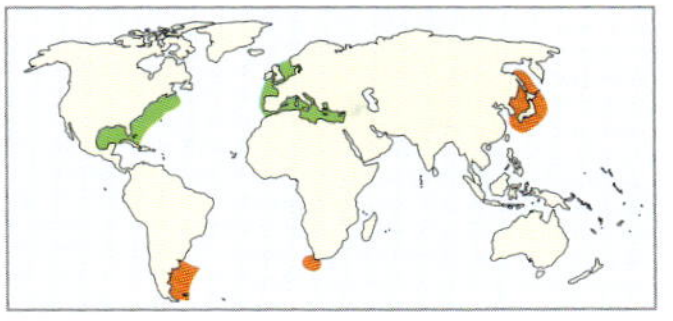

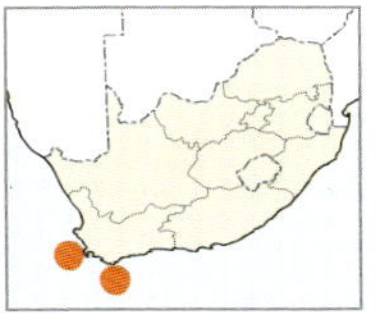

Habitat: Marine, on rocky reefs, harbours and other hard surfaces to over 100 m depth. **Identification and biology:** Attain 100 mm in height. Easily recognised by the lobed oral disc and hundreds of short tentacles, which give the species a characteristic fluffy appearance. The column is smooth and uniform white, pink or orange; flared at the base with a distinct parapet at the top. Often found in closely packed groups and can reproduce by basal laceration, whereby small sections tear free from the base of the column and develop into new individuals. Also reproduces sexually via eggs and sperm shed into the water. As a result of asexual reproduction it is common to see one large individual surrounded by smaller, genetically identical ones of exactly the same colour. Feed on small animals captured by the stinging cells borne on the tentacles. **Introduction and spread:** Native to the North Atlantic and first reported in Table Bay docks in 1995. Recent photographic evidence points to another dense offshore population associated with oil installations on the Agulhas Bank, one of very few deep-water marine invasions in South Africa. **Impacts:** May be significant as a fouling species and compete with local species for food. **Control:** No established techniques other than mechanical removal.

C. Griffiths

Key reference: Griffiths, C.L., Kruger, L.M. & Ewart-Smith, C. 1999. First record of the sea anemone *Metridium senile* from South Africa. *South African Journal of Zoology*, 31:157-158.

Brooding anemone

Sagartia ornata

Group: Actiniaria: Sagartiidae

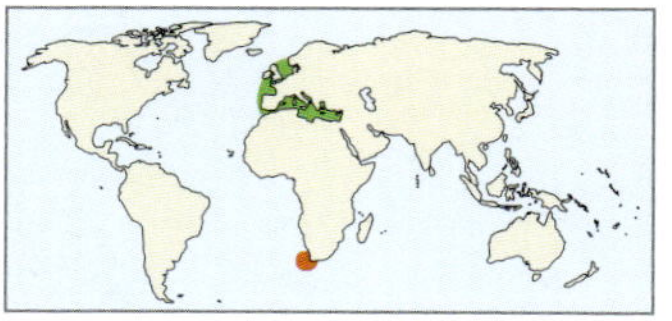

Habitat: Marine, on intertidal rocks and stones buried in sand. **Identification and biology:** Attach to intertidal stones and rocks shallowly buried in sand, the expanded oral disc lying flush with the sand surface. Oral disc up to 20 mm diameter, brownish, with paler centre and radial stripes; up to 200 tentacles, brown with white spots, some white. Column tapering to narrow base, lower part cream, upper part darker with adhesive suckers to which sand and stones adhere. Readily releases sticky white threads when disturbed. Young brooded in body cavity, emerging as fully formed anemones. When exposed to air withdraws, leaving characteristic pits in sand. **Introduction and spread:** Native to Europe and known only from Langebaan Lagoon, where they occur in hundreds per square metre. Probably introduced with dry ballast or shipping decades or centuries ago. **Impacts:** Unknown, but as an abundant predator must impact prey species. **Control:** Not attempted and no appropriate technique known.

C. Griffiths

Key reference: Acuna, F.H., Excoffon, A.C. & Griffiths, C.L. 2004. First record and redescription of the introduced sea anemone *Sagartia ornata* (Holdsworth, 1855) (Cnidaria: Actiniaria: Sagartiidae) from South Africa. *African Zoology*, 39:314-318.

Freshwater jellyfish

Craspedacusta sowerbyi

Group: Hydrozoa: Limnomedusae

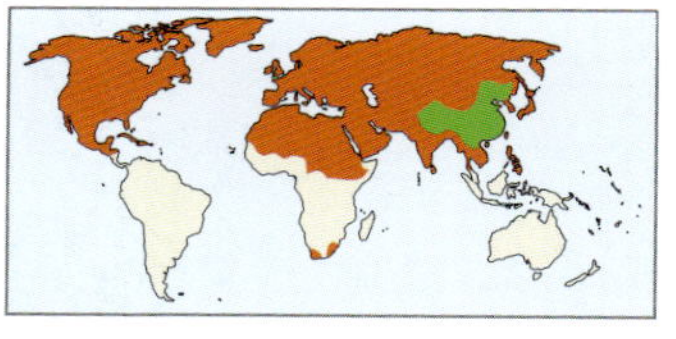

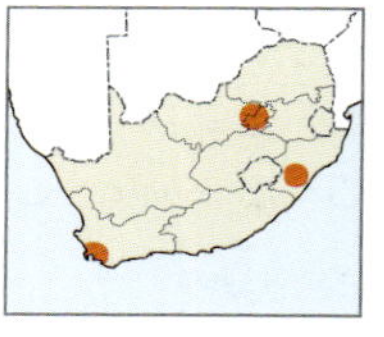

Habitat: Freshwater lakes and dams, free swimming. **Identification and biology:** Life cycle alternates between small (2 mm) sessile polyp, attached to hard substrata, and bell-shaped medusae, developing from polyp via asexual budding. Medusae derived from a particular polyp are all the same sex, and 3–6 mm diameter, with a ring of short tentacles around the margin. Swim by rhythmic muscular contractions of the bell, feeding on water fleas, rotifers and other small zooplankton caught by stinging cells along the tentacles. Sperm and eggs shed into water and develop into free-swimming larvae, which settle and develop into polyp stage. **Introduction and spread:** Origins China. First recorded in Midmar Dam, probably introduced there with aquatic plants in late 1970s. Probably much more widespread, with anecdotal reports as far afield as Cape Town. **Impacts:** May impact zooplankton populations, competing with other predatory species. **Control:** No known means of control, spread can be limited by controlling distribution of aquatic plants.

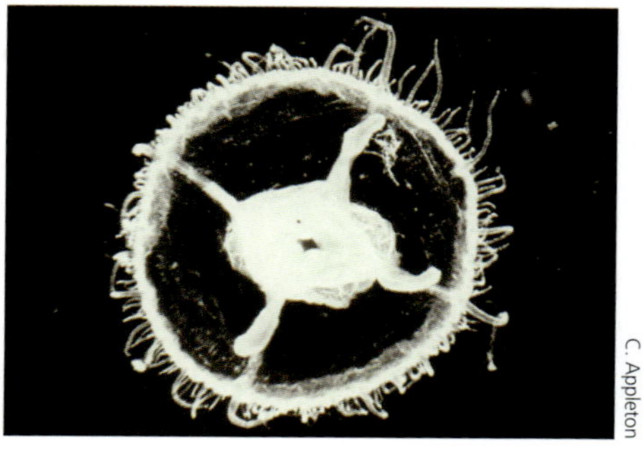

C. Appleton

Key reference: Rayner, N.A. 1988. First record of *Craspedacusta sowerbyi* Lankester (Cnidaria: Limnomedusae) from Africa. *Hydrobiologia*, 162:73-77.

Sea fern hydroid

Pennaria disticha

Group: Hydrozoa: Pennariidae

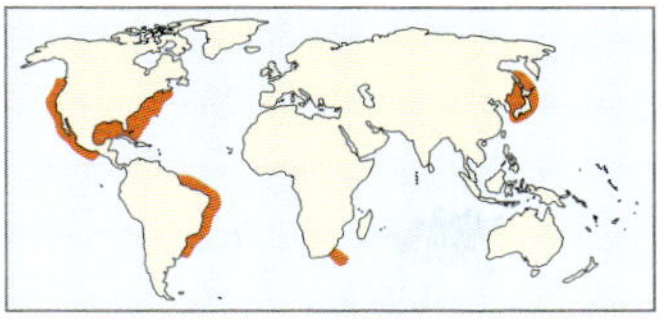

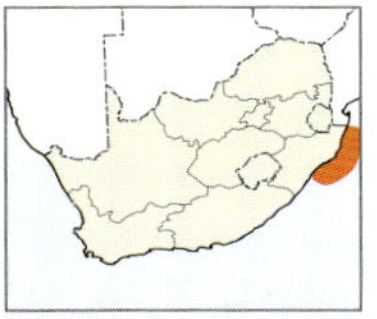

Habitat: Marine, attached to rocks, ship hulls, pier pilings and similar hard surfaces. **Identification and biology:** Fern-like colony up to 60 mm tall, with a central stem that gives off branches alternately to left and right sides. Tiny pink polyps are arranged in a single, evenly spaced row along upper surface of each branch. If examined under the microscope each polyp has a ring of outer tentacles surrounding a knobbed central cone. **Introduction and spread:** So widely dispersed that native range uncertain. First recorded in KwaZulu-Natal in 1906 (when erroneously described as a new indigenous species, *Halocordyle cooperi*). Current range from Durban northwards, on rocky shores and ship hulls. **Impacts:** A minor nuisance as a fouling species, but few other impacts. **Control:** None proposed.

C. Griffiths

Key reference: Millard, N.A.H. 1975. Monograph on the Hydroida of southern Africa. *Annals of the South African Museum*, 68:1-513.

Zigzag hydroid

Obelia geniculata

Group: Hydrozoa: Campanulariidae

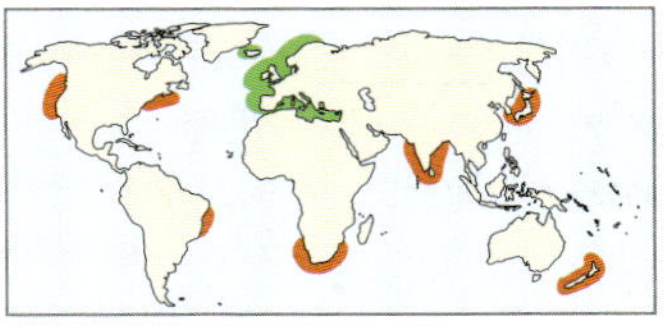

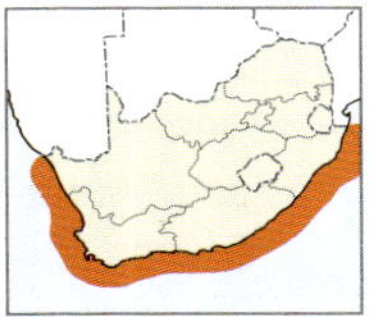

Habitat: Marine, typically growing on surfaces of large algae, such as kelps. **Identification and biology:** A colonial form consisting of a tubular creeping stem, or stolon, from which erect stems about 10 mm tall grow at intervals. Each hollow, vertical stem is bent into a zigzag pattern and bears a series of tiny polyps arranged alternately on either side of the upright. When disturbed, polyps withdraw into cup-shaped capsules. Club-shaped reproductive structures are found at base of polyps in breeding season. One of several introduced *Obelia* species – this species is distinguished by marked thickenings on the sides of stem and by never being branched. **Introduction and spread:** Probably introduced from Europe as a fouling species via shipping. **Impacts:** Minor fouling species, common throughout region but unlikely to have significant impacts due to small size. **Control:** No control system known or proposed.

C. Griffiths

Key reference: Millard, N.A.H. 1975. Monograph on the Hydroida of southern Africa. *Annals of the South African Museum*, 68:1-513.

AUTHOR: C. GRIFFITHS

SPONGES Phylum Porifera

Sponges are simple, sedentary animals that can adopt a wide variety of shapes, including flat, lobed, tubular and cup-shaped. They lack conventional organs and are made up of just a few types of cells, supported by a skeleton of silica, calcium carbonate or flexible fibres. Flagellated collar cells suck water in through numerous tiny pores all over the body surface, trap and digest food particles floating in the water, then expel filtered water through larger, more conspicuous openings called oscula. Most sponges are marine and they can form an important component of the fauna of hard substrata, often reaching a large size. A few relatively small species also occur in freshwater. Only one marine species is currently recognised as introduced to South Africa, although the taxonomy of the group is not well known here, and the discovery of additional introduced species can be expected.

Sulphur sponge

Suberites ficus

Group: Demospongiae: Suberitidae

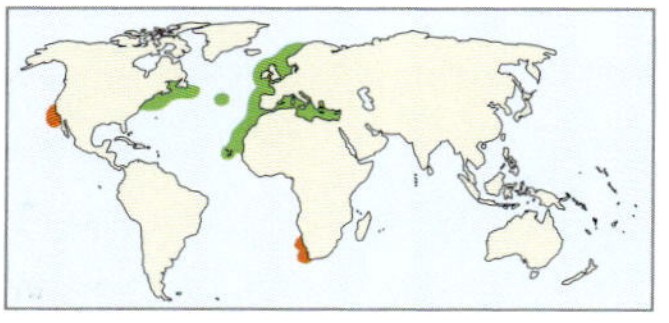

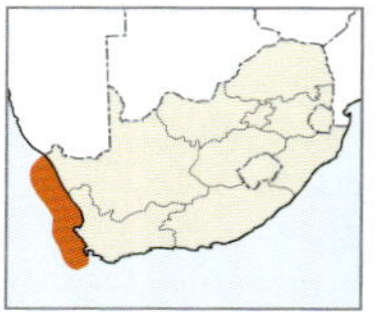

Habitat: Marine, on rocky reefs, harbour walls, pier pilings and the like. **Identification and biology:** An irregularly rounded, lobed, orange sponge, typically 10 cm across, always firm to the touch. The oscula are large, rounded openings that are flush with the upper surface and not raised on elevated turrets. Grows attached firmly to hard substrata, particularly in docks. **Introduction and spread:** Native to Europe. First reported from specimens collected in 1998 and recorded from Luderitz to Table Bay docks. Probably introduced with shipping. **Impacts:** Can form significant fouling growths and these form habitat for other smaller animals, but otherwise unlikely to result in important impacts. **Control:** Established specimens can easily be removed manually, but no systematic eradication or control system is in place, or proposed.

Key reference: Samaai, T. & Gibbons, M.J. 2005. Demospongiae taxonomy and biodiversity of the Benguela region on the west coast of South Africa. *African Natural History*, 1:1-96.

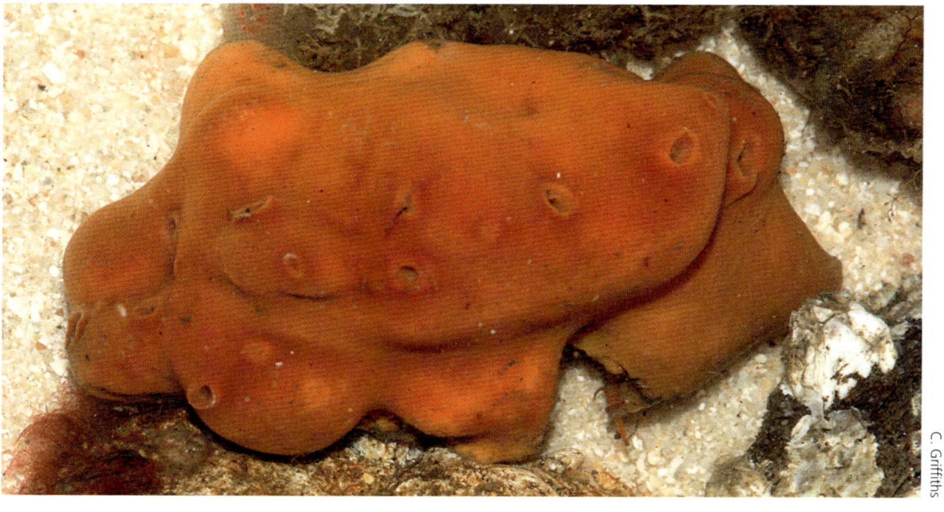

C. Griffiths

FURTHER READING

Chown, S.L., Spear, D., Lee, J.E. & Shaw, J.D. 2009. Animal introductions to southern systems: lessons for ecology and for policy. *African Zoology*, 44:248-262.

Henderson, L. 2001. *Alien weeds and invasive plants: a complete guide to declared weeds and invaders in South Africa*. Pretoria: Agricultural Research Council.

Hulme, P.E. & Roy, D.B. 2010. DAISIE and arthropod invasions in Europe. *BioRisk*, 4(1):1-3.

Joubert, L. 2009. *Invaded – the biological invasion of South Africa*. Johannesburg: University of the Witwatersrand Press.

Lach, L., Picker, M.D., Allsopp, M., Colville, J. & Griffiths, C.L. 2002. Alien invertebrate animals in South Africa. In: Pimental, D. (ed.). *Biological invasions. Economic and environmental costs of alien plant, animal and microbe species*. Florida: CRC Press.

Lockwood, J.L., Hoopes, M.F. & Marchetti, R. 2007. *Invasion ecology*. Malden, USA: Blackwell.

Macdonald, I.A.W., Kruger, F.J. & Ferrar, A.A. 1986. *The ecology and management of biological invasions in southern Africa*. Cape Town: Oxford University Press.

Macdonald, I.A.W., Reaser, J.K., Bright, C., Neville, L.E., Howard, G.W., Murphy, S.J. & Preston, G. (eds). 2003. *Invasive alien species in southern Africa: national reports and directory of resources*. Cape Town: Global Invasive Species Programme.

Macdonald, I.A.W. & Richardson, D.M. 1986. Alien species in terrestrial ecosystems of the fynbos biome. In: Macdonald, I.A.W., Kruger, F.J. & Ferrar, A.A. (eds). *The ecology and management of biological invasions in southern Africa*. Cape Town: Oxford University Press, pp. 77-91.

Perrings, C., Mooney, H. & Williamson, M. 2010. *Bioinvasions and globalisation: ecology, economics, management and policy*. New York: Oxford University Press.

Pimental, D. (ed.). 2001. *Biological invasions: economic and environmental costs of alien plant, animal and microbe species*. Florida: CRC Press.

Preston, G., Brown, G. & van Wyk, E. 2000. *Best management practices for preventing and controlling invasive alien species*. Cape Town: Working for Water Programme.

Rilov, G. & Crooks, J.A. 2009. *Biological invasions in marine ecosystems: ecological, management and geographic perspectives*. Berlin: Springer-Verlag.

Ruis, G.M. & Carlton, J.T. (eds). 2003. *Invasive species: vectors and management strategies*. Washington, D.C.: Island Press.

Visser, D. 2009. *A complete guide to vegetable pests in South Africa*. Pretoria: Agricultural Research Council.

Wittenberg, R. & Cock, M.J.W. (eds). 2001. *Invasive alien species: a toolkit of best prevention and management practices*. Wallington, Oxon, UK: CABI International.

WEBSITES

Global Invasive Species Programme:
www.gisp.org

Global Invasive Species Database:
www.issg.org/database

National Invasive Species Information Center (USA):
www.invasivespeciesinfo.gov

Centre of Excellence for Invasion Biology:
www.sun.ac.za/cib

BioRisk journal home page:
www.pensoft.net/journal_home_page.php?journal_id=2&page=article&type=show&article_id=606

GLOSSARY

Acontia – thread-like defensive organs armed with stinging cells, ejected by certain anemones when irritated

Aestivation – cessation or slowing down of activity during hot or dry season

Annulate – consisting of rings or ring-like segments

Arthropod – one of a huge range of invertebrates having an external, hardened skin

Benthic – living on the sea floor, bottom-dwelling

Byssal threads – strong, silky fibres used by mussels to attach to rocks

Carapace – hard, protective shell covering dorsal part of crab or other arthropod

Caste – specialised group of individuals, such as soldiers or workers, found in colonies of social insects

Cephalothorax – fused head and thorax in arachnids and crustaceans

Chelicerae – first pair of fang-like appendages of arachnids

Cilia – fine, hair-like projections extending from surface of cells, commonly beat in rhythmic waves for locomotion

Clitellum – thickened glandular section of body wall of earthworms, secretes egg capsule

Commensal – relationship between two different species where one receives benefits without damaging the other

Congeneric – belonging to the same genus

Crepuscular – active at dawn and dusk

Cryptic – concealed, difficult to see

Cryptogenic – a species whose origin (native range) and therefore introduced status is uncertain

Dextral – shell coiled in a right-handed spiral, turning clockwise when viewed from the apex

Diapause – state of inactivity or arrested development

Emarginated – having notched margin or apex

Endemic – confined to a particular region (e.g. country, province)

Epibiont – organism living harmlessly on the surface of another

Eutrophication – process where water bodies receive excess nutrients, stimulating excessive plant growth

Extant – still surviving

Extralimital – occurring outside its normal range

Furcula – springing organ on underside of abdomen of springtails (Order Collembola)

Gonopodium – anal fin of certain live-bearing fishes used as copulatory organ

Haploid – cell of organism having a single set of chromosomes

Hermaphroditic – having both male and female reproductive organs in the same individual

Hybrid – offspring of animals or plants resulting from interbreeding of two different species (or varieties)

Instar – stage of arthropod development between successive moults

Intromittent organ – external organ of male adapted to deliver sperm during copulation

Invertebrate – any animal lacking a spinal column

Macrophyte – macroscopic plant, particularly one in aquatic environment (as opposed to microscopic algae)

Mantle – outer skin of molluscs lining inner surface of shell and responsible for secreting and repairing shell

Medusa – jellyfish; tentacled, bell-shaped form of a cnidarian

Notochord – flexible, rod-like structure forming primitive 'backbone' of lower chordates and embryos of higher vertebrates

Ocelli – small, simple eyes found in many invertebrates

Operculum – horny or calcareous plate on foot of gastropods, used to close shell opening when animal retracts

Osculum/a – mouth-like opening/s in a sponge through which water is expelled

Ovipositor – tubular structure projecting from abdomen of females in some arthropods and used to deposit eggs

Ovoviviparous – producing living young from eggs that hatch within body

Parasitoid – insect whose larvae are parasitic on other insects, usually killing them

Parthenogenetic – method of reproduction in which female can generate offspring (clones) without fertilisation by a male

Pedicel – stem or stalk of an animal or organ, or 'waist' of an ant or wasp

Peduncle – stalk of Brachiopoda or goose barnacle

Phytosanitary – inspection of plants or their derivatives to ensure that plant pests are not transported across national boundaries

Pneumostome – respiratory opening in mantle of terrestrial snail or slug

Polyp – body form of anemone, or an individual within a colony of corals or hydroids

Polyphagous – feeding on many different types of food

Proboscis – tubular feeding organ of mollusc or insect

Pronotum – upper plate of first thoracic segment of arthropod

Protist – one-celled organism of Kingdom Protista

Pubescent – covered in soft hair or down

Pulmonate – air-breathing land snail with lung or lung-like organ

Pupate – transformation of mature larva to the immobile pupal or 'chrysalis' stage of the life cycle of an insect

Radula – flexible, ribbon-like and toothed feeding structure in snails

Reticulate – resembling or forming a net or network pattern

Rhinophore – paired club or tentacle-like sensory projections on the head of seaslug

Rodent – mammal of order Rodentia, including rats, mice and squirrels, characterised by large, gnawing incisors

Sessile – sedentary or attached organisms, permanently fixed in one spot

Seta/e – hair-like bristle/s on surface of an organism

Sinistral – shell coiled in a left-handed spiral, turns anti-clockwise when viewed from apex

Solanaceous – plant belonging to family Solanaceae; including potatoes, tomatoes and peppers

Spermatophore – capsule or compact mass of spermatozoa produced by males of certain invertebrates

Stolon – creeping, underground stem

Stridulation – production of sound by rubbing two rough surfaces together, typically in arthropods

Suture line – groove where two whorls, or turns, of snail shell join

Synanthropic – found in association with humans and their habitation

Terminalia – terminal abdominal segments forming genitalia of insects

Test/tunic – hardened outer body wall, as in ascidians

Translocate – migrate/move from one place to another

Trophic – relating to nutrition

Umbilicus – hole penetrating central axis at base of gastropod shell

Ungulates – hoofed mammals

Uropods – posterior abdominal appendages of crustaceans

Vector – mode of transportation by which organism is moved from place to place

Vermiculate – decorated with wavy lines

Vivipary – giving birth to living young, rather than eggs

Whorl – single coil or turn on shell of snail

Zooid – organism forming part of colony, such as a coral, bryozoan or colonial ascidian

INDEX TO SCIENTIFIC NAMES

D

E

Q

R

S

T

C. Griffiths

The Asian Harlequin lady beetle out-competes and even feeds on other lady beetles.

INDEX TO COMMON NAMES

T

W

Y

Z

APPENDIX

This appendix lists introduced animals recorded from South Africa, excluding species translocated from one part of the country to another. Also excluded are poorly supported 'cryptogenic' species (those whose origin remains uncertain, but that are suspected to be introduced). It is currently complete for vertebrates, but remains preliminary for certain invertebrate groups, many of which are incompletely documented and surveyed.

This list of South African alien fauna includes 601 species, of which 487 (81%) are terrestrial, 38 (6.3%) freshwater, and 76 (12.7%) marine. Included in the list are 80 biocontrol agents (13.3% of alien species on the list); of these, 76 (12.7%) are terrestrial, and 4 (0.7%) freshwater.

Major resources used in compiling this Appendix are shown at the foot of the list. In addition to the inclusion of all species covered in the main text of this book, the list has been supplemented from various published sources, the most important of which are referenced at the end of the Appendix. Unpublished lists of alien Arachnida were provided by A. Dippenaar (Araneae) and E. Uekermann (Acari). J. Hoffmann provided distribution data for the biocontrol agents that have successfully established on target invasive plants.

Key for 'Habitat' column: T = terrestrial, M = marine, F = freshwater, BIO = biocontrol agent.

Species and group	Origin	SA distribution	Habitat
MAMMALIA			
Bovidae			
Capra hircus Feral domestic goat	Iran	Widespread, patchy	T
Hemitragus jemlahicus Himalayan tahr	Central Asia to China	Table Mountain	T
Rusa unicolor Sambar deer	SE Asia	Table Mountain	T
Cervidae			
Dama dama European fallow deer	Iran, Iraq, Turkey	Widespread, patchy	T
Equidae			
Equus asinus Donkey	Egypt, Somalia	Widespread, arid parts	T
Equus ferus caballus Feral horse	Central Asia	Kaapsehoop, Kleinmond	T
Felidae			
Felis catus Feral domestic cat	Egypt	Widespread	T
Leporidae			
Oryctolagus cuniculus European rabbit	Europe	Offshore islands	T
Muridae			
Mus musculus House mouse	Eurasia	Widespread	T
Rattus norvegicus Brown rat	China, Russia, Japan	Coastal cities	T
Rattus rattus House rat	S Asia	Widespread (not Karoo/savanna)	T
Sciuridae			
Sciuris carolinensis Grey squirrel	USA	W Cape	T
Suidae			
Sus scrofa Feral domestic pig	Eurasia	W Cape, Gauteng	T
AVES			
Anatidae			
Anas platyrhynchos Mallard	Nearctic	Widespread, Gauteng	F
Columbidae			
Columba livia Rock dove	Mediterranean, Asia	Widespread urban	T

Species and group	Origin	SA distribution	Habitat
Corvidae			
Corvus splendens House crow	S Asia	Cape Town, Durban, Richards Bay	T
Fringillidae			
Fringilla coelebs Common chaffinch	Europe	Cape Peninsula	T
Passeridae			
Passer domesticus House sparrow	Eurasia, N Africa	Widespread, urban	T
Phasianidae			
Alectoris chukar Chukar partridge	Central Asia, China	Robben Island	T
Pavo cristatus Common peacock	S Asia	Widespread, patchy	T
Psittadiae			
Psittacula krameri Rose-ringed parakeet	S Asia	Durban, Gauteng	T
Sturnidae			
Acridotheres tristis Common myna	S Asia	Widespread	T
Sturnus vulgaris Common starling	Europe	Widespread	T
REPTILIA			
Typhlopidae			
Ramphotyphlops braminus Brahminy blind snake	S Asia	SW Cape, Durban	T
TELEOSTEI			
Centrarchidae			
Lepomis macrochirus Bluegill sunfish	E USA	Widespread	F
Micropterus dolomieu Smallmouth bass	USA	Widespread	F
Micropterus punctulatus Spotted bass	USA	W Cape to KZN, coastal	F
Micropterus salmoides Largemouth bass	USA, Mexico	Widespread	F
Cichlidae			
Oreochromis niloticus Nile tilapia	Sahel, Africa	Widespread	F
Cyprinidae			
Carassius auratus Goldfish	China	Widespread, patchy	F
Ctenopharyngodon idella Grass carp	China, Russia	Highveld and KZN	F
Cyprinus carpio Common carp	E Europe, Russia	Widespread	F
Hypophthalmichthys molitrix Silver carp	China	Mpumalanga	F
Tinca tinca Tench	Europe, Russia	W Cape	F
Locariidae			
Pterygoplichthys disjunctivus Vermiculated sailfin	Amazon	Kruger Park, Richard's Bay	F
Perciidae			
Perca fluviatilis Perch	Europe, Russia	Patchy, W to E Cape, Free State	F
Poeciliidae			
Gambusia affinis Mosquitofish	USA, Mexico	Widespread	F
Poecilia reticulata Guppy	S America/Caribbean	KZN	F
Xiphophorus helleri Swordtail	Mexico	Tropical parts	F
Salmonidae			
Onchorhynchus mykiss Rainbow trout	W North America	W Cape, Gauteng	F
Salmo trutta Brown trout	Europe, W Russia	Widespread	F
ASCIDIACEA			
Ascidiidae			
Ascidia sydneiensis Crevice ascidian	Asia	False Bay to Bushman's River	M
Ascidiella aspersa Dirty sea squirt	North Sea	Saldanha Bay to Table Bay	M

Species and group	Origin	SA distribution	Habitat
Cionidae			
Ciona intestinalis Vase tunicate	Europe	Durban to Alexander Bay	M
Clavelinidae			
Clavelina lepadiformis Light-bulb sea squirt	Europe	Durban to Saldanha Bay, harbours	M
Didemnidae			
Diplosoma listerianum Jelly crust tunicate	Europe	Durban to Alexander Bay	M
Pyuridae			
Microcosmus squamiger Blunt-spined microcosmus	Australia	Mossel Bay to Richards Bay	M
Styelidae			
Botryllus schlosseri Star sea squirt	NE Atlantic	Durban to Alexander Bay	M
Cnemidocarpa humilis Leathery sea squirt	Unknown	Alexander Bay to Cape Agulhas	M
Styela plicata Pleated sea squirt	W Pacific	Saldanha Bay to Richards Bay	M
ECHINODERMATA			
Arbaciidae			
Tetrapygus niger Black sea urchin	Chile, Peru	Alexander Bay	M
Ophiactidae			
Ophiactis savignyi Little six-arm brittle star	N Indian Ocean, Pacific Ocean	Durban northwards	M
COLLEMBOLA			
Brachystomellidae			
Brachystomella parvula	Europe		T
Entomobryidae			
Entomobrya nivalis	N hemisphere	W Cape	T
Hypogastruridae			
Hypogastrura armata	Cosmopolitan?		T
Hypogastrura manubrialis Mushroom springtail	N Europe	Scattered	T
Hypogastrura purpurescens	Europe?		T
Mesogastrura libyca	Europe?		T
Xenylla maritima	Eurasia?		T
Isotomidae			
Isotomurus palustris Marsh springtail	Europe	Widespread	T
Parisotoma notabilis	Europe		T
Proisotoma minuta	N hemisphere?		T
Neanuridae			
Neanura muscorum Moss springtail	N Europe	Localised, E & W Cape	T
Sminthuridae			
Sminthurus viridis Clover springtail	Europe	SW Cape	T
Tomoceridae			
Tomocerus minor	Palaearctic?	W Cape	T
THYSANURA			
Lepismatidae			
Ctenolepisma longicaudata Grey silverfish	Caribbean	Widespread, in buildings	T
Ctenolepisma urbana Urban silverfish	? (type USA)	Widespread, in buildings	T
Lepisma saccharina Silverfish	Unknown	SW Cape, in buildings	T

Species and group	Origin	SA distribution	Habitat
BLATTODEA			
Blatellidae			
Blattella germanica German cockroach	SE Asia	Widespread	T
Blattidae			
Blatta orientalis Oriental cockroach	S Russia/Tropical Africa?	Johannesburg	T
Periplaneta americana American cockroach	Tropical Africa	Widespread	T
ISOPTERA			
Kalotermitidae			
Cryptotermes brevis House termite	West Indies	Coastal regions	T
Rhinotermitidae			
Coptotermes formosanus Formosan termite	China, Japan	Simon's Town, Komatipoort	T
DERMAPTERA			
Anisolabididae			
Euborellia annulipes Ring-legged earwig	S Europe	Widespread	T
Labiduridae			
Labidura riparia Sand earwig	Siberian Russia	Widespread	T
Nala lividipes Black field earwig	SE Asia	Widespread	T
Spongiphoridae			
Labia minor Lesser earwig	Palaearctic Europe, N Africa	Widespread	T
PHASMATODEA			
Phasmatidae			
Carausius morosus Indian stick insect	India	Cape, scattered	T
EMBIOPTERA			
Oligotomidae			
Oligotoma saundersii Saunders embiid	India	Durban and surrounds	T
PSOCOPTERA			
Liposcelidae			
Liposcelis bostrychophila Domestic booklouse	Tropical Africa?	Widespread	T
HEMIPTERA			
Aleyrodidae			
Aleurocanthus woglumi Citrus blackfly	Asia		T
Aleurothrixus floccosus Woolly whitefly	Caribbean	W, E Cape, Gauteng	T
Bemisia tabaci Tobacco whitefly	Asia?		T
Siphoninus phillyreae Pomegranate whitefly	S Europe, N Africa	W parts, Gauteng	T
Trialeurodes vaporariorum Greenhouse whitefly	Likely USA	Widespread	T
Aphididae			
Acyrthosiphon kondoi Blue alfafa aphid	Asia		T
Acyrthosiphon pisum Pea aphid	Palaearctic		T
Aphis amoraciae	USA		T
Aphis chloris St. John's wort aphid	Eurasia	W Cape	T (BIO)
Aphis craccivora Groundnut aphid	Europe		T
Aphis fabae Black bean aphid	Europe	Widespread	T
Aphis gossypii Cotton aphid	Unknown	Widespread	T
Aphis nasturtii Buckthorn aphid	Europe?		T
Aphis spiraecola Green citrus aphid	E Asia		T
Aulacorthum circumflexum Mottled arum aphid	SE Asia		T
Aulacorthum solani Greenhouse potato aphid	Europe		T
Brachycaudus amygdalinus	Europe, N Africa		T

Species and group	Origin	SA distribution	Habitat
Brachycaudus helichrysi Leaf curling plum aphid	Palaearctic		T
Brevicoryne brassicae Cabbage aphid	Europe	Widespread	T
Capitophorus elaeagni Artichoke aphid	Mediterranean		T
Cedrobium laportei	N Africa		T
Cerataphis orchidearum Orchid aphid	SE Asia		T
Cerataphis variabilis	SE Asia		T
Chaetosiphon fragaefolii Strawberry aphid	USA?		T
Chaitophorus leucomelas	Palaearctic		T
Chaitophorus populialba	Eurasia, N Africa		T
Cinara cronartii Black pine aphid	E USA	Summer-rainfall parts	T
Cinara cupressivora Cypress aphid	E Greece?	Widespread	T
Diuraphis noxia Russian wheat aphid	Russia, Palaearctic	W, central parts	T
Dysaphis apiifolia Hawthorn parsley aphid	Eurasia		T
Dysaphis foeniculus Carrot aphid	Eurasia		T
Dysaphis tulipae Tulip bulb aphid	Eurasia, N Africa?		T
Eriosoma lanigerum Woolly apple aphid	USA?	Scattered, temperate parts	T
Eulachnus rileyi Pine needle aphid	USA, Europe		T
Hyadaphis coriandri Coriander aphid	Central Asia		T
Hydaphis foeniculi Honeysuckle aphid	Palaearctic		T
Hyperomyzus lactucae Blackcurrent aphid	Palaearctic		T
Hysteroneura seteriae Rusty plum aphid	USA		T
Illinoia azalea	USA		T
Lipaphis pseudobrassicae False cabbage aphid	Palaearctic		T
Macrosiphon euphorbiae Potato aphid	USA		T
Macrosiphoniella sanborni Chrysanthemum aphid	E Asia		T
Monelliopsis pecanis Yellow pecan aphid	USA		T
Myzocallis castanicola	Europe		T
Myzus persicae Green peach aphid	Asia?	Widespread	T
Neotoxoptera oliveri	Asia?		T
Pemphigus populitransversus Poplar gall aphid	USA		T
Pentalonia nigronervosa Banana aphid	SE Asia?		T
Pineus pini Pine woolly aphid	Eurasia?		T
Prociphilus fraxinifolii	USA		T
Rhodobium porosum	USA?		T
Rhopalosiphum maidis Maize aphid	Pakistan?		T
Rhopalosiphum padi Bird cherry aphid	Palaearctic		T
Schizaphis graminum Wheat aphid	Palaearctic, Asia?		T
Schizaphis minuta	Asia, Australia, Iran		T
Schizaphis rotundiventris	S Europe, Asia, Australia		T
Sipha maydis	Eurasia, India, Middle East		T
Sitobium avenae Brown wheat ear aphid	Eurasia?		T
Smynthurodes betae Bean root aphid	Mediterranean?		T
Takecallis taiwanus	Taiwan, Japan		T
Therioaphis trifolii Yellow clover aphid	Mediterranean		T
Toxoptera odinae	India, SE Asia		T
Cicadellidae			
Circulifer tenellus Beet leafhopper	Europe, N Africa	Scattered	T
Opsius stactogalus Tamarix leafhopper	Europe	Scattered	T
Orosius albicinctus Orosius leafhopper	Unknown	Scattered, mainly E parts	T
Coccidae			
Coccus hesperidum Soft brown scale	Afro-Ethiopian?	Widespread	T
Pseudaulacaspis pentagona White peach scale	Tropical Asia?		T
Pulvinaria psidii Guava scale	Asia		T
Dactylopiidae			
Dactylopius austrinus	S USA, Central America	SE parts	T (BIO)
Dactylopius ceylonicus	S USA, Central America	Coastal, Cape Town to E parts	T (BIO)
Dactylopius opuntiae Prickly pear cochineal	USA, Mexico	Cape provinces and NE parts	T BIO)
Dactylopius tomentosus	S USA, Central America	NW parts	T (BIO)
Delphacidae			
Perkinsiella saccharicida Sugarcane planthopper	Australasia	E parts	T

Species and group	Origin	SA distribution	Habitat
Diaspididae			
Aonidiella aurantii Red scale	Likely SE Asia	Widespread	T
Chrysomphalus aonidium Circular purple scale	Indian subcontinent		T
Cornuaspis beckii Citrus mussel scale	Indian subcontinent		T
Diaspidiotus perniciosus Pernicious scale	Likely N China	Widespread	T
Diaspis bromeliae Pineapple scale	S America		T
Parlatoria pergandii Chaff scale	Asia		T
Miridae			
Eccritotarsus catarinensis	Brazil	NE parts	F (BIO)
Falconia intermedia	S USA, Caribbean	NE parts	T (BIO)
Monophlebidae			
Icerya purchasi Australian bug	Australia	Widespread	T
Ortheziidae			
Orthezia insignis	S, Central America	W parts	T (BIO)
Pentotamidae			
Nezara viridula Green stinkbug	Ethiopia?	Widespread	T
Phylloxeridae			
Moritziella corticalis	Europe		T
Viteus vitifoliae Grapevine phylloxera	E USA		T
Pseudococcidae			
Ferrisia malvastra White-tailed mealybug	Americas		T
Hypogeococcus pungens	S America	NE parts	T (BIO)
Phenacoccus parvus	S, Central America	E parts	T (BIO)
Planococcus citri Citrus mealybug	China, Japan?	Widespread	T
Planococcus ficus Grapevine mealybug	Unknown	Widespread	T
Psyllidae			
Blastopsylla occidentalis Eucalypt shoot psyllid	Australia	W parts, Gauteng	T
Ctenarytaina eucalypti Blue gum psyllid	Australia	All save arid central parts	T
Thaumastocoridae			
Thaumastocoris peregrinus Eucalyptus thaumastocorid	Australia	Widespread, save dry parts	T
Tingidae			
Carvalhotingis hollandi	S, Central America	NE parts	T (BIO)
Carvalhotingis visenda	S, Central America	NE parts	T (BIO)
Gargaphia decoris	Argentina	E parts	T (BIO))
Teleonemia elata	Brazil	W parts	T (BIO)
Teleonemia scrupulosa Lantana lace bug	Mexico	W parts	T (BIO)
THYSANOPTERA			
Thripidae			
Fulmekiola serrata Sugarcane thrips	Mauritius	NE parts	T
Heliothrips haemorrhoidalis Greenhouse thrips	S America	NE parts	T
Thrips tabaci Onion thrips	Unknown	Widespread	T
PHTHIRAPTERA			
Boopiidae			
Heterodoxus spiniger Dog-chewing louse	Australia	Widespread	T
Haematopinidae			
Haematopinus eurysternus Short-nosed cattle louse	Unknown	Widespread	T
Haematopinus suis Pig-louse	Unknown	Widespread	T

Species and group	Origin	SA distribution	Habitat
Linognathidae			
Linognathus setosus Dog-sucking louse	Unknown	Widespread	T
Linognathus vituli Long-nosed cattle louse	Unknown	Widespread	T
Menoponidae			
Goniocotes gallinae Poultry-fluff louse	Asia?	Widespread	T
Lipeurus caponis Poultry-wing louse	Asia?	Widespread	T
Menacanthus stramineus Chicken body louse	Unknown	Widespread	T
Menopon gallinae Chicken louse	Asia?	Widespread	T
Polyplacidae			
Polyplax spinulosa Spined rat louse	Asia?	Widespread	T
Trichodectidae			
Damalinia bovis Cattle-biting louse	Unknown	Widespread	T
Damalinia caprae Common goat louse	Unknown	Widespread	T
Damalinia ovis Sheep-biting louse	Unknown	Widespread	T
COLEOPTERA			
Anobiidae			
Lasioderma serricorne Cigarette beetle	Unknown	Widespread	T
Stegobium paniceum Drug store beetle	Unknown		T
Apionidae			
Coelocephalapion camarae	Mexico	E parts	T (BIO)
Trichapion lativentre	S America	Widespread	T (BIO)
Bostrichidae			
Prostephanus truncatus Larger grain borer	Central America	N Mpumalanga	T
Rhyzopertha dominica Lesser grain borer	Possibly tropical Asia	Widespread	T
Buprestidae			
Hylaeogena jureceki	Korea	NE parts	T (BIO)
Cerambycidae			
Aphanasium australe	SE Australia	W to E Cape	T (BIO)
Chlorophorus annularis Bamboo longhorn beetle	Indian subcontinent	Hluhluwe	T
Lagocheirus funestus	Mexico	E Cape	T (BIO)
Nealcidion cereicola	Argentina	N parts	T (BIO)
Phoracantha recurva Eucalypt borer	Australia	SE parts	T
Phoracantha semipunctata Zig-zag eucalypt borer	Australia	SE parts	T
Chrysomelidae			
Acanthoscelides macrophthalmus	W Africa	E parts	T (BIO)
Acanthoscelides obtectus Bean weevil	Central America	Widespread	T
Algarobius prosopis Prosopis seed beetle	SW USA	NW parts	T (BIO)
Charidotis auroguttata	Costa Rica	NE parts	T (BIO)
Chrysolina quadrigemina	N Africa, Europe	W Cape?	T (BIO)
Gratiana spadicea	S America	Gauteng	T (BIO)
Leptinotarsa defecta Satansbos leaf beetle	N America	Scattered, E parts	T (BIO)
Leptinotarsa texana Texan Satansbos leaf beetle	N America	Scattered, E parts	T (BIO)
Longitarsus bethae	Mexico	E parts	T (BIO)
Lysathia sp.	S America	Widespread	T (BIO)
Neltumius arizonensis Prosopis seed beetle	SW USA	NW parts	T (BIO)
Octotoma scabripennis	Mexico, Central America	NE parts	T (BIO)
Oulema bilineata Tobacco slug	S America		T
Oulema trilineata	S America		T
Phenrica guerini	S America	E Cape	T (BIO)
Sulcobruchus subsuturalis	Asia		T (BIO)
Trachymela tincticollis	Australia	W Cape	T
Uroplata girardi	S America	E parts	T (BIO)

Species and group	Origin	SA distribution	Habitat
Coccinellidae			
Harmonia axyridis Harlequin lady beetle	Central and E Asia	Widespread, scattered	T
Hippodamia variegata Variegated lady beetle	Palaearctic	Widespread	T
Rodolia cardinalis Vedalia beetle	Australia	Widespread	T
Curculionidae			
Anthonomus santacruzi	S America	E parts	T (BIO)
Cosmopolites sordidus Banana root borer	SE Asia?		T
Cydmaea binotata	SE Australia	W to E Cape	T (BIO)
Cylas formicarius Sweet potato weevil	Indonesia	KZN	T
Cyrtobagous salviniae	S, Central America	Widespread	T (BIO)
Dicomada rufa	SE Australia	W Cape	T (BIO)
Erytenna consputa	SE Australia	W Cape	T (BIO)
Gonipterus scutellatus	S Australia, Tasmania		T
Listroderes costirostris Vegetable weevil	Brazil		T
Melanterius acaciae	Australia	W to E Cape, KZN	T (BIO)
Melanterius compactus	Australia	W Cape	T (BIO)
Melanterius maculatus	Australia	Widespread	T (BIO)
Melanterius servulus	Australia	W Cape	T (BIO)
Melanterius ventralis	Australia	W to E Cape	T (BIO)
Metamasius spinolae	Neotropics	E Cape	T (BIO)
Naupactus leucoloma Whitefringed weevil	S America	SW parts	T
Neochetina bruchi	S, Central America	Widespread	T (BIO)
Neochetina eichhorniae	S, Central America	Widespread	T (BIO)
Neodiplogrammus quadrivittatus	S America	Widespread	T (BIO)
Neohydronomus affinis Waterlettuce weevil	S, Central America	Widespread	F (BIO)
Pantomorus cervinus Fuller's rose beetle	Americas?		T
Pissodes nemorensis Northern pine weevil	SE USA	All save arid NW parts	T
Rhinocyllus conicus	Eurasia	E Cape	T (BIO)
Rhyssomatus marginatus	S America	Widespread	T (BIO)
Sitophylus granarius Granary weevil	Palaearctic		T
Stenopelmus rufinasus Azolla weevil	Americas	Widespread, scattered	T (BIO)
Sternochetus mangiferae Mango weevil	SE Asia		
Dermestidae			
Anthrenus verbasci Varied carpet beetle	California?		T
Dermestes maculatus Hide beetle	Unknown	Widespread	T
Trogoderma granarium Khapra beetle	Asia		T
Trogoderma inclusum Larger cabinet beetle	USA		T
Trogoderma variabile Warehouse beetle	Asia		T
Melyridae			
Astylus atromaculatus Spotted maize beetle	S America	Scattered, E parts	T
Nitidulidae			
Carpophilus dimidiatus Corn-sap beetle	Neotropics		T
Ptinidae			
Trigonogenius globulus Globular spider beetle	Americas		T
Scolytidae			
Hylastes angustatus Pine bark beetle	Europe		T
Silvanidae			
Ahasverus advena Foreign grain beetle	Americas		T
Oryzaephilus mercator Merchant grain beetle	Unknown		T
Oryzaephilus surinamensis Saw-toothed grain beetle	Unknown		T
Staphylinidae			
Cafius xantholoma	Europe	W Cape beaches	T/M
Habrocerus capillaricornis Verticillate rove beetle	Europe, N Africa	Scattered	T

Species and group	Origin	SA distribution	Habitat
Tenebrionidae			
Alphitobius diaperinus Lesser mealworm beetle	African?	Widespread	T
Gnathocerus cornutus Broad-horned flour beetle	Americas		T
Latheticus oryzae Long-headed flour beetle	Asia		T
Tenebrio molitor Yellow mealworm	Eurasia	Widespread	T
LEPIDOPTERA			
Arctiidae			
Pareuchaetes insulata Yellow-winged pareuchaetes	Cuba, Jamaica	E parts	T (BIO)
Carposinidae			
Carposina autologa	Australia	W Cape	T (BIO)
Crambidae			
Niphograpta albiguttalis Waterhyacinth moth	Amazon	NE parts	F (BIO)
Salbia haemorrhoidalis Lantana leaftier	Caribbean, Central America	E parts	T (BIO)
Gelechiidae			
Pectinophora gossypiella Pink bollworm	Australasia		T
Phthorimaea operculella Potato tuber moth	Bolivia	Widespread, save dry parts	T
Sitotroga cerealella French grain moth	Australia	Widespread	T
Gracillariidae			
Aristaea thalassias	Australia	W to E Cape	T (BIO)
Noctuidae			
Spodoptera exigua Beet armyworm	SE Asia		T
Pieridae			
Pieris brassicae Large cabbage white butterfly	Eurasia	W parts	T
Plutellidae			
Plutella xylostella Diamondback moth	Likely Mediterranean	Widespread	T
Pterophoridae			
Lantanophaga pusillidactyla Lantana plume moth	S USA to Central America	E parts	T (BIO)
Pyralidae			
Achroia grisella Lesser wax moth	Europe	Widespread	T
Cactoblastis cactorum Prickly pear moth	S America	Widespread	T (BIO)
Chilo partellus Sorghum stem borer	Likely Asia	E, S parts	T
Ephestia elutella Tobacco moth	Europe	Widespread	T
Ephestia kuehniella Mediterranean flour moth	India	Widespread	T
Galleria mellonella Greater wax moth	Eurasia?	Widespread	T
Hellula undalis Cabbage web worm	S Europe, Asia?		T
Plodia interpunctella Indian meal moth	S America?	Widespread	T
Tineidae			
Tinea pellionella Case-making clothes moth	Europe	Widespread	T
Tortricidae			
Crocidosema lantana Lantana tortricid moth	Mexico, S USA	E parts	T (BIO)
Cydia pomonella Codling moth	Temperate Asia	Widespread	T
DIPTERA			
Agromyzidae			
Calycomyza eupatorivora	Neotropics	E parts	T (BIO)
Calycomyza lantanae	Americas	E parts	T (BIO)
Chromatomyia horticola Pea leafminer	Eurasia?		T
Liriomyza huidobrensis Potato leafminer	S America?		T

Species and group	Origin	SA distribution	Habitat
Liriomyza trifolii American leafminer	USA		T
Ophiomyia camarae	Neotropics, S USA	E parts	T (BIO)
Ophiomyia lantanae	S USA, S America	E parts	T (BIO)
Psila rosae Carrot rust fly	Temperate Eurasia		T
Anthomyiidae			
Delia platura Bean seed maggot	Europe		T
Fucellia tergina Common kelp fly	Europe	W coast beaches	M/T
Fucellia maritima	Europe	W coast beaches	M/T
Calliphoridae			
Calliphora vicina European bluebottle	Europe, N America	Scattered	T
Chrysomya megacephala Oriental latrine fly	SE Asia, E Africa	Widespread	T
Lucilia sericata Common greenbottle	Europe, N America	Widespread	T
Cecidomyiidae			
Contarinia sorghicola Sorghum midge	Asia?		T
Dasineura dielsi	Australia	W to E Cape	T (BIO)
Dasineura rubiformis	W Australia	W to S Cape	T (BIO)
Zeuxidiplosis giardi	France	W Cape	T (BIO)
Culicidae			
Aedes albopictus Asian tiger mosquito	SE Asia	Cape Town	T
Faniidae			
Fannia albitarsis White-footed lesser house fly	S America		T
Phoridae			
Megaselia scalaris Common coffin fly	Americas	Widespread	T
Piophilidae			
Piophila casei European cheese fly	Europe	Widespread	T
Psychodidae			
Clogmia albipunctata Moth fly	Unknown	Widespread (urban)	T
Stratiomyidae			
Hermetia illucens Window-waisted soldier fly	Americas	Widespread	T
Syrphidae			
Eristalis tenax European drone fly	N Europe, Asia	Widespread	T
Tephritidae			
Bactrocera invadens Invasive fruit fly	Sri Lanka	NE parts	T
Ceratitis capitata Mediterranean fruit fly	Mediterranean	Widespread	T
Procecidochares utilis	Mexico	W Cape, KZN	T (BIO)
SIPHONAPTERA			
Pucilidae			
Ctenocephalides felis Cat flea	Egypt	Widespread	T
Echidnophaga gallinacea Stick-tight flea	SE Asia	Widespread except in N	T
Pulex irritans Human flea	S, Central America	Widespread	T
Xenopsylla cheopis Oriental rat flea	Egypt, Sudan	Widespread except in N Cape	T
Tungidae			
Tunga penetrans Jigger flea	Tropical S America	KZN coast/Kruger Park	T
HYMENOPTERA			
Aphelinidae			
Aphytis coheni	Asia		T (BIO)
Aphytis holoxanthus	Indian subcontinent		T (BIO)
Aphytis lepidosaphes	Indian subcontinent		T (BIO)

Species and group	Origin	SA distribution	Habitat
Aphytis lingnanensis *Aphytis melinus* Red scale parasite	Asia Asia		T (BIO) T (BIO)
Braconidae *Apanteles subandinus* Potato tuber moth parasitoid *Cotesia plutellae* Diamondback moth parasitoid	 Bolivian Andes Europe	 Widespread, potato-growing areas Widespread	 T T
Encyrtidae *Comperiella bifasciata* *Copidosoma koehleri*	 Asia S America		 T (BIO) T (BIO)
Eulophidae *Leptocybe invasa* Eucalyptus gall wasp	 Australia	 Upington, Polokwane	 T
Evaniidae *Evania appendigaster* Ensign wasp	 S China, SE Asia	 Widespread	 T
Formicidae *Linepithema humile* Argentine ant	 S America	 Scattered, mainly SW parts	 T
Ibaliidae *Ibalia leucospoides* Sirex ibaliid wasp	 Holarctic	 W Cape	 T (BIO)
Megalyridae *Megalyra fasciipennis* Megalyrid wasp	 Australia	 W Cape, Tzaneen	 T
Mymaridae *Patasson nitens*	 S Australia		 T (BIO)
Pteromalidae *Trichilogaster acaciaelongifoliae* Acacia gall wasp *Trichilogaster signiventris* Acacia gall wasp	 Australia Australia	 W, E and N Cape, KZN SW and S Cape	 T T
Siricidae *Sirex noctilio* Sirex woodwasp	 Europe, Asia, N Africa	 W and E Cape, KZN	 T
Tenthridinidae *Caliroa cerasi* Pear slug *Fenusa dohrnii* Alder leafminer	 Europe Europe, USA	 Widespread, scattered W Cape	 T T
Vespidae *Polistes dominulus* European paper wasp *Vespula germanica* German wasp	 Europe, Asia, N Africa Europe, Asia, N Africa	 Cape Peninsula SW Cape	 T T
MYRIOPODA			
Blaniulidae *Proteroiulus fuscus*	 Europe	 Stellenbosch	 T
Julidae *Brachyiulus pusillus* *Cylindroiulus brittanicus* *Cylindroiulus truncorum* *Ommattoiulus moreleti* Portuguese millipede	 Europe Europe N Africa W Europe	 Stellenbosch SW Cape Cape Town SW Cape	 T T T T
Lithobiidae *Lithobius obscurus* Purple stone centipede *Lithobius peregrinus* Peregrine's stone centipede	 W Mediterranean Central Europe	 Cape Town Cape Town	 T T
Paradoxosomatidae *Orthomorpha gracilis* Hothouse millipede	 East Indies	 Scattered	 T

Species and group	Origin	SA distribution	Habitat
Scutigeridae			
Scutigera coleoptrata House centipede	S Europe	Cape Town, KZN, urban	T
PYCNOGONIDA			
Ammotheidae			
Ammothella appendiculata	Pacific	Durban Bay	M
ARACHNIDA			
Agelenidae			
Tegenaria domestica Barn funnel-web spider	Europe	Cape Town, Witwatersrand	T
Anystidae			
Anystis wallacei Wriggling mite	? (type France)	W Cape	T
Araneidae			
Cyrtophora citricola Tropical tent-web spider	Unknown	Widespread	T
Bdellidae			
Bdellodes lapidaria Snout mite	? (type Europe)	W Cape	T
Dysderidae			
Dysdera crocata Long-fanged six-eyed spider	Mediterranean	Cape Town, Witwatersrand	T
Eriophyidae			
Aceria aloinis Aloe gall mite	USA	W parts	T
Aceria cynodoniensis Grass rosette mite	Egypt	N and E parts	T
Aceria ficus Fig bud mite	France	Gauteng	T
Aceria lantanae Lantana gall mite	S USA to S America	E parts	T (BIO)
Aceria mangiferae Mango bud mite	Egypt	E parts	T
Aceria oleae Olive bud mite	Europe	W Cape	T
Aceria sheldoni Citrus bud mite	USA	W Cape, E parts	T
Aculops lycopersici Tomato rust mite	Japan	Cape Town, Limpopo	T
Galumnidae			
Orthogalumna terebrantis Waterhyacinth mite	Americas	E parts	F (BIO)
Ixodidae			
Rhipicephalus microplus Asiatic cattle tick	Australasia, Madagascar, Neotropics	Widespread	T
Linyphiidae			
Ostearius melanopygius Black tailed sheet-web spider	New Zealand	Widespread	T
Oecobiidae			
Oecobius navus House ant-eater	Unknown	Widespread	T
Penthaleidae			
Halotydeus destructor Black sand mite	? (type locality South Africa)	W Cape and coast of N Cape	T
Pholcidae			
Crossopriza lyoni Humped back daddy-long-legs	SE Asia	Pretoria	T
Pholcus phalangiodes House daddy-long-legs	Europe	Widespread	T
Phytoseiidae			
Neoseiulus californicus Predatory mite	? (type California)	W Cape	T
Salticidae			
Hasarius adansoni Adanson's house jumping spider	Egypt	Widespread	T
Scytodidae			
Scytodes thoracica Spitting spider	N Africa?	Widespread	T

Species and group	Origin	SA distribution	Habitat
Sparassidae			
Heteropoda venatoria Brown huntsman spider	Unknown	Widespread?	T
Tarsonemidae			
Polyphagotarsonemus latus Citrus silver mite	? (type USA)	Widespread	T
Tarsonemus waitei Peach bud mite	USA?	W Cape	T
Tenuipalpidae			
Brevipalpus californicus Citrus flat mite	? (type California)	Widespread	T
Brevipalpus obovatus Ornamental flat mite	? (tropical?)	Widespread	T
Brevipalpus phoenicis Reddish black mite	? (type France)	Widespread except interior	T
Raoiella indica Red palm mite	E Africa, Asia	Not recorded but future threat	T
Tetragnathidae			
Tetragnatha boydi Long-jawed water spider	Mexico	Widespread	T
Tetranychidae			
Bryobia praetiosa Brown clover mite	? (type Germany)	Widespread	T
Panonychus citri Citrus mite	? (type Florida)	Widespread	T
Panonychus ulmi European red mite	? (type Germany)	SW Cape	T
Tetranychus evansi Tobacco spider mite	Brazil, Argentina	Widespread	T
Tetranychus urticae Two-spotted mite	Mediterranean region	Widespread	T
Theridiidae			
Latrodectus geometricus Brown button spider	S America	Widespread	T
Parasteatoda tepidariorum Common house orb-web spider	New World	Widespread	T
Steatoda grossa False widow spider	Greece	Widespread	T
Uloboridae			
Uloborus plumipes Feather-legged lace weaver	Old World	Widespread	T
Varroidae			
Varroa destructor Varroa mite	Asia	Widespread	T
CRUSTACEA			
Acartiidae			
Acartia spinicauda Spinytailed copepod	SE Asia	Durban, Richards Bay	M
Argulidae			
Argulus japonicus Fish louse	SE Asia	Gauteng, NW parts	F
Armadillidiidae			
Armadillidium vulgare Pill bug	Europe	Cape Town	T
Balanidae			
Amphibalanus venustus	Tropical N Atlantic	Hermanus to Mozambique	M
Balanus glandula Pacific barnacle	N American Pacific	W coast	M
Cheluridae			
Chelura terebrans Wood boring amphipod	Pacific N America	Saldanha Bay to Port Elizabeth	M
Coropiidae			
Apocorophium acutum	N Atlantic	Durban	M
Erichthonius brasiliensis	N Atlantic	Widespread coastal	M
Monocorophium acherusicum Stout-antenna amphipod	N Atlantic	Alexander Bay to Durban	M
Ischyroceridae			
Cerapus tubularis Hermit amphipod	Atlantic N America	Saldanha Bay to Kosi Bay	M
Ischyrocerus anguipes	N Atlantic	Entire coast	M
Jassa marmorata	N Atlantic	Table Bay to KZN	M

Species and group	Origin	SA distribution	Habitat
Jassa morinoi	N Pacific	False Bay to KZN	M
Jassa slatteri Hitchhiker amphipod	Pacific N America	Saldanha Bay to Knysna	M
Limnoriidae			
Limnoria quadripunctata Gribble	Unknown	Table Bay to Port Elizabeth	M
Limnoria tripunctata Gribble	Unknown	Table Bay	M
Oniscidae			
Philoscia elongata	Europe, Middle East	Cape Town	T
Philoscia muscorum Fast woodlouse	Europe, N Africa	Durban, Pietermaritzburg	T
Parastacidae			
Cherax destructor Yabby	Australia	Free State (in culture)	F
Cherax quadricarinatus Australian redclaw	Australia	Crocodile River	F
Cherax tenuimanus Marron	SW Australia	W Cape, previously E Cape	F
Procambarus clarkii North American red swamp crayfish	N America	Crocodile River	F
Pinnotheridae			
Pinnixa occidentalis Pea crab	Pacific N America	Saldanha Bay	M
Porcellionidae			
Porcellio laevis Smooth woodlouse	British Isles	Cape Town	T
Porcellio scaber Rough woodlouse	S and W Europe	Cape Town	T
Porcellionides pruinosus	Europe, N America	Cape Town	T
Portunidae			
Carcinus maenas European shore crab	Europe, Mediterranean	Cape Peninsula	M
Sphaeromidae			
Dynamene bidentata Horned isopod	Europe	Port Elizabeth	M
Paracerceis sculpta Sponge isopod	NE Pacific	Port Elizabeth	M
Sphaeroma serratum	Europe	Durban Bay	M
Sphaeroma walkeri Pill isopod	N Indian Ocean	KZN coastline	M
Talitridae			
Orchestia gammarella Beach Hopper	Europe, Mediterranean	Scattered harbours	M
Platorchestia platensis	Origin unknown	Gansbaai only	M
Talitroides alluaudi	? (type Paris)	Cape Town	T
Talitroides topitotum Land hopper	Origin unknown	Cape Town and Pretoria	T
Xanthidae			
Xantho incisus Black-fingered crab	Europe, Mediterranean	Kleinsee	M
NEMATODA			
Beddingiidae			
Beddingia siricidicola Woodwasp nematode	Europe	E and W Cape, KZN	T
Heteroderidae			
Globodera rostochiensis Potato cyst nematode	Bolivia and Peru	E and W Cape, Gauteng	T
Meloidogyne javanica Root-knot nematode	S America	Widespread	T
Tylenchulidae			
Radopholus similis Burrowing nematode	Australasia	E Cape, KZN	T
Tylenchulus semipenetrans Citrus nematode	Unknown	E, W and N Cape	T
ANNELIDA			
Acanthodrilidae			
Dichogaster affinis	? (type Zanzibar)	Rare, Soutpansberg, KZN	T
Dichogaster annae	Java	Pietermaritzburg	T
Dichogaster bolaui	? (type, Germany)	E and N parts	T
Dichogaster modiglianii	Sumatra	KZN	T

Species and group	Origin	SA distribution	Habitat
Dichogaster saliens	W Africa	Drakensberg, KZN	T
Microscolex dubius	S America	E and W Cape, Free State	T
Microscolex phosphoreus	S America	Widespread	T
Cirratulidae			
Dodecaceria fewkesi Black coral worm	Pacific N America	Table Bay docks	M
Eudrilidae			
Eudrilus eugeniae	W Africa	Potchefstroom, Durban	T
Glossoscolecidae			
Pontoscolex corethrurus Quincunx earthworm	Amazonian Brazil	Widespread in E and N parts	T
Lumbricidae			
Allolobophoridella eiseni	Europe, Americas	Scattered in plantations	T
Allolobophoridella parva	USA	Widespread but scattered	T
Aporrectodea caliginosa	Palaearctic	Widespread	T
Aporrectodea longa	? (type Germany)	Roodepoort, Heidelberg	T
Aporrectodea rosea Pink soil earthworm	Temperate Eurasia	Widespread	T
Aporrectodea trapezoides	? (type France)	Widespread	T
Dendrobaena cognettii	? (type Sardinia)	Cape Peninsula	T
Dendrobaena hortensis	? (type Germany)	Port Elizabeth	T
Dendrobaena octaedra	? (type France)	W and E Cape, KZN	T
Dendrodrilus rubidus Bark-eating earthworm	Temperate Eurasia	Widespread	T
Eisenia fetida Composting earthworm	Eurasia?	Widespread	T
Eiseniella tetraedra	? (type France)	Widespread	T
Lumbricus castaneus	Unknown	Karkloof (KZN)	T
Lumbricus rubellus	Unknown	Scattered (W Cape, KZN)	T
Octolasion cyaneum	? (type France)	Stutterheim	T
Octolasion lacteum	Unknown	Scattered, widespread	T
Megascolecidae			
Amynthas aeruginosus	? (type Guam)	KZN, Pretoria	T
Amynthas corticis	? (type Hawaii)	Widespread	T
Amynthas diffringens Snake earthworm	China?	Widespread	T
Amynthas gracilis	? (type Argentina)	Mpumalanga, E Cape	T
Amynthas hawayanus	? (type Hawaii)	Widespread	T
Amynthas minimus	? (type Java)	Subtropical zone	T
Amynthas morrisi	? (type Malaysia)	Pretoria, Sabie, Potchefstroom	T
Amynthas rodericensis	? (type Rodrigues)	E parts	T
Metaphire californica	? (type California)	KZN	T
Perionyx excavatus	Asia	Potchefstroom,KZN	T
Pontodrilus litoralis	? (type France)	Coastal KZN	T
Nereidae			
Neanthes succinea Pileworm	Europe	Mossel Bay to Durban	M/E
Ocnerodrilidae			
Eukerria saltensis	S America	Widespread, littoral	T/F
Nematogenia lacuum	Unknown	KZN	T
Ocnerodrilus occidentalis	Unknown	Parys, Free State, rare	T
Opheliidae			
Travesia forbesii Pink spindle worm	Palaearctic, Japan	Cape Town harbour	M
Serpulidae			
Ficopomatus enigmaticus Estuarine tube-worm	Australia	Cape Town to Kosi Bay	M
Hydroides elegans	Indo-Pacific	False Bay	M
Janua pagenstecheri	Europe	Cape Town to Durban	M
Neodexiospira brasiliensis Spiral fanworm	West Indies, Brazil	Cape Town to Port Elizabeth	M
Spionidae			
Boccardia proboscidea Shell worm	N Pacific	Abalone aquaculture, W Cape	M
Polydora hoplura Mudworm	Europe, Mediterranean	Saldanha Bay to Plettenberg Bay	M

Species and group	Origin	SA distribution	Habitat
MOLLUSCA			
Agriolimacidae			
Deroceras laeve Marsh slug	Europe	Widespread	T
Deroceras panormitanum Brown field slug	Europe	Widespread	T
Deroceras reticulatum Grey field slug	Europe	Widespread	T
Ampullariidae			
Pomacea diffusa Mystery snail	Brazil	W parts, except in N Cape	F
Arionidae			
Arion hortensis aggregate Garden arion	W Europe	W Cape	T
Arion intermedius Hedgehog slug	W Europe	Widespread	T
Bradybaenidae			
Bradybaena similaris Asian tramp snail	E Asia	KZN	T
Cochlicellidae			
Cochlicella barbara Small conical snail	Mediterranean	Widespread, scattered	T
Cochlicopidae			
Cochlicopa cf. *lubrica* Slippery moss snail	Europe, N America	Widespread	T
Discidae			
Discus rotundatus Spotted disc	Europe	E and W Cape	T
Gastrodontidae			
Zonitoides arboreus Orchid snail	N America	Widespread, scattered	T
Helicidae			
Cornu aspersum European brown garden snail	W Europe, Mediterranean	Widespread	T
Eobania vermiculata Vermiculate snail	Mediterranean	Widespread, scattered	T
Otala puncata White-lipped milk snail	W Mediterranean	W Cape	T
Theba pisana Dune snail	Mediterranean	Coastal N, W and E Cape	T
Lauriidae			
Lauria cylindracea Common chrysalis snail	W Europe, Mediterranean	W Cape	T
Limacidae			
Lehmannia nyctelia Striped garden slug	E Europe	Widespread	T
Lehmannia valentiana Three-banded garden slug	E Europe	Widespread	T
Limacus flavus Yellow garden slug	Mediterranean	Widespread, scattered	T
Limax maximus Giant garden slug	Europe	Cape Peninsula	T
Littorinidae			
Littorina saxatilis Lagoon snail	Europe, Mediterranean W Atlantic	Langebaan, Knysna	M
Lymnaeidae			
Lymnaea columella Reticulate pond snail	N America	Orange-Vaal river systems	F
Radix rubiginosa Rust-coloured pond snail	SE Asia	Durban, Amatikulu	F
Milacidae			
Milax gagates Black keeled slug	Mediterranean	Scattered	T
Muricidae			
Thais blanfordi Blanford's whelk	Tropical Indo-Pacific	Durban	M
Thais tissoti Tissot's rock-shell	Tropical Indo-Pacific	Durban	M
Mytilidae			
Mytilus galloprovincialis Mediterranean mussel	Mediterranean, NE Atlantic	Alexander Bay to East London	M
Perna viridis Green mussel	SE Asia	East London harbour	M
Semimytilus algosus Bisexual mussel	Pacific S America	Alexander Bay to Cape Town	M

Species and group	Origin	SA distribution	Habitat
Ostreidae			
Crassostrea gigas Japanese oyster	Japan, NW Pacific	S Cape estuaries	M
Ostrea edulis European oyster	Europe, Mediterranean	Alexander Bay	M
Oxychilidae			
Aegopinella nitidula Smooth glass snail	Europe	W Cape	T
Oxychilus alliarius Garlic glass snail	W Europe	Widespread	T
Oxychilus cellarius Cellar glass snail	Europe	Widespread	T
Oxychilus draparnaudi Draparnaud's glass snail	W Europe, Mediterranean	Widespread	T
Physidae			
Aplexa marmorata Slender bladder snail	S America	E parts	F
Physa acuta Sharp-spired bladder snail	S America, Caribbean	Widespread	F
Planorbidae			
Gyraulus chinensis Chinese ram's-horn snail	SE Asia	N KZN	F
Helisoma duryi Dury's ram's-horn snail	SE USA	W parts, except N Cape	F
Pristilomatidae			
Hawaiia minuscula Minute gem	N America	E Cape	T
Vitrea contracta Milky crystal snail	Europe, Middle East, N Africa	W Cape	T
Vitrea crystallina Common crystal snail	Europe	W Cape	T
Subulinidae			
Rumina decollata Decollate snail	Mediterranean	Cape Town	T
Subulina octona Wandering awl snail	Caribbean	Cango Caves	T
Teredinidae			
Lyrodus pedicellatus Shipworm	Unknown	Simon's Town	M
Teredo navalis Shipworm	Europe, Mediterranean	E coast	M
Tergipedidae			
Catriona columbiana Columbian nudibranch	N Pacific	Cape Town	M
Testacellidae			
Testacella maugei Maug's shelled slug	W Europe, Mediterranean	W Cape	T
Thiaridae			
Tarebia granifera Quilted melania	SE Asia	KZN coastal strip	F
Valloniidae			
Vallonia costata Ribbed grass snail	Europe, E USA	E Cape, Gauteng, NW parts	T
Vallonia pulchella Smooth grass snail	Europe, E North America	Widespread	T
BRACHIOPODA			
Discinidae			
Discinisca tenuis Disc lampshell	Namibian coast	Saldanha Bay, Port Elizabeth	M
BRYOZOA			
Bugulidae			
Bugula dentata Blue dentate moss animal	Indo-Pacific	Table Bay to Sodwana Bay	M
Bugula flabellata	Unknown	Port Nolloth to Plettenberg Bay	M
Bugula neritina Purple dentate moss animal	Europe?	Port Nolloth to Durban	M
Cryptosulidae			
Cryptosula pallasiana	Europe	Saldanha Bay to False Bay	M
Membraniporidae			
Conopeum seurati	Europe	Saldanha Bay, False Bay	M
Watersiporidae			
Watersipora subtorquata Red-rust bryozoan	Caribbean	Saldanha Bay to Knysna	M

Species and group	Origin	SA distribution	Habitat
PLATYHELMINTHES			
Bothriocephalidae			
Bothriocephalus acheilognathi Asian tapeworm	China	E parts	F
Dactylogyridae			
Pseudodactylogyrus anguillae Eel gill fluke	China, Japan, Taiwan	E Cape	F
Geoplanidae			
Bipalium kewense Spade-headed planarian	SE Asia	Cape Town, E coast?	T
Kontikia ventrolineata Kontikia flatworm	Australia	Gauteng	T
Gyrodactylidae			
Gyrodactylus kherulensis Fish skin fluke	E Asia	W Cape	F
Gyrodactylus kobayashii Fish gillworm	Asia	Widespread, aquaculture	F
CNIDARIA			
Campanulariidae			
Gonothyraea loveni	N Atlantic	Table Bay harbour	M
Laomedea calceolifera	N Atlantic	Table Bay harbour	M
Obelia bidentata Wine-glass hydroid	Unknown	Table Bay harbour, Durban	M
Obelia dichotoma Bushy wine-glass hydroid	Unknown	Lambert's Bay to Algoa Bay	M
Obelia geniculata Zigzag hydroid	Europe, Mediterranean	Entire coastline	M
Corynidae			
Coryne eximia	N Atlantic, Pacific	Table Bay harbour	M
Limnomedusae			
Craspedacusta sowerbyi Freshwater jellyfish	Nearctic, Paleoarctic	W Cape, E parts	F
Metridiidae			
Metridium senile Plumose anemone	N Atlantic	Table Bay, Agulhas Bank	M
Moerisiidae			
Moerisia maeotica	Black Sea region	KZN lagoons	M
Oceanidae			
Pachycordyle navis	Europe, Mediterranean	Table Bay harbour	M
Pennariidae			
Pennaria disticha Sea fern hydroid	Unknown	Durban northwards	M
Sagartiidae			
Sagartia ornata Brooding anemone	Europe, Mediterranean	Saldanha Bay	M
Tubulariida			
Pinauay larynx	N Atlantic	Cape Peninsula	M
Pinauay ralphi	N Atlantic	Table Bay harbour, Durban	M
PORIFERA			
Demospongiae – Suberitidae			
Suberites ficus Sulphur sponge	NE Atlantic and Mediterranean	W coast	M

MAJOR RESOURCES USED FOR THE APPENDIX

Marine groups
Mead, A., Carlton, J.T., Griffiths, C.L. & Rius, M. Introduced and cryptogenic marine and estuarine species of South Africa. *Journal of Natural History* (in press).

Terrestrial molluscs
Herbert, D.G. 2010. The introduced terrestrial Mollusca of South Africa. *SANBI Biodiversity Series* 15. Pretoria: South African National Biodiversity Institute.

Collembola
Coates, T.J. 1970. Check-list of the Collembola of South African Parks (Part 1). *Koedoe*, 13:181-184.

Earthworms
Plisko, J.D. 2010. Megadrile earthworm taxa introduced to South African soils (Oligochaeta: Acanthodrilidae, Eudrilidae, Glossoscolecidae, Lumbricidae, Megascolecidae, Ocnerodrilidae). *African Invertebrates,* 51:289-312.

Crop pests
Annecke, D.P. & Moran, V.C. 1982. *Insects and mites of cultivated plants in South Africa.* Durban: Butterworths.

Visser, D. 2009. *A complete guide to vegetable pests in South Africa.* Pretoria: ARC-Roodeplaat Vegetable and Ornamental Plant Institute.

Terrestrial biocontrol agents
Klein, H. Biological control of invasive alien plants in South Africa: a catalogue of candidate agents and ultimately released insect and pathogenic agents and their target weeds, since 1913. *African Entomology* (in press).

Aphids
Millar, M. 1994. A catalogue of the aphids (Homoptera: Aphidoidea) of sub-Saharan Africa. *Plant Protection Research Handbook* No. 4. Pretoria: Plant Protection Research Institute, Biosystematics Division, ARC.

Nematodes
Heyns, J. 1971. *A guide to the plant and soil nematodes of South Africa.* Cape Town: A.A. Balkema.

Top left: German wasp
Top right: European shore crab
Bottom left: Window-waisted soldier fly larvae
Bottom right: Green vegetable bug nymph

C. Griffiths

C. Griffiths